Biopesticides for sustainable agriculture

It is widely recognised that agriculture is a significant contributor to global warming and climate change. Agriculture needs to reduce its environmental impact and adapt to current climate change whilst still feeding a growing population, i.e. become more 'climate-smart'. Burleigh Dodds Science Publishing is playing its part in achieving this by bringing together key research on making the production of the world's most important crops and livestock products more sustainable.

Based on extensive research, our publications specifically target the challenge of climate-smart agriculture. In this way we are using 'smart publishing' to help achieve climate-smart agriculture.

Burleigh Dodds Science Publishing is an independent and innovative publisher delivering high quality customer-focused agricultural science content in both print and online formats for the academic and research communities. Our aim is to build a foundation of knowledge on which researchers can build to meet the challenge of climate-smart agriculture.

For more information about Burleigh Dodds Science Publishing simply call us on +44 (0) 1223 839365, email info@bdspublishing.com or alternatively please visit our website at www.bdspublishing.com.

Related titles:

Critical issues in plant health: 50 years of research in African agriculture
Print (ISBN 978-1-78676-232-0); Online (ISBN 978-1-78676-235-1, 978-1-78676-234-4)

Integrated management of diseases and insect pests of tree fruit
Print (ISBN 978-1-78676-256-6); Online (ISBN 978-1-78676-258-0, 978-1-78676-259-7)

Pesticides and agriculture
Print (ISBN 978-1-78676-276-4); Online (ISBN 978-1-78676-279-5, 978-1-78676-278-8)

Chapters are available individually from our online bookshop:
https://shop.bdspublishing.com

climate•SMART•publishing
IN AGRICULTURAL SCIENCE

BURLEIGH DODDS SERIES IN AGRICULTURAL SCIENCE
NUMBER 89

Biopesticides for sustainable agriculture

Edited by Professor Nick Birch, formerly The James Hutton Institute, UK; and Professor Travis Glare, Lincoln University, New Zealand

burleigh dodds
SCIENCE PUBLISHING

Published by Burleigh Dodds Science Publishing Limited
82 High Street, Sawston, Cambridge CB22 3HJ, UK
www.bdspublishing.com

Burleigh Dodds Science Publishing, 1518 Walnut Street, Suite 900, Philadelphia, PA 19102-3406, USA

First published 2020 by Burleigh Dodds Science Publishing Limited

Library of Congress Control Number: 2020931823

British Library Cataloguing in Publication Data
A catalogue record for this book is available from the British Library

ISBN 978-1-78676-356-3 (Print)
ISBN 978-1-78676-359-4 (PDF)
ISBN 978-1-78676-358-7 (ePub)
ISSN 2059-6936 (print)
ISSN 2059-6944 (online)

DOI 10.19103/AS.2020.0073

Typeset by Deanta Global Publishing Services, Dublin, Ireland

Contents

Series list

Title	Series number
Achieving sustainable cultivation of maize - Vol 1 From improved varieties to local applications Edited by: Dr Dave Watson, CGIAR Maize Research Program Manager, CIMMYT, Mexico	001
Achieving sustainable cultivation of maize - Vol 2 Cultivation techniques, pest and disease control Edited by: Dr Dave Watson, CGIAR Maize Research Program Manager, CIMMYT, Mexico	002
Achieving sustainable cultivation of rice - Vol 1 Breeding for higher yield and quality Edited by: Prof. Takuji Sasaki, Tokyo University of Agriculture, Japan	003
Achieving sustainable cultivation of rice - Vol 2 Cultivation, pest and disease management Edited by: Prof. Takuji Sasaki, Tokyo University of Agriculture, Japan	004
Achieving sustainable cultivation of wheat - Vol 1 Breeding, quality traits, pests and diseases Edited by: Prof. Peter Langridge, The University of Adelaide, Australia	005
Achieving sustainable cultivation of wheat - Vol 2 Cultivation techniques Edited by: Prof. Peter Langridge, The University of Adelaide, Australia	006
Achieving sustainable cultivation of tomatoes Edited by: Dr Autar Mattoo, USDA-ARS, USA & Prof. Avtar Handa, Purdue University, USA	007
Achieving sustainable production of milk - Vol 1 Milk composition, genetics and breeding Edited by: Dr Nico van Belzen, International Dairy Federation (IDF), Belgium	008
Achieving sustainable production of milk - Vol 2 Safety, quality and sustainability Edited by: Dr Nico van Belzen, International Dairy Federation (IDF), Belgium	009
Achieving sustainable production of milk - Vol 3 Dairy herd management and welfare Edited by: Prof. John Webster, University of Bristol, UK	010
Ensuring safety and quality in the production of beef - Vol 1 Safety Edited by: Prof. Gary Acuff, Texas A&M University, USA & Prof. James Dickson, Iowa State University, USA	011
Ensuring safety and quality in the production of beef - Vol 2 Quality Edited by: Prof. Michael Dikeman, Kansas State University, USA	012
Achieving sustainable production of poultry meat - Vol 1 Safety, quality and sustainability Edited by: Prof. Steven C. Ricke, University of Arkansas, USA	013
Achieving sustainable production of poultry meat - Vol 2 Breeding and nutrition Edited by: Prof. Todd Applegate, University of Georgia, USA	014

Introduction

With increasing concern about the environmental impact of synthetic pesticide use, including their impact on beneficial insects, the problem of insect resistance and the lack of new products, there has been an increasing interest in developing alternative biopesticides to control insect and other pests. This collection reviews the wealth of research on identifying, developing, assessing and improving the growing range of biopesticides. Part 1 reviews research on developing new biopesticides in such areas as screening new compounds, ways of assessing effectiveness in the field and improving regulatory approval processes. Part 2 summarises advances in different types of entomopathogenic biopesticide including entomopathogenic fungi and nematodes and the use of Bt genes in insect-resistant crops. The final part, Part 3, assesses the use of semiochemicals such as pheromones and allelochemicals, peptide-based and other natural substance-based biopesticides.

Part 1 General

The first chapter of the book begins with a discussion of improving methods for developing new microbial biopesticides. Chapter 1 highlights the role and economics of biopesticides, analyses the strategic framework for biopesticides development and provides an overview of the biopesticide development process. The chapter then focuses on microbial exploration and discovery and draws attention to the biological characterisation and development of biopesticides. A section on fermentation and formulation is also provided, focusing on various fermentation methods and a selection of formulations useful to biopesticide production. The chapter concludes by providing an overview of the future of biopesticides as an emerging industry as well as giving recommendations for future research.

Chapter 2 reviews the implementation of biopesticides as part of an integrated pest management (IPM) programme. It begins by highlighting the way biopesticides and other biocontrol agents can be integrated into IPM programmes. The chapter then discusses regulatory assessment of when it is safe and appropriate to use biopesticides and the use of biopesticides in different crop production systems. The chapter also provides a case study of the French government's programme to move its agricultural sector from reliance on synthetic pesticides to the use of more sustainable biocontrol techniques.

Moving on to the final chapter of Part 1, Chapter 3, the subject is improving regulatory approval processes for biopesticides and other new biological technologies in agriculture. The chapter begins by discussing how a regulatory framework for biopesticides can be established. It moves on to review the regulation of pesticides in the European Union (EU) and how the biopesticide sector and new regulatory requirements have been developed. The chapter also highlights the challenges faced in improving the regulatory framework for biopesticides and discusses current EU regulation and global national initiatives in biopesticides. The chapter concludes by reviewing how good regulatory practice can be developed.

Part 2 Microbial biopesticides, entomopathogenic nematodes and mites

Part 2 begins with Chapter 4, a discussion of the advances in the use of entomopathogenic fungi as biopesticides in suppressing crop pests. Hypocrealean ascomycetes have a set of properties such as their unique mode of action by direct penetration through the cuticle, potential for mass production, and their newly discovered ecological roles as endophytes and/or plant growth promoters which put them at the forefront of the global development of alternative control strategies. The chapter first addresses general aspects related to the natural occurrence and biodiversity, life cycle and delivery methods of the entomopathogenic fungi. It uses case studies to review the use of entomopathogenic fungi as biopesticides for the control of a wide range of insect and mite pests including locust and grasshoppers, soil dwelling insect pests, piercing and sucking insect and mite pests, stored-grain pests, forestry pests, invasive pests, and medical and veterinary pests. Finally, the advantages and limitations of entomopathogenic fungi as biopesticides together with a market overview are provided, with a concluding section indicating possible future directions in research on these fungi.

The next chapter, Chapter 5, addresses the advances in the use of entomopathogenic bacteria/microbial control agents (MCAs) as biopesticides in suppressing crop insect pests. The chapter first provides an overview of commercialised entomopathogenic bacteria and then discusses commercialised spore-forming bacterial entomopathogens. It also draws attention to non-spore forming entomopathogenic bacteria with potential for commercialisation and provides a section on microbial control agents (MCAs). A section on the advantages and drawbacks on the use of bacterial biopesticides is included before the chapter concludes by discussing research needs and emerging initiatives and providing future research trends in the subject.

Chapter 6 focuses on the advances in the use of Bt genes in insect-resistant crops. Insecticidal toxins derived from the soil bacterium Bacillus thuringiensis (Bt) have been widely used as microbial pesticides to selectively control juveniles of Lepidoptera, Coleoptera, Diptera and soil-dwelling nematodes. Synthetic genes derived from different strains of Bt have been used to produce genetically-modified (GM) plants resistant to insects. The chapter begins with an introduction to *Bacillus thuringiensis* (Bt) insecticidal toxins and moves on to discuss how Bt genes can incorporated into integrated pest management (IPM). It also reviews the importance of effective insecticide resistance programs before providing a section on potential areas for future research and sources for further information.

The subject of Chapter 7 is on plant growth-promoting bacteria (PGPBs) and how they can be used as biocontrol agents against invertebrate pests. Plant growth promoting bacteria (PGPBs) represent an important resource for agricultural crops, providing several benefits to the plant such as optimizing the use of environmental resources, improving plant health and resistance to biotic and abiotic factors, and directly acting against plant pathogens and invertebrate pests. The chapter begins by reviewing plant growth promotion and how plant growth-promoting bacteria (PGPBs) can be used against plant pathogens. It also addresses their potential against invertebrate pests and applications in agriculture and forestry. The chapter provides a case study on the soil-dwelling bacterium *Pseudomonas protegens* as a PGPB with dual action of biocontrol against both plant pathogens and invertebrate pests. It concludes by providing potential areas for future research.

Chapter 8 reviews advances in the use of entomopathogenic viruses as biopesticides in suppressing crop insect pests. It begins by discussing insect pests and their impact on the environment and how insect pest control has developed over time. It also highlights how viruses have been used as potential insect control agents. The chapter first analyses RNA viruses, specifically focusing on tetraviruses and reoviruses. It then looks at DNA viruses, focusing on Parvoviridae, Poxviridae, Nudiviridae and Baculoviridae. As baculoviridae is the most studied group of insect-specific viruses, the chapter also provides information on attributes of baculoviruses, an analysis of baculovirus genomics and molecular biology, a discussion on biopesticide development and reviews baculovirus production and formulation. It also provides a paragraph on the development of host-resistance to baculoviruses. The chapter concludes by providing an overview of how insect-specific viruses have potential for development as biological control agents of insect pests in integrated pest management (IPM) systems.

The next chapter, Chapter 9, discusses advances in the use of entomopathogenic nematodes (EPNs) as biopesticides in suppressing crop insect pests. Entomopathogenic nematodes have been commercialized as biopesticides since the 1980s. Since the 1990s, research on the application, biology, and ecology of these biocontrol agents has seen exponential growth, and since the mid-2000s basic research on EPNs has further expanded due to the use of the nematodes and their symbiotic bacteria as tractable model systems for many basic biological and ecological questions. The emphasis of the chapter is on the use of EPNs as biopesticides in a wide range of agricultural and other commodities. However, the chapter also provides overviews on EPN biology and ecology and mass production and application technology and interactions with other management tools as they are necessary for a better understanding of the strengths and limitations of EPNs as biopesticides. The chapter also provides a section on potential areas for future research.

Chapter 10, the final chapter of Part 2, reviews advances in the use of entomopathogenic oomycetes as biopesticides in suppressing crop insect pests. It begins by discussing how oomycetes have evolved from invertebrate pathogens to plant pathogens, what they are commonly known as today. The chapter also provides a case study on *Lagenidium giganteum* as a biocontrol agent against insects. The chapter moves on to review how the mode of infection can be established by entomopathogenic oomycetes, and how oomycetes can be used as sources of novel biopesticidal metabolites. It concludes by providing potential sources for further information and directs researchers to key conferences in the field.

Part 3 Natural substance-based biopesticides

The final part of the book begins by addressing the advances in the use of semiochemicals in integrated pest management, specifically focussing on pheromones. Semiochemicals enable intra- and interspecific chemical communication in insects. Among them, pheromones are species-specific and have been used in the management tactics for control of many insect pests for the last 50 years. During this time, important advancements have been made in understanding pheromones and their uses, particularly pheromone receptor neurons and pheromone inhibitors. Chapter 11 presents the latest developments of pheromones and their applications as key elements to specifically control insect pests within IPM programs. It also includes a section which provides researchers with potential sources for further information on the subject.

The next chapter, Chapter 12, discusses the possible use of allelochemicals in integrated pest management (IPM). Allelochemicals are produced and released into the environment by one species and have an effect on another species. They include attractants, repellents and toxic defence compounds which could provide valuable tools for use in IPM. The chapter begins by reviewing allelochemicals for pest management and also discusses botanical pesticides. Sections on repellents, attractants and defence activators are also provided. The chapter concludes by discussing the importance of overcoming barriers to improve the use of allelochemicals in pest management (IPM).

Chapter 13 reviews the use of peptides as novel biopesticides. Venom peptides are small toxic proteins originated from animal venoms. They have attracted great interest as a route to develop the next generation of novel, effective and safe biopesticides. The chapter begins by reviewing the characteristics and discovery of a range of venom peptides. It shows how inhibitory cysteine knot (ICK) type peptides, characterized by strong selectivity/specificity, stability and capacity for production at high volume, were used to develop the first commercialised peptide-based insecticide: SPEAR®. It also discusses other commercialised peptides and techniques to incorporate venom into plants to protect against insect and fungal attack.

The final chapter of the book, Chapter 14, reviews the development of plant-derived compounds as biopesticides. It begins by providing an overview of the legal framework and authorization of plant-derived plant protection products, then goes onto discuss how these products are developed. The chapter focuses on how plant-derived substances with activity against plant pathogens can be identified, as well as the efficacy of botanicals and novelty of invention and IP protection. It also highlights the importance using raw material of a high quality for plant-derived plant protection products. The chapter also provides a section on the extraction process that is used to produce extracts suitable for plant protection products. The chapter concludes by highlighting the growing interest in environmentally-friendly produced agricultural products, including plant-derived plant protection products, and also provides potential sources of further information on the subject.

Part 1

General

Chapter 1

Improving methods for developing new microbial biopesticides

Susan M. Boyetchko, Agriculture and Agri-Food Canada, Canada

1 Introduction

Biopesticides are attracting global attention as pest management tools that have a role in sustainable pest management. They are broadly defined as the deliberate use of living organisms, such as bacteria, fungi, viruses and nematodes, and their natural products (bioactive compounds produced as secondary metabolites) that directly or indirectly harm weeds, reduce the impact of plant diseases, or control the insect pests. Although the scientific community has been involved in biopesticide technology development for decades, it is only in recent years that it has been recognized as a viable emerging technology that is acceptable for mainstream use, rather than niche market products (Vincent et al., 2007; Bailey et al., 2009; Glare et al., 2012, 2016; Morán-Diez and Glare, 2016). They have generally been accepted as environmentally friendly technology because they are naturally occurring and thus considered attractive alternatives to synthetic pesticides. This is not to say that all naturally derived microbes are safer than chemicals because some of the mycotoxin-producing fungi can be considered very toxic to humans and animals.

2 The role of biopesticides

Several factors have precipitated the necessity for developing biopesticides as pest control strategies to synthetic pesticides. These include the deregistration or

http://dx.doi.org/10.19103/AS.2020.0073.02
Published by Burleigh Dodds Science Publishing Limited, 2020.

phasing out of older chemistries for agricultural and horticultural crop protection by the regulators and phasing out of numerous synthetic pesticides in lawns and gardens (Bailey et al., 2009; Boyetchko, 2017). Government legislation was enacted to ban synthetic pesticides for cosmetic use in urban municipalities in Canada through the Pest Management Regulatory Agency (PMRA). For example, Halifax, Nova Scotia, became the first Canadian city to ban chemicals within city limits in 2000 (Bailey et al., 2009). More than 100 urban municipalities in Canada followed suit, while the provinces of Ontario, Quebec, and Manitoba have ratified similar bans for cosmetic use of pesticides within urban municipalities. These regulations have thus generated a demand for biopesticides. However there are not sufficient biopesticide products to replace those that have been forbidden for use by the public and/or city employees. Therefore there is an even more urgent need to conduct future research for the discovery, development, and registration of new biopesticide products. The legislation accordingly supported policy changes within the Canadian regulatory body to register lower-risk pest control products. Biopesticides were thus classified under the Reduced-Risk Products Initiative (http://www.hc-sc.gc.ca/cps-spc/pest/index-eng.php; Bailey et al., 2009). This program helped expedite the registration of currently registered biopesticides in the United States which led to the harmonization of regulations with the US EPA and Canadian PMRA. During this time, there was a significant increase in PMRA registrations because registrants were allowed to undergo a pre-submission consultation process and the length of time it took to register biopesticides was reduced. In addition to the changes in federal legislation, the federal government created the Pest Management Centre (PMC) in 2003, an entity that helped in the delivery of new biological control products by jointly funding research to aid in their registration. PMC also provided support to encourage the adoption of these new pest management tools.

The general public has a greater awareness for environmentally friendly and safe products in agriculture and forestry, and for use in home gardens.(Ott et al., 1991; Magnusson and Cranfield, 2005; Bailey et al., 2009; Glare et al., 2012) . Therefore there is a greater expectation for organic and pesticide-free products. Agricultural practices such as the continued use of chemicals that pollute soil, water, and food supplies are putting more pressure on industry to provide suitable products for pest management and this practice has forced the industry to evaluate current pest management practices and how chemical pesticide loads have affected our environmental performance. Changes in consumer acceptance of biopesticide technology are thus driving the need for such products and are leading the initiatives into the development of 'green' technologies for food production in grocery stores and farmer's markets.

Another major issue of concern related to traditional chemical pest control products is the development of pesticide-resistance to chemicals making them ineffective to the pests they were intended to control (Boyetchko and Rosskopf,

2006; Lazarovits et al., 2007; Bailey, 2010; Bailey et al., 2009; Glare et al., 2016). Repeated use of synthetic pesticides has created this problem, but researchers and the industry involved in their manufacture were skeptical to believe that pesticide-resistance could occur until the scientific community reported that such problems were occurring. For example, there are numerous examples of (i) insecticide-resistance to pyrethroids, organophosphates, and carbamates with insects; (ii) herbicide-resistance to acetyl-CoA carboxylase (ACCase), acetolactate synthase (ALS), and 5-enolpyruvylshkimate-3-phosphate (EPSPS or glyphosate) with weeds; and (iii) fungicide-resistance to QoI and azole fungicides with plant pathogens (Hawkins et al., 2019).

3 The economics of biopesticides

The discovery of new synthetic chemicals with novel modes of action is becoming more expensive and difficult. It is estimated that their discovery is decreasing by 1–2% per annum, while the biopesticide industry is expected to grow by over $1 billion by 2010, and projections indicate the biopesticide market to exceed $4 billion by 2023 (Oerke, 2006; Thakore, 2006; Bailey et al., 2009; Glare et al., 2012). The cost to develop a new synthetic pesticide is more than $250 million and takes at least 10 years for a product launch (Oerke, 2006; Kelly and Allen, 2011; Glare et al., 2012). Conversely, it takes fewer microbial strains to discover a new biopesticide and can cost approximately $3–5 million to commercialize in the United States and take 3 years to get to market. In the EU, it typically takes 5–7 years to get a biopesticide to market and there are fewer biopesticide-active substances registered in the EU versus the United States, mainly due to greater complexity of EU-based biopesticide regulations (Balog et al., 2017).

In the past, the multinational companies did not invest in many biopesticide products. The commercialization of the majority of biopesticides was engaged by small to medium-sized companies (Bailey, 2010; Bailey et al., 2009; Boyetchko, 2017). Investment in infrastructure for the discovery of synthetic chemicals is expensive and facilities and equipment can be specialized for their discovery. Large firms invested in discovery and development of synthetically derived compounds whereas a change in infrastructure investment is required for microbial-based technology, which requires access to fermentation capabilities. Moreover, formulation of chemical pesticides may or may not be compatible with biopesticide technology. Therefore, small- to medium-sized enterprises (SMEs) were often responsible for biopesticide development. It was anticipated that multinational companies would tackle the screening of microorganisms and their secondary metabolites once they acquired small to medium-sized biopesticide companies (Glare et al., 2016). For example, Becker Underwood was acquired by BASF, and Bayer CropScience purchased

Agraquest and Prophyta, in addition, Syngenta purchased Valent BioSciences and Pasteuria BioScience. This investment by multinationals demonstrated their recognition of the growing value of the biopesticide market. Nonetheless, multinationals have the resources to devote to R&D however, their focus has been on major crops such as corn, soybean, rice, potato, and cotton being the main emphasis. Other markets that remain overlooked are pulse crops, small and large fruit crops and high value field and greenhouse vegetable crops and other horticultural crops. This therefore creates an opportunity for biopesticides to fill this niche.

4 Strategic framework for biopesticide development

One of the major challenges to the development and commercialization of biopesticide products is how to manufacture them at an affordable cost and in a consistent manner. Fermentation systems, formulation technology, and spray application or delivery of biopesticides has often been ignored in the R&D phase by researchers. It should be borne in mind that the initial screening of microbial organisms for biopesticides represents the first step in the discovery phase. The perception of finding a suitable organism that effectively controls the target pest by senior managers and the general public unfamiliar with the complexities of biopesticide development often assumes that once the organism has been discovered, that is the end of the process. However, the microbe must be mass-produced through fermentation systems, subsequently formulated into a stable form and applied using spray application systems to facilitate their delivery to the soil, seed, and/or foliage (Boyetchko and Peng, 2004; Hynes and Boyetchko, 2006). Developing a suitable fermentation system requires an understanding of the microbial physiology of the organism and a suitable formulation often involves the knowledge of the microbe's mode of action in order to facilitate delivery of the biopesticide. Contrary to the popular belief that once a microbial organism has been discovered with biopesticidal properties in the laboratory, the final biopesticide product is imminent, a biopesticide product is one where all the processes related to its manufacture are considered, including platform technologies related to fermentation, formulation, and spray application (Fig. 1; Boyetchko and Peng, 2004). The taxonomy, biological characterization, mode of action, and efficacy are among the factors that are core to the selection of the biopesticide organism. The platform technologies are interrelated and closely linked and their complexities have been underestimated (Boyetchko and Peng, 2004; Boyetchko et al., 2007; Bailey, 2014). They are integral factors to consider with the biopesticide product development. The goal is to find 'nature's best' with respect to the microbe selected for biological control. However, fermentation or formulation may not necessarily improve a mediocre microbe, but a superior microbial agent may

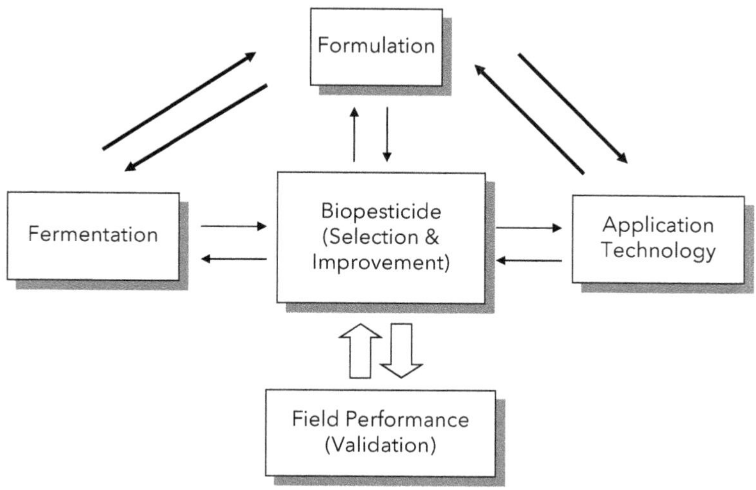

Figure 1 Strategic framework for evaluation and development of biopesticides. Source: adapted from Boyetchko and Peng (2004).

have to be sacrificed if it cannot be readily grown in suitable fermentation media or it cannot be readily formulated. Compromises are sometimes made in the selection process.

The earlier prototype models of biopesticides were manufactured using simple formulations but the intricacies between the platform technologies are much more complicated (Auld and Morin, 1995; Fravel, 1999; Auld et al., 2003). Lack of appropriate formulation may have led to ineffectual or inconsistent field performance and fermentation of the microbial organism may be too expensive or scale-up unacceptable that the product commercialization was abandoned which led to an 'orphaned' biopesticide product (Butt et al., 2001; Hallett, 2005; Hynes and Boyetchko, 2006).

4.1 The process of biopesticide development

Biopesticide discovery and development follows a process of incremental steps that are unique for each target pest-biopesticide system. Different research or industry groups follow different stepwise protocols to facilitate in the selection of promising biopesticide candidates (Boyetchko and Svircev, 2009; Bailey et al., 2009; Bailey, 2010; Köhl et al., 2011; Ravensberg, 2011; Glare et al., 2012). The 'Biopesticide Innovation Chain' (Fig. 2) was developed in response to a need for several researchers in the disciplines of weed science, entomology, and plant pathology to follow a similar blueprint to develop microbial biopesticides. The innovation chain depicts nine critical stages for developing a biopesticide product using a series of 'Go vs. No-Go' criteria in order to make decisions on

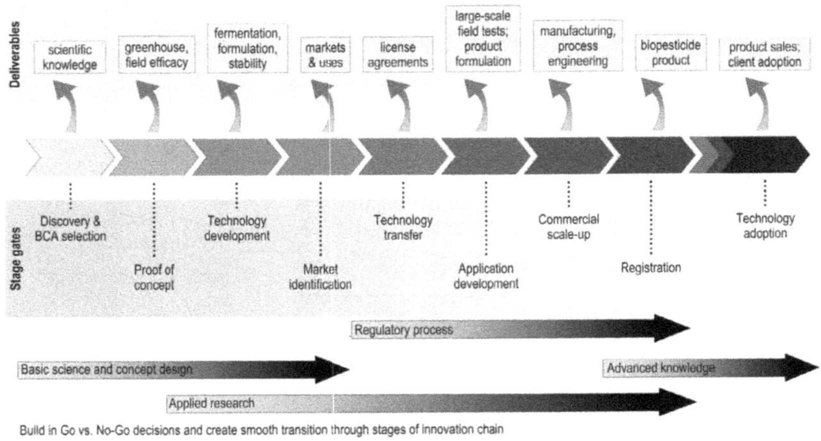

Figure 2 Solution for delivery of biopesticides: AAFC biopesticide science innovation chain. Source: adapted from Boyetchko and Svircev (2009).

the feasibility of the organisms and the target pest. This concept is applicable to any type of biopesticide (i.e. bacteria, fungi, viruses, natural products) and is appropriate for any type of crop pest (i.e. weed, invertebrate/insect pest, plant pathogen). The stages of the biopesticide innovation chain include (i) discovery and selection; (ii) proof of concept; (iii) technology development; (iv) market identification; (v) technology transfer; (vi) application development; (vii) commercial scale-up; (viii) registration; and (ix) technology adoption. The innovation chain describes the process for developing a biopesticide without considering the scientific terminology specific to each crop protection discipline; everyone will be describing the process using similar terminology. This road-map forces the researcher to proceed to the subsequent steps while not being fixated on one or two stages, thus preventing him/her from advancing toward the end goal: developing a biopesticide.

The early stages of discovery involve characterization of the biopesticide agent, assessing growth characteristics, basic understanding of the biology, genomics, and initial in vitro and in vivo screening bioassays that measure efficacy on the target pest (Bailey et al., 2009; Bailey, 2010). Proof of concept entails evaluation of biopesticide performance on live plants under controlled-environmental conditions where light, temperature, and relative humidity/moisture conditions can be monitored. It is then essential to bring the evaluation to the field, where possible, in order to demonstrate the adequacy of the biopesticide performance. Technology development involves the bioprocesses of fermentation/mass-production, formulation, and application technology for delivery of an efficacious, stable biopesticide with qualities that contribute to an 'adequate' shelf life for a microbe that is easy to grow and deliver to the target pest. As previously indicated, these platform technologies that are critical to biopesticide development are often linked. Recent advances have been publicly available, but, unfortunately, industry has often kept fermentation and formulation as proprietary or trade secret information. Several fermentation and formulation may tested before a final process is selected (Bailey, 2014).

The importance of market considerations cannot be underestimated since they will dictate the success or failure for commercialization. Although this area is not necessarily a research consideration to scientists, it is important because it defines the field of use and the market the biopesticide is being developed for. It will pre-determine the experimentation and scientific data required to register the product. For example, will the biopesticide be developed for use in the horticultural or greenhouse market, in large acreage field crops, or for use in forestry or for the home gardener? Another key feature of the innovation chain is that it encourages additional applications and crop pests into other potential use patterns, thus broadening to other target pests and production systems (Boyetchko and Rosskopf, 2006; Hynes and Boyetchko, 2006). A great deal of understanding of the microbial physiology and biochemistry is also required, but the business side of the research is an essential factor.

The later stages in the innovation chain (i.e. technology development, technology transfer) test the robustness of the earlier decisions by actively working with various stakeholders involved in the development of biopesticides. These include the industry partner, other collaborators, and regulators necessary to develop the data required for the registration package. Moreover, involvement of additional stakeholders generates 'buy-in' of the new technology which will help with the adoption of the biopesticide into the field of use. For example, the innovation chain can be used as a tool for the training of farmers and crop protection consultants in the safe and effective use of biopesticides. Moving into commercial scale-up can still affect the final stages of product development and is more often led by industry, often requiring significant investment back into technology development to registration, and technology adoption, which is usually directed by the industry partner and extension agrologists. The champion of the technology is usually the lead scientist or inventor who should work collaboratively with industry until the technology adoption has occurred. The corporate memory or 'know-how' and intuition of the lead scientist, along with his/her scientific expertise, can ease in the transition phases toward commercialization. Do not assume that the technology will be successful once it has been handed off to industry, without input from the scientist who originally discovered the biopesticide. The lead scientist can provide guidance into the overall idea and direction of the scientific discovery, along with a number of unsuccessful trials that had been attempted but failed during the process of developing the biopesticide.

5 Microbial exploration and discovery

If we are to identify the 'nature's best', a systematic approach is essential during the exploration and discovery phase to thoroughly evaluate the microbial biodiversity. This biodiversity provides excellent opportunities for finding microbial strains with suitable traits for biocontrol potential (Boyetchko and Peng, 2004).

Numerous discoveries of biopesticides have taken place in university and government institutions resulting in the licensing of the technology to industry. Many of these institutions may be in possession of undervalued but extensive microbial culture collections which can be exploited for the discovery of future biopesticide candidates. To add value to these collections, it is essential to conduct additional mass-screening of these same microbes against additional target pests, whether they are for insecticidal, fungicidal, or herbicidal activity using established in vitro and in vivo screening protocols. Over the years, a researcher can accumulate numerous future biopesticide candidates which can be stored under −80°C conditions until required. Future collections can thus access potential microbes based on the taxonomy of related pests they are known to be inhibitory to. Different strains to the same genus species may perform differently from one another and should not be readily discarded. Glare et al. (2012) described how Trichobank™, a culture collection of over 2000 isolates of 21 *Trichoderma* spp. has been effectively utilized to select for promising biopesticide candidates with various growth characteristics for control of several soil-borne diseases. Similarly, Bailey et al. (2009) explained their philosophy behind screening for microorganisms as bioherbicides. Once the target weed is selected, several hundreds to thousands of fungi and bacteria that have been collected through surveys were isolated from a broad range of environmental conditions and evaluated to select those microbes that possess a list of desirable characteristics. Therefore, a systematic approach for selection of appropriate biopesticide candidates is conducted. For example, the fungal pathogen, *Phoma macrostoma*, was initially discovered to be a foliar pathogen, but its bioheribicidal potential to control Canada thistle and chickweed was only discovered when it was found to be effective when applied to the soil (Bailey, 2014). The potential for *P. macrostoma* as a bioheribicide was only discovered when the researchers understood what was the best method to apply the fungal pathogen and what was its mode of action. No longer are potential biopesticide candidates developed by a fortuitous discovery, as was the case with *Colletotrichum gloeosporioides* f.sp. *malvae* to control round-leaved mallow (Boyetchko et al., 2007; Bailey, 2010). Experienced researchers are developing their own road-map and selection process for screening biopesticide organisms.

Endophytic bacteria and fungi found in vascular plants are an underexplored bioresource as promising biological control agents (Vega,

2008; Glare et al., 2012; Eljounaidi et al., 2016; Rybakova et al., 2016; De Silva et al., 2019). Endophytes grow inside the plant for part or all of its life cycle but do not cause visible symptoms. Almost all vascular plants examined to date have been found to harbor bacterial and/or fungal endophytes. It is believed that many endophytic microbes may have originated in the rhizosphere and phylloplane and they gained entry through natural openings or plant wounds. Due to the diversity of species discovered to be endophytic, it is postulated that they may prove to be very suitable biopesticides and plant-growth promoters. Endophytes can be introduced as biocontrol agents through seeds and propagated plant material such as those introduced via tissue culture (Eljounaidi et al., 2016). Once endophytes enter the plant, they will not be subjected to environmental stresses that many other nonendophytic biocontrol agents can succumb to. This may mitigate the need to develop formulations to protect them from unfavorable environmental conditions. Some endophytes may demonstrate their biocontrol potential by secreting antifungal and antibacterial metabolites, thereby inhibiting the competition of pathogens, or they may exhibit mycoparasitic activity (i.e. parasitism of one microorganism by another microorganism) (de Silva et al., 2019). Those endophytic microbes that are entomopathogenic are surmised to act as a feeding deterrent or by antibiosis. Commonly isolated bacteria that can become endophytic belong to the genera *Bacillus, Pseudomonas, Agrobacterium, Paenibacillus* and *Enterobacter.* Common endophytic fungal microbes belong to *Fusarium oxysporum, Trichoderma* spp., *Colletotrichum gloeosporioides*, and *Phomopsis* spp.

Many biopesticide products were developed because they targeted a single major pest (Glare et al., 2012; Bailey, 2014; Boyetchko, 2017). Narrow host-range possessed by these biopesticides was initially considered an asset, if not essential. A narrow host-range contributes to environmental safety and limits the risk of a biopesticide affecting nontarget crops and other beneficial microbiota and microfauna. This trait has also restricted its success to a niche market product. However, it was later considered a hurdle to commercialization because industry was hesitant to invest in technology with such a small market potential. It was recognized that it is important to select a product that has activity against multiple pests in order for industry to get a return on their investment. Moreover, it is not practical or economical to develop a crop protection product for use by farmers on a single target pest when agriculture and forestry ecosystems are comprised of multiple crop pests. For example, bioinsecticide products containing *Bacillus thuringiensis* have had remarkable success in the marketplace because they kill several invertebrate pest species (Côté, 2007). Similarly, the fungus *Chondrostereum purpureum* was developed as a broad-spectrum bioherbicide to control a number of woody tree species and shrubs for vegetative management in forests and riparian lands and utility-rights-of-way (Hintz, 2007). Control of a broad spectrum of plant diseases is also possible with the commercial product Serenade®, hence adding to its commercial value (Marrone, 1999, 2002).

5.1 Biological characterization and development

The discovery of new and promising microbial candidates as active ingredients for biopesticides has far out-paced the knowledge and related technology required to bring them to commercialization. Although there are numerous biopesticides registered for use globally (Vincent et al., 2007), including over 100 biopesticide-active ingredients in the United States and more than four different microbial-based active ingredients registered in Canada since 1972 (Fravel, 1999; Hynes and Boyetchko, 2006; Bailey et al., 2009; Kabaluk and Gazdik, 2011), the public still wonders why more biopesticides are not commercially available. It has been pointed out that scientific efforts in biological control have not contributed to the lack of commercial success of biopesticides, but that researchers have failed to focus on resolving critical technical elements such as formulation and methods for stabilization of living organisms (Zorner et al., 1993). It is true that a great deal of research effort has focused on the search for microbial agents with pest control potential, but technological challenges that include economical methods for mass production and scale-up, viable formulation and application strategies, and shelf life have been encountered and have contributed to the lack of commercial success (Auld and Morin, 1995; Boyetchko and Peng, 2004). However, the long-term commitment toward biological control has been predominantly supported by public research institutions such as government and universities (Boyetchko, 2017), while investment by industry to commercialize this technology has been sorely lacking. Access to public funding has been one constraint to this area; biopesticide research requires long-term commitment to funding, while many funding agencies provide financial support for projects that are 3–4 years in duration. In addition, the teams of researchers dedicated in understanding the fundamentals of pest biology and ecology and the agronomists necessary to implement pest management practices, have not been active participants in the application and implementation of biopesticide technology. It should be recognized that the success of biopesticides will depend on the involvement of multidisciplinary research teams consisting of crop protection specialists in either entomology, weed science, or plant pathology, microbiologists, natural products chemists, agronomists, molecular biologists, and economists with a vested interest in biopesticide adoption (Fig. 3). Biopesticides should not be considered a stand-alone pest management tool. A holistic approach using various tools for controlling pests can often lead to an additive or

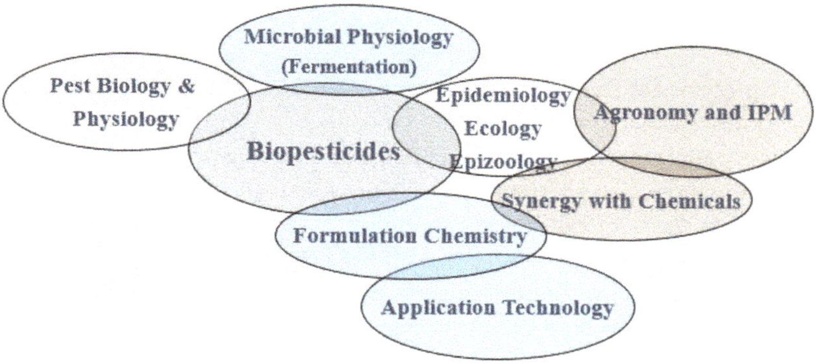

Figure 3 Use of integrated pest management (IPM) tools will lead to successful adoption of biopesticides.

synergistic effect of the pest control. Using a single method for controlling a pest may eventually lead to the pest control method becoming effective or the pest developing resistance to the practice or product. Different aspects of biopesticide discovery, including microbial physiology and genomics, target pest biology, ecology and knowledge of farming systems are also essential.

Biological characterization of the biopesticide involves knowledge of its taxonomic designation to genus and species. Early identification alerts the researchers to the general traits that the microbe may possess, including how to grow it in culture. Identification to species level is required for registration and increasingly involves molecular characterization, in addition to traditional cultural and morphological studies. The nomenclature of the potential biopesticide provides the researcher a heads-up if they are dealing with a potential human or animal pathogen, and thus alerts them to avoid further evaluation of a microbe that the regulators will not consent to develop into a future biopesticide product (Bailey et al., 2009). Early biological assessments involve extensive efficacy testing under controlled environments and in the field, assessing consistency of response to the target pest. Specific data requirements included in the assessment are (i) the origin of microorganism, (ii) derivation and identification, (iii) biological properties, (iv) manufacturing methods, (v) estimate of potency and product guarantee, (vi) human health and safety, (vii) environmental fate, and toxicology, and (viii) efficacy (US EPA does not require this) (Bailey et al., 2009; Bailey, 2014).

6 Fermentation and formulation

Some of the technological challenges predominantly encountered in the development of successful biopesticides include (i) effective methods for scale-up mass production, (ii) formulations and (iii) application technology

(Auld and Morin 1995; Boyetchko and Peng, 2004; Boyetchko and Rosskopf, 2006). Cost-effective fermentation methods have been one reason cited for not pursuing commercialization of a biopesticide product. The microbe must be mass produced in large quantities as viable, highly efficacious and genetically stable propagules in a cost-effective manner. Submerged (liquid) fermentation has been well-developed for the production of bacteria and some filamentous fungi and appears to be highly economical (Stowell, 1991; Rosskopf et al., 1999; Wraight et al., 2001). Solid-substrate fermentation is highly applicable for mass production of fungi that do not produce spores or other infective propagules readily in liquid culture. However, it is not considered as economical due to other challenges such as higher labor costs, spore harvest, maintenance of sterile conditions and inability to control cultural conditions, including temperature and moisture (Churchill, 1982). A bi-phasic system has been employed for some fungi by using liquid fermentation to generate mycelium followed by sporulation using shallow trays that are exposed to various light cycles, depending on the requirement of the microbial agent (Rosskopf et al., 1999). Further, optimization of microbial biomass production requires that a suitable nutritional medium be developed (Wraight et al., 2001). A variety of nutrients from carbon and nitrogen sources, trace elements, and carbon-to-nitrogen ratios will affect propagule yield, stability, and ultimately biopesticidal activity. Often, these elements must be custom-designed on an individual basis for each biopesticide agent and low-cost ingredients substituted for the lab-grade ingredients.

Early formulations of biopesticides were comprised of simple ingredients such as surfactants which facilitated the uniform distribution of the biopesticide across the target surface. However, the lack of future biopesticides were partially attributed to the lack of suitable formulations and resulted in several orphaned biopesticide technologies (Boyetchko et al., 2005; Hallett, 2005; Hynes and Boyetchko, 2006). It was assumed that the industry could fill the void in knowledge for formulation technology, but the ingredients available for use in chemical pesticide formulation development were not always compatible with the living organisms used in biopesticides. This resulted in products that were prematurely considered as failures. The fact that biopesticides are not analogous to chemical pesticides, lacking in some features of chemicals, has perhaps led to the opinion by some critics that this technology has failed to deliver the goods.

Availability and selection of suitable formulations that address critical issues of product stability and shelf life as well as environmental factors including desiccation, temperature, and availability of high humidity or free water, will have a major impact on the field performance of the biopesticide (Auld et al., 2003; Boyetchko et al., 2005; Burges, 1998). Formulation ingredients can be used to reduce the rate of evaporation and/or improve the rate of infection of the biopesticide and increase the moisture-retaining properties. Non-ionic surfactants such as Tween are commonly used-in formulations as wetting agents while humectants, oil emulsions and hydrophilic polymers have been used to alleviate the impact of desiccation and reduce the dew period requirement (Auld et al., 2003). Formulations can also enhance the stability, shelf life, and survival of the microbial agent. Germinated fungal spores are also sensitive to desiccation and UV irradiation. Under ideal conditions, a formulation acts as a buffer against environmental extremes that are encountered by the biopesticide agent, thus promoting disease development on the target pest. This holds true for soil-applied microbial inoculants as well. Ideally, shelf life of a microbial-based product should be at least one to two years, particularly at room temperature (Auld et al., 2003; Wraight et al., 2001), allowing for production, transport, and distribution of the product. Stabilization of microbial agents in order to extend shelf life has been a major challenge for many biopesticides, yet more efforts in formulation have focused on the environmental constraints, especially on reducing dew period requirements. Finally, the final product must be formulated in a manner that is safe and easy to use with existing application technology and equipment.

7 Conclusion and future trends

The future of biopesticides as an emerging industry and as a viable tool for sustainable crop production is very promising. We must encourage researchers, industry, and the regulatory bodies to work together to create a smooth transition from basic scientific discovery, to application, development, scale-up, registration and adoption. Having an effective roadmap to guide researchers through the process is critical. This forces them through the various stages of biopesticide development (Boyetchko and Svircev, 2009; Bailey et al., 2009; Bailey, 2010)

It is easy for researchers to feel overwhelmed by the complexities of investing in biopesticide research. However, it takes a holistic and multidisciplinary approach and a team of scientists with complementary expertise in entomology/plant pathology/weed science, microbiology, pest biology and ecology/epidemiology/epizoology, natural products chemistry, molecular biology (plant and microbial genomics), and economics. Weak links for improving biopesticides include lack of knowledge of pest biology, physiology, and ecology. Fundamental knowledge of agronomic concepts for the application of biopesticides into crop production is very essential for improving adoption of the technology - inclusion of agrologists for demonstration and implementation of biopesticides will likely speed up acceptance of its importance as a pest management tool (Boyetchko and Rosskopf, 2006).

Critics and sceptics alike could claim that the lack of more biopesticides in the marketplace is the proof that they are not a viable technology. However, it can be argued that there is a need for more research dollars into publicly funded institutions to advance the development of biopesticide technology. Lack of investment and diminishing resources for research have resulted in less rapid progress in this area. Long-term funding is essential, while 3–4 years financial support only allows researchers to conduct early screening for biopesticide candidates before progressing to detailed biological assessments into the mode of action, natural products chemistry and development of formulation and fermentation that will prove the applicability of the technology. As industry becomes more aware of the promise of biopesticides, partnerships with academic scholars and researchers should result in more rapid advancement in the area.

Industry partners must be willing to take the risks of commercializing an under-developed and unfamiliar technology and work closely with the researchers in academic and government institutions. Scientists may be rudely awakened by the realities of commercializing the technology and what economic costs and marketing factors are required to make biopesticides a reality. But these partnerships appear to be more essential because it will lead to more success and progress toward biopesticide technology.

Choice of the target pest is very important when attempting to attract interest by industry to invest in the technology. Selection of target pest for control in a high-value market or of significance over a large geographic region is always preferred. In addition, the biopesticide candidate with a broad spectrum activity will make the product more appealing to industry to commercialize. Consider the business side of science and how to make decisions to the selection process of the target (what is the market size for resolving the problem? how many acres are affected? is it a global pest or is it restricted to a particular region? is it a high-value crop?). The industry has access to market and economic assessments by economists within their own company structure and the information may be very fruitful for attracting funding.

The perception that biopesticides are replacements for synthetic pesticides and that they are silver bullets leads to unrealistic expectations for the performance of biopesticides. Scientists need to explore how biopesticides fit into an integrated pest management (IPM) program where we utilize several pest management tools. Using more than one pest control option can delay the onset of pesticide-resistance, including possible resistance occurring to biopesticides. We need to resist the urge to use simplistic crop management tools and be judicious in our use of biopesticides, as well as synthetic pesticides and breeding for resistance to plant diseases and insect pests. Therefore, inclusion of agronomists and other field specialists on research teams is important in order to put biopesticides into practice and make them a reality. Pesticide-resistance to several insect pests, weeds, and plant pathogens is a problem that continues unless we deal with it by developing a number of pest management tools and implement them in a thoughtful manner. Development of resistance to biopesticides has been reported with *Bt* spp. in diamondback moth (Tabashnik et al., 1990), with entomopathogenic fungi on greater wax moth (Serebrov et al., 2006), and with viral biopesticdes such as granulovirus (CpGV) on codling moth (Sauer et al., 2017). Excessive use of any one pest management tool, including biopesticides, can lead to the development of resistance. Therefore, integrated use of a variety of pest management tools is always a desirable approach.

Glare et al. (2012) asked the question 'Have biopesticides come of age?'. The answer is emphatically Yes! First, the general public has become much more aware and educated about the technology. Food safety with regards to chemical pesticide residues and environmental health issues related to soil and water quality and contamination are significant social factors driving the technology. Acquisition of small- to medium-sized biopesticide companies by multinationals is improving R&D investment into biopesticide products. Moreover, government legislative changes that are affecting the synthetic pesticide industry are generating an urgent need to create more environmentally friendly products. The future of biopesticides is reaching a reality, but all stakeholders must be united and work toward a common goal. Simplicity and ease-of-use of chemicals has often been used as a quick-fix for pest control, but we all have an obligation as environmental stewards to utilize chemicals and other pest-control strategies judiciously and responsibly.

8 References

Ash, G. J. 2010. The science, art, and business of successful bioherbicides. *Biological Control* 52(3), 230–40. doi:10.1016/j.biocontrol.2009.08.007.
Auld, B. A. and Morin, L. 1995. Constraints in the development of bioherbicides. *Weed Technology* 9(3), 638–52. doi:10.1017/S0890037X00023964.
Auld, B. A., Hetherginton, S. D. and Smith, H. E. 2003. Advances in bioherbicide formulation. *Weed Biology and Management* 3(2), 61–7. doi:10.1046/j.1445-6664.2003.00086.x.
Backman, P. A. and Sikora, R. A. 2008. Endophytes: an emerging tool for biological control. *Biological Control* 46(1), 1–3. doi:10.1016/j.biocontrol.2008.03.009.
Bailey, K. L. 2010. Canadian innovations in microbial biopesticides. *Canadian Journal of Plant Pathology* 32(2), 113–21. doi:10.1080/07060661.2010.484195.
Bailey, K. L. 2014. The bioherbicide approach to weed control using plant pathogens. In: Abrol, D. P. (Ed.), *Integrated Pest Management*. Elsevier, Inc, pp. 245–65.
Bailey, K. L., Boyetchko, S. M., Peng, G., Hynes, R. K., Taylor, W. G. and Pitt, W. M. 2009. Developing weed control technologies with fungi. In: Rai, M. (Ed.), *Advances in Fungal Biotechnology*. I.K. International Pvt. Ltd., New Delhi, pp. 1–44.
Balog, A., Hartel, T., Loxdale, H. D. and Wilson, K. 2017. Differences in the progress of the biopesticide revolution between the EU and other major crop-growing regions. *Pest Management Science* 73(11), 2203–8. doi:10.1002/ps.4596.

Boyetchko, S. M. 2005. Biological herbicides in the future. In: Ivany, J. A. (Ed.), *Weed Management in Transition. Topics in Canadian Weed Science* (vol. 2). Canadian Weed Science Society - Societe Canadienne De Malherbologie, Sainte-Anne-de-Bellevue, Quebec, pp. 29–47.

Boyetchko, S. M. 2017. The use of microorganisms in integrated weed management. In: Zimdahl, R. (Ed.), *Integrated Weed Management for Sustainable Agriculture*. Burleigh Dodds Science Publishing, Cambridge, UK, pp. 401–16. doi:10.19103/AS.2017.0025.21.

Boyetchko, S. M. and Peng, G. 2004. Challenges and strategies for development of mycoherbicides. In: Arora, D. K., Bridge, P. and Bhatnagar, D. (Eds), *Fungal Biotechnology in Agricultural, Food, and Environmental Applications* (vol. 21). Marcel Dekker Inc., pp. 111–21.

Boyetchko, S. M. and Rosskopf, E. N. 2006. Strategies for developing bioherbicides for sustainable weed management. In: Singh, H. P., Batish, D. R. and Kohli, R. K. (Eds), *Handbook of Sustainable Weed Management*. The Haworth Press, Inc., New York, pp. 393–430.

Boyetchko, S. M. and Svircev, A. M. 2009. Biopesticides: strategies for discovery, development, and adoption. AAFC 10733, Cat. No. A52-120/2009E-PDF, ISBN 978-1-100-11640-2.

Boyetchko, S. M., Bailey, K. L., Hynes, R. K. and Peng, G. 2007. Development of an inundative mycoherbicide: BioMal®. In: Vincent,C., Goettel, M. S. and Lazarovits, G. (Eds), *Biological Control: A Global Perspective*. CABI Publishing, Wallingford, UK, pp. 274–83.

Burges, H. D. (Ed.). 1998. *Formulation of Microbial Biopesticides: Beneficial Microorganisms, Nematodes, and Seed Treatments*. Kluwer Academic Publishers, Dordrecht, the Netherlands.

Butt, T. M., Jackson, C. and Magan, N. 2001. Introduction - fungal biological control agents: progress, problelms and potential. In: Butt, T. M., Jackson, C. and Magan, N. (Eds), *Fungi as Biocontrol Agents: Progress, Problems, and Potential*. CABI Publishing, Wallingford, UK, pp. 1–8.

Churchill, B. W. 1982. Mass production of microorganisms for biological control. In: Charudattan, R. and Walker, H. L. (Eds), *Biological Control of Weeds with Plant Pathogens*. Wiley, New York, NY, pp. 209–22.

Côté, J.-C., Otvos, I., Schwartz, J.-L. et al. (Eds) 2007. *Proceedings of the 6th Pacific Rim Conference on the Biotechnology of Bacillus thuringiensis and its Environmental Impact*, Biocontrol Network of Canada, Montreal, Canada.

De Silva, N. I., Brooks, S., Lumyong, S. and Hyde, K. D. 2019. Use of endophytes as biocontrol agents. *Fungal Biology Reviews* 33(2), 133–48. doi:10.1016/j.fbr.2018.10.001.

Eljounaidi, K., Lee, S. K. and Bae, H. 2016. Bacterial endophytes as potential biocontrol agents of vascular wilt diseases - review and future prospects. *Biological Control* 103, 62–8. doi:10.1016/j.biocontrol.2016.07.013.

Fravel, D. 1999. Hurdles and bottlenecks on the road to biocontrol of plant pathogens. *Australasian Plant Pathology* 28(1), 53–6. doi:10.1071/AP99007.

Glare, T., Caradus, J., Gelernter, W., Jackson, T., Keyhani, N., Köhl, J., Marrone, P., Morin, L. and Stewart, A. 2012. Have biopesticides come of age? *Trends in Biotechnology* 30(5), 250–8. doi:10.1016/j.tibtech.2012.01.003.

Glare, T. R., Gwynn, R. L., Morán-Diez, M. E. 2016. Development of biopesticides and future opportunities. In: Glare, T. R. and Morán-Diez, M. E. (Eds), *Microbial-Based Biopesticides: Methods and Protocols, Methods in Molecular Biology* (vol. 1477). Springer Science+Business Media, New York, NY, pp. 211–21.

Hallett, S. G. 2005. Where are the bioherbicides? *Weed Science* 53(3), 404–15. doi:10.1614/WS-04-157R2.

Hawkins, N. J., Bass, C., Dixon, A. and Neve, P. 2019. The evolutionary origins of pesticide resistance. *Biological Reviews* 94(1), 135–55. doi:10.1111/brv.12440.

Hintz, W. 2007. Development of *Chondrostereum purpureum* as a mycoherbicide for deciduous brush control. In: Goettel, C. and Lazarovits, G. (Eds), *Biological Control: a Global Perspective Vincent*. CABI Publishing, Wallingford, UK, pp. 284–90.

Hubbard, M., Hynes, R. K., Erlandson, M. and Bailey, K. L. 2015. The biochemistry behind biopesticide efficacy. *Sustainable Chemical Processes* 2, 18–25.

Hynes, R. K. and Boyetchko, S. M. 2006. Research initiatives in the art and science of biopesticide formulations. *Soil Biology and Biochemistry* 38(4), 845–9. doi:10.1016/j.soilbio.2005.07.003.

Kabaluk, T. and Gazdik, K. 2011. *Directory of Biopesticides for Agricultural Crops in OECD Countries*, Agriculture and Agri-Food Canada, Ottawa, Canada.

Kelly, I. D. and Allen, R. 2011. An industry perspective on challenges and hurdles faced in the development of agrochemicals. In: Harker, K. N. (Ed.), *The Politics of Weeds. Topics in Canadian Weed Science* (vol. 7). Canadian Weed Science Society - Societe Canadienne De Malherbologie, Pinawa, Manitoba, pp. 3–12.

Köhl, J., Postma, J., Nicot, P., Ruocco, M. and Blum, B. 2011. Stepwise screening of microorganisms for commercial use in biological control of plant pathogenic bacteria. *Biological Control* 57(1), 1–12. doi:10.1016/j.biocontrol.2010.12.004.

Lazarovits, G., Goettel, M. S. and Vincent, C. 2007. Adventures in biocontrol. In: Vincent, C., Goettel, M. S. and Lazarovits, G. (Eds), *Biological Control: a Global Perspective*. CABI Publishing, Wallingford, UK, pp. 1–6.

Magnusson, E. and Cranfield, J. A. L. 2005. Consumer demand for pesticide free food products in Canada: a probit analysis. *Canadian Journal of Agricultural Economics/Revue Canadienne d'Agroeconomie* 53(1), 67–81. doi:10.1111/j.1744-7976.2005.00354.x.

Marrone, P. G. 1999. Microbial pesticides and natural products as alternatives. *Outlook on Agriculture* 28(3), 149–54. doi:10.5367/000000099101292917.

Marrone, P. G. 2002. An effective biofungicide with novel modes of action. *Pesticide Outlook* 13(5), 193–4. doi:10.1039/b209431m.

Morán-Diez, M. E. and Glare, T. 2016. What are microbial-base biopesticides? In: Glare, T. R. and Diez-Morán, M. E. (Eds), *Microbial-Based Biopesticides: Methods and Protocols, Methods in Molecular Biology* (vol. 1477). Springer Science+Business Media, New York, NY, pp. 1–10.

Oerke, E.-C. 2006. Crop losses to pests. *Journal of Agricultural Science* 144(1), 31–43. doi:10.1017/S0021859605005708.

Ott, S. L., Huang, C. L. and Misra, S. K. 1991. Consumer's perceptions of risks from pesticide residues and demand for certification of residue-free produce. In: Caswell, J. A. (Ed.), *Economics of Food Safety*. Elsevier Science Publishing Co., Amsterdam, the Netherlands, pp. 175–87.

Ravensberg, W. J. 2011. *A Roadmap to the Successful Development and Commercialization of Microbial Pest Control Products for Control of Arthropods*. Springer, New York, NY.

Rosskopf, E. N., Charudattan, R. and Kadir, J. B. 1999. Use of plant pathogens in weed control. In: Bellows, T. S. and Fisher, T. W. (Eds), *Handbook of Biological Control*. Academic Press, New York, NY, pp. 891–918.

Rybakova, D., Cernava, T., Köberi, M., Liebminger, S., Etemadi, M. and Berg, G. 2016. Endophytes-assisted biocontrol: novel insights in ecology and the mode of action of *Paenibacillus*. *Plant and Soil* 405(1–2), 125–40. doi:10.1007/s11104-015-2526-1.

Sauer, A. J., Schulse-Bopp, S., Fritsch, K., Undorf-Spahn, K. and Jehle, J. A. 2017. A third type of resistance of codling moth against Cydia pomonella granulovirus (CpGV) shows a mixture of a Z-linked and autosomal inheritance pattern. *Applied and Environmental Microbiology* 83(17). doi:10.1128/AEM.01036-17.

Serebrov, V. V., Gerber, O. N., Malyarchuk, A. A., Martemyanov, V. V., Alekseev, A. A. and Glupov, V. V. 2006. Effect of entomopathogenic fungi on detoxification enzyme activity in greater wax moth *Galleria mellonella* L. (Lepidoptera, Pyralidae) and role of detoxification enzymes in development of insect resistance to entomopathogenic fungi. *Biology Bulletin* 33(6), 581–6. doi:10.1134/S1062359006060082.

Stowell, L. J. 1991. Submerged fermentation of biological herbicides. In: TeBeest, D. O. (Ed.), *Microbial Control of Weeds*. Chapman and Hall, New York, NY, pp. 225–61.

Tabashnik, B. E., Cushing, N. L., Finxon, N. and Johnson, M. W. 1990. Field development of resistance to *Bacillus thuringiensis* in diamondback moth (Lepidoptera: Plutellidae). *Journal of Economic Entomology* 83(5), 1671–6. doi:10.1093/jee/83.5.1671.

Thakore, Y. 2006. The biopesticide market for global agricultural use. *Industrial Biotechnology* 2(3), 194–208. doi:10.1089/ind.2006.2.194.

Vega, F. E. 2008. Insect pathology and fungal endophytes. *Journal of Invertebrate Pathology* 98(3), 277–9. doi:10.1016/j.jip.2008.01.008.

Vincent, C., Goettel, M. S. and Lazarovits, G. (Eds). 2007. *Biological Control: a Global Perspective*. CABI Publishing, Wallingford, UK.

Wraight, S. P., Jackson, M. A. and de Kock, S. L. 2001.Better control of weeds with enhanced bioherbicides. In: Butt, T. M., Jackson, C. and Magan, N. (Eds), *Fungi as Biocontrol Agents. Progress, Problems, and Potential*. CABI Publishing, Oxon, UK, pp. 253–87.

Zorner, P. S., Evans, S. L. and Savage, S. D. 1993. Perspectives on providing a realistic technical foundation for the commercialization of bioherbicides. In: Duke, S. O., Menn, J. J. and Plimmer, J. R. (Eds), *Pest Control with Enhanced Environmental Safety*. American Chemical Society, Washington DC, pp. 79–86.

Published by Burleigh Dodds Science Publishing Limited, 2020.

Chapter 2

Implementing biopesticides as part of an integrated pest management (IPM) programme

József Kiss, Szent István University, Hungary; and Marc Delos, Académie d'Agriculture de France, France

1 Introduction

Humans have co-existed with pests of cultivated plants for at least 9000 years, seeking to reduce their impact on yields and the quality of agricultural products. The tools and methods for achieving these goals have varied, depending on regional conditions, the developmental stage of societies and priorities and available technologies. Historically, there has been a focus on optimising food production. However, in the late 1980s, global agricultural development strategies were broadened from commodity production towards a wider focus on sustainable agriculture and rural development, defined by the FAO as 'the management and conservation of the natural resource base, and the orientation of technological change in such a manner as to ensure the attainment of continued satisfaction of human needs for present and future generations. Sustainable agriculture conserves land, water, and plant and animal genetic resources, and is environmentally non-degrading, technically appropriate, economically viable and socially acceptable' (FAO, 1988). The different elements of sustainable agriculture highlight the multiple roles of farming and the need for system-based approaches with the integration

http://dx.doi.org/10.19103/AS.2020.0073.03

of different components in agriculture, including pest management. In this chapter we use the widely accepted definition of pest management where 'pest' refers to any type of plant, animal or pathogenic agent injurious to plants or plant products (ISPM No. 05, 2010).

The need for an integrated approach to controlling pests was highlighted by the scientific community more than 50 years ago, focussing on integrated pest control (IPC), later referred to as integrated pest management (IPM). IPM used the same approach as IPC but diverged in the relative importance of specific details (for the historical overview of the development of definitions, see Bajwa and Kogan, 2002). In the International Code of Conduct on the Distribution and Use of Pesticides (FAO, 2002), 'Integrated Pest Management (IPM) means the careful consideration of all available pest control techniques and subsequent integration of appropriate measures that discourage the development of pest populations and keep pesticides and other interventions to levels that are economically justified and reduce or minimize risks to human health and the environment. IPM emphasizes the growth of a healthy crop with the least possible disruption to agro-ecosystems and encourages natural pest control mechanisms'.

The EU definition in the EU Framework Directive 2009/128/EC on the sustainable use of pesticides is almost identical: 'Integrated Pest Management means careful consideration of all available plant protection methods and subsequent integration of appropriate measures that discourage the development of populations of harmful organisms and keep the use of plant protection products and other forms of intervention to levels that are economically and ecologically justified and reduce or minimise risks to human health and the environment. Integrated pest management emphasises the growth of a healthy crop with the least possible disruption to agro-ecosystems and encourages natural pest control mechanisms'.

An agro-ecological approach has also been identified by the International Organisation of Integrated and Biological Control (IOBC), established in 1955 (http://www.iobc-global.org) as the basis of integrated production (IP) as follows: *IP is an agro-ecological approach*: 'That starts from the notion that a farm is an agro-ecosystem. IP focusses on managing ecological processes and their interactions within the farm and its related context (landscape and region), to optimise the use of internal farm resources, to minimise the need for external inputs and to avoid or minimise the effect of farming practices on the direct environment, future generations or conditions elsewhere in the world'.

IP is based on a system approach: 'Taking the farming system as a basic unit: the full potential of an agro-ecology approach can only be realized when all interactions between different elements of farming are utilized. Synergy between objectives might be realized'.

From this perspective, farming and associated plant protection measures are situated within and are designed to support agro-ecosystems.

Agro-ecosystems involve cropped fields, pastures and non-cropped areas such as semi-natural or protected habitats (Natura 2000 areas, COUNCIL DIRECTIVE 92/43/EEC). In crop fields complex food webs are established where different functional groups of living organisms (the cultivated crop, other plants including weeds, herbivores including arthropod pests and pathogens, decomposers, pollen and nectar feeders, predators, parasitoids, etc.) are linked by trophic interactions. Agriculture aims at reducing the potential of damage by pests to crops while supporting those functional groups that provide important ecosystem services (pest regulation, pollination, etc.). This is achieved by reducing external interventions (including synthetic pesticides) which have unacceptable side effects on those beneficial organisms that allow the agro-ecosystem to function (IOBC WPRS Pesticide Side Effect Database). The protection of the environment and the use of natural pest control mechanisms, emphasised in the IPM definitions, are seen as part of this broader ecosystem approach. Ecosystems and the ecosystem services they provided are also defined in the Millennium Ecosystem Assessment documents (MAE, 2005) and discussed by Lele et al. (2013). Ecosystem services (pest control and pollination as regulating services), genetic resources (provisioning services) and aesthetic (cultural services) are delivered by agro-ecosystems and therefore support the implementation and success of IPM programmes.

This overview suggests that the use of biopesticides in IPM programmes should be part of a system-based approach, starting from a smaller unit such as crop fields but moving on to include adjacent semi-natural habitats (SNHs) and whole farms up to integration at the landscape scale, as described in the EU Strategic Research Agenda for IPM in Europe (Lamichhane et al., 2016a). A starting point is the recent global overview by van Lenteren et al. (2018) that provides a comprehensive overview of biopesticide use in agriculture and also what is included by terms such as 'biopesticide' and 'biological control'.

2 Biocontrol products and biopesticides: definitions

Biopesticides or biocontrol agents (BCAs) use natural means to control pests with low risk for humans and the environment. They can be used alone or in combination with other plant protection methods to combat crop enemies in IPM programmes. In OECD programmes biological pesticides or biopesticides are defined as various living organisms (microbials, such as bacteria, algae, protozoa, viruses and fungi; macrobials/invertebrates such as insects and nematodes), plant extracts/botanicals as well as pheromones and semiochemicals (http://www.oecd.org/env/ehs/pesticides-biocides/biological-pesticides.htm). Thakore (2006) has also defined biopesticides as 'living organisms (plants, microscopic animals such as nematodes, and microorganisms, including bacteria, viruses, and fungi) or natural products derived from these organisms that are used to suppress pest populations'. He emphasises 'natural products' though some types of biopesticides such as semiochemicals and pheromones are usually chemically synthesised products. Thakore also includes plant-incorporated protectants which can be part of genetically modified plants, though there is disagreement about whether this constitutes a 'natural product'.

Gupta and Dikshit (2010) list three categories of biopesticide:

- microbial pesticides: containing whole microorganisms.
- plant pesticides: pesticidal substances from genetic material that have been added to plants.
- biochemical pesticides: naturally occurring substances that control pests by non-toxic mechanisms. This includes substances that interfere with growth or mating, such as semiochemicals, or substances that repel or attract pests, such as pheromones.

These definitions reflect the natural origin of biopesticides. Depending on the regulatory framework, biopesticides may be divided into various sub-categories. Examples are the categories used in French regulations:

- auxiliary macro-organisms: invertebrates, insects, mites or nematodes used in an integrated approach to protect crops from pests.
- phytopharmaceutical products, including microorganisms to control pests; fungi, bacteria and viruses used to protect crops from pests and diseases or boost plant growth.
- chemical mediators: insect pheromones and kairomones. These can be used to track the flights of insect pests and control insect populations through mating disruption or attracting insects to a trap where they can be killed by an insecticide. Chemical mediators imitate natural substances but may be synthetic (i.e. manufactured).
- natural substances: these are substances obtained from plant, microbial, animal and mineral sources. They are found in the natural environment.

These categories highlight the diverse range of biopesticide products and their potential applications.

A useful definition (given its breadth) is that given by Glare (2015):

> 'Biopesticides are generally regarded as biologically based pesticides. The specific definition however has depended on the researchers, and the term has been used broadly to encompass all biologically derived pesticides, or more often specifically for certain groups, such as microbial biopesticides. The general aim of using the term biopesticide is to suggest both a biological rather than synthetic origin of the active agent, and implicit is a level of environmental and mammalian safety derived

from the use of naturally derived actives. Various subgroups of biopesticides that could be included under a broad definition embrace packaged pesticides based on live microbes, nematodes, microbial and plant extracts, genetically modified plants, semiochemicals, endophytes, invertebrates used inundatively and other compounds derived from animals or minerals. The term biopesticide implies some form of packaging and active application, without the expectation of long term persistence, and often has an element of formulation. The market for biopesticides is increasing rapidly around the world, but regulators in many countries are struggling with providing registration procedures appropriate for these pesticides'.

In this chapter we address the wide-ranging use of biopesticides within IPM programmes, including biopesticides sprayed from a packaged formulation (microbial biopesticides), the release of mass-reared BCAs (predators, parasitoids), supporting existing living endemic populations of natural enemies (augmentative release and/or conservation biological control via habitat management). Natural products, plant extracts and synthetised pheromones or solutions made possible by new breeding techniques will also be considered in the chapter. This broad and flexible approach demonstrates the continuum from living organisms (artificially released or endemic) to their products (whether derived from nature or plant-incorporated gene products), which can all be used in IPM programmes using techniques such as spatial mapping of pest densities, pest population regulation and regional action/economic threshold levels.

3 Regulatory assessment of the use of biopesticides in IPM programmes

Registration and placing on the market of biopesticides requires efficacy and environmental safety assessments. The European Plant Protection Organization (EPPO) 'principles of efficacy evaluation for low-risk plant protection products' (EPPO Guideline, 2017) assess aspects such as:

- effectiveness
- development of resistance
- adverse effects on treated crops
    - phytotoxicity
    - yield quantity and quality
    - effects on plant parts for propagation
- undesirable or unintended side effects
    - damage to succeeding or adjacent crops
    - effects on beneficial and other non-target organisms
- interaction with other crop protection measures

EPPO evaluation also includes the 'contribution of a new product to IPM and sustainable agriculture' as follows: 'The contribution of the proposed use to agricultural sustainability is considered in the evaluation of low-risk products. Any agronomic benefits that are expected to result from the registration of the proposed use may be included in the dossier. A description of the product's fit within a cropping system and its benefits in relation to alternatives may be provided, for example compatibility within an IPM system'. The integration of new biopesticide products in IPM programmes is also reviewed by Matyjaszczyk (2018).

A specific aspect of biopesticide registration is the potential effect of a new product on non-target organisms and therefore on ecosystem services such as pest regulation and pollination. The European Food Safety Authority's (EFSA) scientific opinion on the assessment of the potential impacts of genetically modified plants on non-target organisms provides a useful background on assessing these effects (EFSA Panel on Genetically Modified Organisms (GMO), 2010). Devos et al. (2019) also discusses key protection goals (ecosystem services) when assessing the potential risks from regulated stressors (including biopesticides).

Standard data in assessments include the amount of active substances (kg/ha) or treatment frequency indexes (TFIs; that is usually the registered rate per use × the number of applications per the growing season). These data may be less useful in assessing biopesticides which have typically lower application rates and, as discussed earlier, may be applied more flexibly. Quantitative data may not address the ecotoxicological profile of the product and its broader impact (e.g. effect on non-target organisms). Assessments of biopesticide products need evaluations of longer-term impacts, for example in the reduction of target pest population levels over time (multiple years) using measures such as the reduction in crop pest damage, trapping and monitoring data. This type of impact assessment is particularly relevant for those IPM tools that have an indirect impact on the target pest (e.g. confusion techniques, mating disruption) or where an area-wide pest population level decrease is the aim. The effectiveness of biopesticides in IPM systems may also be assessed using a comparative approach (comparing former or present pesticide use and new candidate biopesticides) using multiple criteria (e.g. social, environmental and economic measures) (Pelzer et al., 2012).

One example of a more flexible approach to biopesticide assessment from the USA is the exemption from the fees of pesticide registration if the novel pesticide fits into current IPM programmes (Farrar et al., 2018). The Western Integrated Pest Management Center, one of the four regional centres funded by the US Department of Agriculture (USDA), has produced an IPM Criteria Guidance Document. This lists 21 specific factors in eight categories (efficacy, economic benefit, non-target effects, resistance concerns, environmental fate, worker risk, compatibility with monitoring and utility as a preventative) with descriptors of affirmative, intermediate and negative compatibility attributes.

An example from Europe is the DEXiPM tool assessing economic, environmental and social impacts of IPM programmes in maize-based and other cropping systems (Vasileiadis et al., 2011, 2013). Examples of the application of DEXiPM can be found in the EU-funded Pesticide Use and Risk Reduction (PURE) in European Farming Systems Project (http://www.pure-ipm.eu/) (Pelzer et al., 2012; Aubertot, 2017). The model makes it possible to assess the overall sustainability of a cropping system using IPM techniques. One application of the model is assessing the performance of demonstration farms using IPM techniques. Examples include the German Model Project 'Demonstration Farms Integrated Plant Protection' (https://www.ble.de) and the Ecophyto programme in France discussed later in this chapter (http://www.ecophytopic.fr/).

Another attempt to adapt the regulatory regime to accommodate biopesticides is the concept of 'minor uses'. The high cost of developing and commercialising new pesticides means that it can be uneconomic to develop them for 'minor uses', for example for speciality crops grown on too small a scale to create a viable market for new pesticide products. Although the definition of 'minor uses' varies globally, it is commonly understood by organisations such as the OECD to cover:

- uses of pesticides on small crop areas or against infrequent pests.
- uses which are too small to warrant a sufficient return for manufacturers to develop pesticides for them.

EU and other regulatory authorities have recognised the need for a simpler and more flexible regulatory framework to allow 'minor use' biopesticide development whilst maintaining safety and product quality. Sources on the definitions and types of minor use include the Global Minor Use Portal (http://gmup.org/) and the European Minor Uses Coordination Facility (http://minoruses.eu/).

4 The use of biopesticides in different crop production systems

There are many different farming systems, in terms of variables such as type of crops, climatic conditions and cultivation methods, which will have implications for the implementation of IPM programmes and the role of biopesticides within those programmes. BCAs involve the use of naturally occurring biopesticidal agents. An advantage of techniques such as conservation and augmentative biological control is avoiding the environmental impact of pesticides based on synthetic chemical compounds, since BCAs are typically already present in the fauna. In addition biopesticidal products (microbial biopesticides, biochemical semiochemicals) are often more selective with less toxic residues. These advantages offset disadvantages (compared to synthetic pesticides) such as a slower kill rate, sensitivity to environmental conditions and the frequent need for repeated applications (Chandler et al., 2011). This section assesses the role of biopesticides in IPM programmes for the following production systems:

- perennial plants (e.g. vineyards, fruit tree orchards).
- protected systems (e.g. greenhouse vegetables, crops under exclusion nets).
- open field vegetables.
- annual arable field crops.

Perennial production systems, for example fruit orchards and vineyards, are established for a longer time period than annual cropping systems. The cultivated plant, together with associated flora and fauna, offers unique possibilities for pest suppression and management due to reduced ecological disruption compared with annual cropping systems. Since animal and plant assemblages in orchards (both species and trophic interactions) are more 'constant' over time compared to arable crops systems in annual rotations, it is possible to provide continuous pest and natural enemy monitoring and early warning systems. This provides a more predictable basis for estimating pest risks and identifying appropriate suppression measures, including those based on biopesticides.

Vineyards and orchards are well suited to conservation biological control measures such as the establishment and maintenance of non-crop habitats, flowering borders and flower strips within plant rows. These attract pollinators as well as pollen and nectar feeders including predators and parasitoids. These naturally occurring beneficial arthropods help to suppress pest populations below economic thresholds, particularly when part of a well-designed IPM system. Novel pest detection and monitoring tools (pheromone and colour-enhanced traps, attractants-based traps, remote digital imaging of pests or pest damage, meteorological data records linked to predictive models) also allow specific 'local' or targeted use of biopesticides to suppress pest 'hot spots' within fields.

A broader landscape approach (including the coordination of IPM practices such as host plants and habitats for beneficial organisms including BCAs and pollinators) will also have an impact on the success of IPM in crop stands. Based on the studies from North America and Europe, Bianchi et al. (2006) concluded that a diversified landscape has most potential for sustainable pest control. A similar conclusion based on over 70 separate studies was reached by Veres et al. (2013). In addition to landscape complexity, crop management intensity may impact natural enemy densities and population dynamics, as demonstrated by Veres et al. (2012) for *Orius niger*, a predatory pirate bug (key generalist predator of Western flower thrips *Frankliniella occidentalis* Pergande). Key issues and research priorities relating to the use of biological control measures in perennial, arable and vegetable crops are reviewed in Lamichhane et al. (2017).

Protected systems (e.g. glasshouses, polytunnels) are generally more isolated from the surrounding external flora and fauna, while production conditions (temperature, humidity, water and nutrient supply, pest access to crops) are generally highly regulated. In these controlled conditions, pests and their natural enemies can be more precisely monitored, which makes it easier to identify and implement an effective IPM programme, including the use of BCAs. Protected systems are often used for higher-value crops, which make it possible for farmers to use automated, remote monitoring tools (e.g. web-linked camera traps). Using this precision monitoring, together with well-defined action thresholds, commercially developed equipment can then be used for large-scale inundative release of predators and parasitoids. Since other IPM tools such as rotations are not suitable for protected cultivation systems, biopesticides are a particularly important tool in protected cultivation IPM.

Open field vegetables are, in many cases, value-added, specialty products grown on a smaller scale than many arable crops. As 'minor' crops, there may be fewer synthetic pesticides available if the market is not deemed sufficiently attractive to develop and register specific pesticides for a particular crop. Other IPM measures such as rotation are possible but may be limited by farm conditions. This makes non-chemical pest management techniques particularly important, for example intelligent mechanical weed control techniques, biopesticides, flower strips and push-pull strategies (http://www.pure-ipm.eu/publication/29).

Annual arable crops cover the largest acreages and are 'open' systems which are particularly exposed to pests over the growing season. Since pests can be more difficult to predict in such open systems, larger acreages are more challenging to manage and arable crops can be lower-value commodity crops, arable farmers prefer cheap, simply applied (e.g. fewer applications) and robust, season-long crop protection solutions against often less-predictable pest attacks. This makes them potentially more reliant on synthetic pesticides, which are cheap and relatively easy to apply. However, the large cultivation areas of these crops also makes it possible to implement IPM measures such as carefully planned rotations (which break the life cycle of pests by replacing host plants with non-hosts). These techniques can be combined with the use biopesticides, triggered at a lower action threshold to prevent pest population build-up earlier in the infection cycle. A reduced reliance on synthetic pesticides may also be driven by consumer concerns about safety and the environment, as well as by the government policy to restrict conventional pesticide use and promote lower-risk bioactive substances, for example as in the case of the EU (https://eur-lex.europa.eu/legal-content/EN/ALL/?uri=celex%3A32009L0128).

5 Integrating biopesticides into IPM programmes

In assessing how best to integrate biopesticides into IPM programmes, a good starting point is the EU Framework Directive 2009/128/EC (European Union 2009) on the sustainable use of pesticides and the eight principles listed in Annex III of the Framework Directive (see Barzman et al., 2015, for further details). These principles are:

- Principle 1 - prevention and suppression
- Principle 2 - monitoring
- Principle 3 - decision-making
- Principle 4 - non-chemical methods
- Principle 5 - pesticide selection
- Principle 6 - reduced pesticide use
- Principle 7 - anti-resistance strategies
- Principle 8 - evaluation

Underpinning these principles is the focus on growing a healthy crop with the least possible disruption to agro-ecosystems, including preserving and supporting natural pest control mechanisms. Supporting natural pests provides good opportunities for complementary biopesticide use. In this section we will discuss the principles of prevention and suppression and the use of sustainable biological, physical and other non-chemical methods.

Successful use of biopesticides with an IPM programme requires a more flexible approach in the use of economic and action thresholds. For instance, some arthropod pests like the population of Western corn rootworm (WCR, *Diabrotica virgifera virgifera*), a well-known pest of maize in the USA and Europe, can efficiently be managed with population monitoring in maize fields and primarily (in Europe) with rotation of maize with non-maize crops to disrupt the pest's life cycle (Kiss et al., 2005). To optimise maize production, farmers cultivate maize for more than 1 year on the same field, rotating the crop only when the population of WCR increases to a potentially economically damaging level (Kiss, 2008; Szalai et al., 2014).

However, WCR population growth can be decreased and larval damage to maize root system can be reduced in continuous maize production with the application of entomopathogenic nematodes and fungi (Pilz et al., 2009), although the efficacy of such control tools is influenced by local soil and weather conditions, as demonstrated in Central European field studies (Toepfer et al., 2010). The efficacy of entomopathogenic nematodes and fungi in reducing the density and subsequent damage by root-feeding WCR larvae may vary over fields and regions and may not reach the level of control given by specific synthetic pesticides. However, biopesticide intervention is a practical option in the early phase of WCR population increase (in advance economically damaging WCR population levels) where variations in performance are less significant. This approach demonstrates how to manage a formerly invasive pest and develop 'a peaceful co-existence' with the pest (Boriani et al., 2006; Kiss, 2008). Combinations of various non-chemical control tactics based on WCR monitoring and precautionary WCR population suppression can result in more sustainable maize production. The management of WCR population levels ideally needs to occur at a broader spatial scale (over several fields, several farms or a whole region) within an 'area-wide IPM' approach for pest regulation.

Pest population monitoring and flexible interpretation of action thresholds for applying biopesticides may also be appropriate for those pests with a life cycle that extends to more than 1 year and one crop, such as wireworms (larvae of elaterids) or *Melolontha* species. The same approach is valid for a pest such as white mould (*Sclerotinia sclerotiorum*), a polyphagous multi-host pathogen of various dicotyledonous crops, such as sunflower, oilseed rape and soybean. The surviving form of the fungus, the sclerotia, remains in the soil for several years. Therefore, its parasite *Coniothyrium minitans*, which survives in the soil for at least 6 months (Jones and Stewart, 2011), can be efficiently used regardless of specific crop or time period. The setting and use of economic thresholds for key pests is a complex task and needs flexible approach, depending on factors such as the region, farm and given year. This flexible approach is more relevant for arable systems than for perennial or protected production systems.

It is important to take account of the fact that pests, their natural enemies and other beneficial organisms (e.g. parasitoids and pollinators) occur in agricultural crops in large species numbers (e.g. several thousand species in an annual crop such as maize or a perennial such as apple) (Mészáros et al., 1984a,b). A key aspect of IPM programmes is to support these beneficial organisms to keep pests within acceptable thresholds. This involves managing the farm landscape to create a supportive habitat. This includes the need for area-wide IPM strategies. It has been found, for example, that broader 'landscape' approaches successfully suppressed WCR population density levels in the maize and soybean-dominated corn belt in the USA (Chandler et al., 2008). Other examples of the use of area-wide pest management are given in Birch et al. (2011).

Flowering strips within/among fields support beneficial organisms by providing them with additional feed sources (e.g. pollen, nectar, alternative prey or hosts) and shelter for overwintering. The potential of non-cropped habitats, specifically of SNHs, to contribute to pollination, pest control regulation and provision of other ecosystem services has been widely discussed in recent years (Figs. 1–3). It has been found, for example, that carabid and plant species

Figure 1 Agricultural landscape with crop fields and semi-natural habitats in South Hungary.

Figure 2 Flowering field margin in Hungary.

Figure 3 Syrphid flies on field margin flower.

diversity depends on the type and number of SNHs within a broader 'habitat mosaic' (Duflot et al., 2015).

Research has provided differing results on the effectiveness of SNHs, with some studies suggesting only a most contribution to improved pest control (Holland et al., 2016, 2017). Martin et al. (2019) have synthesised data from 49 studies (1515 landscapes) across Europe with contrasting results. For arable-dominated landscapes with high-edge densities, the provision of SNHs resulted in improved pest control services. Conservation biological control, therefore, could be enhanced by the establishment of flowering strips, field edges and other SNHs. However, the impact of SNHs is likely to be different in various landscape contexts.

All these methods of biological control require farmers to have the appropriate knowledge and skills to recognise the value of these tools and to be able to implement them effectively. The consequences of dependence on a narrow range of synthetic chemical pesticides with little flexibility to adapt to changing agronomic conditions and pest problems have been demonstrated in rice pest outbreaks in Asia in the 1980s and 1990s (Ooi, 1996). This highlighted the need to educate farmers about alternative IPM strategies as well as equip them with new biopesticide and biocontrol products and techniques. Participatory training of farmers was developed by the FAO in 1970s in the Philippines, based on Farmer Field Schools (FFS) and then expanded into other regions such as Central Europe in the late 1990s (Papp-Komáromi et al., 2007) (Figs. 4 and 5). A

Figure 4 Adviser with farmers evaluating maize root damage by Western corn rootworm larvae in FAO Farmers Field School, Hungary.

Figure 5 Grandfather and grandson searching for Western corn rootworm larvae in FAO Farmers Field School, Hungary.

recent global overview of farmer field school approaches towards sustainable and resilient food systems is provided by FAO (2019), which has a Global FFS Network Platform. The development of IPM in Asia using FFS and other community approaches is discussed by Pontius et al. (2002). A detailed discussion of these approaches in helping to disseminate new IPM techniques to combat the fall armyworm (*S. frugiperda*) in Africa is provided by Toepfer et al. (2019). Although FFS focus mainly on supporting resource poor farmers in developing countries, their use in more than 90 countries shows their broader potential, including promoting the greater use of biopesticides in IPM programmes in developed as well as developing countries. A similar networking approach is demonstrated by CABI in its global Plantwise program, which has a network of plant clinics run by trained 'plant doctors' in 30 countries where farmers can find practical advice on pests, diseases and their management. In Europe one of the most promising ways of disseminating information about IPM is the establishment and operation of IPM farm networks (Lamichhane et al., 2016b). These are particularly useful given the value of area-wide implementation of IPM programmes.

6 Using biopesticides in IPM programmes in practice: a case study

France's Ministry of Agriculture launched its Agroecology Project in 2012 with the ambition of developing a path for the future of French agriculture. Its central aims include guaranteeing food security and increasing production while preserving natural resources and biodiversity. The reduction in the use of synthetic pesticidal chemicals, as well as a reduction in fossil fuel use and in tillage practices, is one of the top priorities of this project (which are seen as damaging the soil health). In a series of plans implemented from 2010 onwards (Ecophyto, Ecophyto II and now Ecophyto II+), the government aims to reduce the use of pesticides and other chemical products by half on farms and in other areas by 2025. After a slow start, there has been a steady decrease since 2012 in the use of key pesticides targeted in the programme (Note de suivi Écophyto, 2017). The Ecophyto plans aim to shift the use from synthetic pesticides to biocontrol products, 'basic substances' and low-risk plant production products as defined by the EU (https://ec.europa.eu/food/plant/pesticides/sustainable_use_pesticides_en). The French government has its own list of approved biopesticides (https://agriculture.gouv.fr/quest-ce-que-le-biocontrole).

A network of 1900 French test farms (DEPHY farms) has also shown that it is possible to reduce the use of pesticide products without having an impact on productivity or farm profits. This network has seen decreases of 18% in the overall pesticide use, with a 25% reduction in vegetable crops, 12% in wine production and 33% in horticultural production. With the Ecophyto II plan, published in October 2015, the French government intends to go from demonstration to action, through the wider implementation of the practices used by the IPM test farms to 30 000 farmers throughout the French farming sector (https://agriculture.gouv.fr/le-plan-ecophyto-quest-ce-que-cest). From 2014, IPM, even though it was already a widely used practice in France (mainly in greenhouse, orchards, vineyards and grain crops), became mandatory. Other measures have included increasing restrictions on the use of synthetic pesticides (e.g. in public spaces such as municipal gardens), limiting consumers to the use of biocontrol products in their gardens and requiring training for those applying pesticides in commercial settings.

Plant-derived biopesticides include macro-organisms such as pyrethrins, as well as Bt biopesticides, *Beauveria bassiana*, Spinosad and avermectines, which affect target pest behaviour. Seventy-five per cent of French glasshouse vegetables are protected by biopesticidal macro-organisms. Twenty-five per cent of the maize area needing protection (120 000 ha on 480 000 ha protected) are protected against European corn borer using *Trichogramma* wasp species. Fifty per cent of apple and pear orchards are protected using insect pheromones, kairomones and other chemicals designed to disrupt behaviour such as mating. More than 70% vineyards in the Champagne region (and 12% of all vineyards in France) are protected using pheromone-based biocontrol methods. The relatively high price of pheromone-based IPM products appears to be the main factor limiting commercial uptake. The use of pheromones for mating disruption has a large adoption rate in French orchards because of the large spectrum of lepidopteran pests controlled (including *Grapholita molesta, Zeuzera pyrina, Synanthedon tipuliformis, Cydia pomonella, Adoxophyes orana, Pandemis heparana, Archips podana, Argyrotaenia pulchellana, Archips rosana, Cacoecimorpha pronubana, Rhopobota naevana*). Pheromones are used to control pests in French vineyards (such as *Lobesia botrana* and *Eupoecilia ambiguella*) and in French and Spanish rice cultivation (e.g. *Chilo suppressalis*).

Other biocontrol techniques include the use of 'push-pull' companion crops and intercrops which either repel insects away from the cropping area or attract them to the companion plant and away from the main crop. One examples is faba bean or alfalfa intercropped with oil seed rape to combat the cabbage stem flea beetle (*Psylliodes chrysocephalus*) and rape winter stem weevil (*Ceutorhynchus picitarsis*) (Ruck et al., 2017). The potential impact of this strategy can be seen in research from East Africa where farmers were able to reduce fall armyworm (FAW, *Spodoptera frugiperda*) infestation and crop damage by up to 86%, with a 2.7-fold increase in yield relative to neighbouring fields that did not implement this approach (Midega et al., 2018). Pest-repellent ('push') plant species (e.g., *Desmodium* spp.) can be used as an intercrop (in conjunction with maize or sorghum as the main crop), while pest-attractive trap ('pull') plant species (usually grasses such as *Brachiaria* spp.) are grown around the main crop (Khan et al., 2018). A push-pull strategy has been implemented in the French Island of Mayotte in the Indian Ocean (PRÉFET DE MAYOTTE. Compte rendu réunion d'information sur la chenille légionnaire d'automne Spodoptera frugiperda du 14 septembre 2018: http://daaf.mayotte.agriculture.gouv.fr/). In mainland France a pull strategy has been implemented by attracting adults of pollen beetles (*Meligethes aeneus*) in oilseed rape fields using early flowering plants, either intercropped with the main crop or in field margins. This approach reduces pest density when the main crop flowers (Frearson et al., 2005; Cook et al., 2002; Délos, 2008).

In terms of biopesticides derived from natural plant extracts, only pyrethrins are still authorised in the EU. Nicotine was banned in 2008 and rotenone was banned in 2009 in the EU (with some exceptions mainly for use in organic vineyards). Azadirachtin (Margosa or neem active extract) has been approved in the EU since 2011 and authorised in organic farming in France since December 2018, but with major restrictions (due to potential toxicological effects on aquatic and other organisms) (https://ephy.anses.fr/ppp/neemazal-ts).

Biopesticides have a limited efficacy in controlling plant diseases. Fenugreek and laminarin, for example, have a limited effect in inducing plant defence responses (Bernard, 2013, 2017). The official French biocontrol list includes mineral-based fungicides such as sulphur and phosphonates, which can be effective against a range of diseases (Maumené, 2018). Biopesticides such as *Coniothyrium minitans* give good control, but the target spectrum is limited to *Sclerotinia sclerotiorum* and *Sclerotinia minor*. Under the French Ecophyto plan, sulphur and phosphonate use has increased greatly, from less than 8000 tons of sulphur in 2010 to more than 12 000 tons per year in 2013. Phosphonate use increased from 200 tons in 2011 to 800 tons in 2017 (Pesticides Sold National Database, 2017). The French cereal institute Arvalis has tested the available biopesticides on the market over the last 3 years to control *Zymoseptoria tritici*, and found only phosphonates and sulphur to be effective when used together with resistant or tolerant cultivars (https://www.semencesdefrance.com/actualite-semences-de-france/septoriose-ble-bonne-surprise-biocontrole/; https://www.terre-net.fr/observatoire-technique-culturale/strategie-technique-culturale/article/le-t1-est-il-toujours-necessaire-sur-cereasles-217-143798.html). A small percentage of oilseed rape (5%) is protected by microbial fungicides (*Bacillus pumilus* QST 2808 or *Bacillus subtilis* QST 713 or *Coniothyrium minitans*). A new bio-fungicide, fenpicoxamide, was approved in October 2018 and may help to improve the performance of biopesticides in the disease control.

Biological control of weeds is practically limited to pelargonic acid. Because of its non-selective action, large amounts are required per hectare to be effective, increasing costs. Its use has increased from 27 tons in 2015 to 290 tons in 2017, following the prohibition of synthetic herbicides in public spaces.

However, there remains a significant reliance on cultural and physical methods of weed control as an alternative to synthetic herbicides.

An International Biocontrol Manufacturers Association (IBMA) Colloquium in 2019 reported the following rates of growth for different types of biopesticide in French agriculture between 2016 and 2019 (https://www.ibmafrance.com/colloque-2019/):

- mineral-based biopesticides (e.g. sulphur and phosphonates): 30%
- 'straight chain Lepidopteran pheromones': 18%
- semiochemicals: 25%
- macro-organisms: 14%
- microorganisms: 15%

An IBMA panel survey in 2019 found that 44% of French growers sampled included biopesticide products in their IPM strategies. Orchards are more comprehensively protected with biopesticide-based products (82% of professional growers) than vineyards (74% of growers) and vegetables (53% of growers). Seventy-six per cent of public green areas (e.g. parks) use biopesticides, whilst only 9% of the overall agriculture sector in France use biopesticides. Biopesticides and BCAs still represent only 5% of the overall pesticide market in France (https://www.ibmafrance.com/). Sixty-five per cent of organic growers consider biopesticides and BCAs as very efficient, compared with only 45% in conventional agriculture. Difficulties in adoption include the cost of biological products (44% of respondents), lack of knowledge and training in use (41%), poor efficacy (37%) and a limited spectrum of pests and disease controlled (27%).

7 Where to look for further information

Further reading:

- Lamichhane, J. M., Bischoff-Schaefer, M., Bluemel, S., Dachbrodt-Saaydeh, S., Dreux, L., Jansen, J. P., Kiss, J., Köhl, J., Kudsk, P., Malausa, Th., Messéan, A., Nicot, Ph. C., Ricci, P., Thibierge, J. and Villeneuve, F. 2016a. Identifying obstacles and ranking common biological control research priorities for Europe to manage most economically important pests in arable, vegetable and perennial crops. *Pest Management Science*. doi:10.1002/ps.4423.
- Lamichhane, J. M., Aubertot, J. N., Begg, G., Birch, A. N. E., Boonekamp, P., Dachbrodt-Saaydeh, S., Hansen, J. G., Hovmøller, M. S., Jensen, J. E., Nistrup Jørgensen, L., Kiss, J., Kudsk, P., Moonen, A. C., Rasplus, J. Y., Sattin, M., Streito, J. C., Messéan, A. 2016b. Networking of Integrated Pest

Management: a powerful approach to address common challenges in agriculture. *Crop Protection* 89, 139–151.

- van Lenteren, J. C., Bolckmans, K., Köhl, K., Ravensberg, W. J. and Urbaneja, A. 2018. Biological control using invertebrates and microorganisms: plenty of new opportunities. *BioControl* 63, 39–59. doi:10.1007/s10526-017-9801-4.

Key journals and weblinks:

- *BioControl*
- *Biological Control*
- *Biocontrol Sience and Technology*
- Daaf.mayotte.agriculture.gouv.fr, 2018 - PRÉFET DE MAYOTTE. Compte rendu réunion d'information sur la chenille légionnaire d'automne Spodoptera frugiperda du 14 septmbre 2018.
- http://www.iobc-global.org/
- https://www.iobc-wprs.org/ip_ipm/IOBC_Pesticide_Side_Effect_Database.html
- http://www.millenniumassessment.org/en/About.html
- http://www.ibma-global.org/
- http://gmup.org/
- http://minoruses.eu/
- http://www.oecd.org/env/ehs/pesticides-biocides/minoruses.htm
- http://www.oecd.org/env/ehs/pesticides-biocides/biological-pesticides.htm http://www.fao.org/farmer-field-schools/en/)
- https://www.plantwise.org/
- https://ec.europa.eu/eip/agriculture/en
- http://www.ecophytopic.fr/
- https://www.semencesdefrance.com/actualite-semences-de-france/septoriose-ble-bonne-surprise-biocontrole/ and https://www.terre-net.fr/observatoire-technique-culturale/strategie-technique-culturale/article/le-t1-est-il-toujours-necessaire-sur-cereasles-217-143798.html
- https://agriculture.gouv.fr/quest-ce-que-le-biocontrole
- IBMA biocontrole colloquium – https://www.ibmafrance.com/colloque-2019/
- IBMA France and ADquation inquieries for IBMA France (July 2018) – https://www.ibmafrance.com/
- Note de suivi Écophyto 2017 in http://www.consultation-ecophyto2plus.gouv.fr/presentation-du-projet-de-plan-ecophyto-ii-a1.html
- Pesticide sold national database-banque nationale de données des ventes des distributeurs de produits phytosanitaires (BNV-D) – https://www.data.gouv.fr/fr/datasets/donnees-de-vente-de-pesticides-par-departement/

Major international research projects and networks:

- ENDURE was originally a Network of Excellence funded by the European Commission from 2007 to 2010. ENDURE brings together some of Europe's leading agricultural research, teaching and extension institutes with a special interest in integrated pest management (IPM) within the general context of environmentally friendly and sustainable agriculture – http://www.endure-network.eu/what_is_endure
- C-IPM (Coordinated Integrated Pest Management in Europe): the project was an ERA-NET project in 2014–2016. Information is available on ENDURE Network website.
- PURE (pesticide use and risk reduction with IPM): the overall objective of PURE is to provide practical IPM solutions to reduce dependence on pesticides in selected major farming systems in Europe – http://www.pure-ipm.eu/project

8 References

Aubertot, J. N. 2017. D1.7. Final versions of DEXiPM assessment tools. Available at: http://www.pure-ipm.eu/node/449.

Bajwa, W. I. and Kogan, M. 2002. Compendium of IPM Definitions (CID) – what is IPM and how is it defined in the Worldwide Literature? IPPC Publication No. 998. Integrated Plant Protection Center (IPPC), Oregon State University, Corvallis, OR.

Barzman, M., Bàrberi, P., Birch, A. N. E., Boonekamp, P., Dachbrodt-Saaydeh, S., Graf, B., Hommel, B., Jensen, J. E., Kiss, J., Kudsk, P., Lamichhane, J. R., Messéan, A., Moonen, A., Ratnadass, A., Ricci, P., Sarah, J. and Sattin, M. 2015. Eight principles of Integrated Pest Management. *Agronomy for Sustainable Development* 35(4), 1199–215. doi:10.1007/s13593-015-0327-9.

Bernard, J. L. (coord.). 2013. *Protection intégrée des cultures. Fiches pour le conseil des techniques utilisables*. ACTA, Editions France Agricole, 256p.

Bernard, J. L. 2017. *Biocontrôle en protection des cultures: Périmètre, Succès, Freins, Espoirs*. L'Harmattan, Paris, 206pp (from Académie of Agriculture of France WG).

Bianchi, F. J. J. A., Booij, C. J. H. and Tscharntke, T. 2006. Sustainable pest regulation in agricultural landscapes: a review on landscape composition, biodiversity and natural pest control. *Proceedings. Biological Sciences* 273(1595), 1715–27. doi:10.1098/rspb.2006.3530.

Birch, A. N. E., Begg, G. S. and Squire, G. R. 2011. How agro-ecological research helps to address food security issues under new IPM and pesticide reduction policies for global crop production systems. *Journal of Experimental Botany* 62(10), 3251–61. doi:10.1093/jxb/err064.

Boriani, M., Agosti, M., Kiss, J. and Edwards, C. R. 2006. Sustainable management of the western corn rootworm (*Diabrotica virgifera virgifera* LeConte, Coleoptera: Chrysomelidae) in infested areas: Italian, Hungarian and USA experiences. *Bulletin OEPP/EPPO Bulletin* 36(3), 531–7. doi:10.1111/j.1365-2338.2006.01055.x.

Council Directive 92/43/EEC of 21 May 1992 on the conservation of natural habitats and of wild fauna and flora. Available at: https://eur-lex.europa.eu/legal-content/EN/ALL/?uri=CELEX%3A31992L0043.

Chandler, L. D., Coppedge, J. R., Edwards, C. R., Tollefson, J. J., Wilde, G. R. and Faust, R. M. 2008. Corn rootworm areawide pest management in the Midwestern USA. Publications from USDA-ARS/UNL Faculty. 652. Available at: https://digitalcommons.unl.edu/usdaarsfacpub/652.

Chandler, D., Bailey, A. S., Tatchell, G. M., Davidson, G., Greaves, J. and Grant, W. P. 2011. The development, regulation and use of biopesticides for integrated pest management. *Philosophical Transactions of the Royal Society of London B* 366, 1987-98.

Cook, S. M., Bartlet, E., Murray, D. A. and Williams, I. H. 2002. The role of pollen odour in the attraction of pollen beetles to oilseed rape flowers. *Entomologia Experimentalis et Applicata* 104(1), 43–50. doi:10.1046/j.1570-7458.2002.00989.x.

Délos, M. 2008. Méligèthes du colza: evolution et gestion des résistances aux pyréthrinoïdes en France et recommandations pour gérer le risque. *Bulletin OEPP/EPPO Bulletin* 38(1), 75–8. doi:10.1111/j.1365-2338.2008.01185.x.

Devos, Y., Munns Jr., W. R., Forbes, V. E., Maltby, L., Stenseke, M., Brussaard, L., Streissl, F. and Hardy, A. 2019. Applying ecosystem services for pre-market environmental risk assessments of regulated stressors. *EFSA Journal* 17 (Suppl.1), e170705. doi:10.2903/j.efsa.2019.e170705.

Duflot, R., Aviron, S., Ernoult, A., Fahrig, L. and Burel, F. 2015. Reconsidering the role of 'semi-natural habitat' in agricultural landscape biodiversity: a case study. *Ecological Research* 30(1), 75–83. doi:10.1007/s11284-014-1211-9.

EFSA Panel on Genetically Modified Organisms (GMO). 2010. Scientific Opinion on the assessment of potential impacts of genetically modified plants on non-target organisms. *EFSA Journal* 8(11), 1877.

EPPO Guideline. 2017. PP 1/296 (1) principles of efficacy evaluation for low-risk plant protection products. *Bulletin OEPP/EPPO Bulletin* 47, 297–304.

European Union. 2009. Directive 2009/128/EC of the European Parliament and of the council of 21 October 2009 establishing a framework for community action to achieve the sustainable use of pesticides. *Official Journal of the European Union* 52, 71-86. Available at: http://eur-lex.europa.eu/legal-content/EN/TXT/?uri=OJ:L:2009:309:TOC.

FAO. 1988. *Report of the FAO Council*, 94th Session, Rome.

FAO. 2002. *International Code of Conduct on the Distribution and Use of Pesticides*, 123. Session of the FAO Council, Rome.

FAO. 2019. Farmers taking the lead – thirty years of farmer field schools. FAO, Rome. Licence: CC BY-NC-SA 3.0 IGO.

Farrar, J. J., Ellsworth, P. C., Sisco, R., Baur, M. E., Crump, A., Fournier, A. J., Murray, M. K., Jepson, P. C., Tarutani, C. M. and Dorschner, K. W. 2018. Assessing compatibility of a pesticide in an IPM program. *Journal of Integrated Pest Management* 9(1), 1–6. doi:10.1093/jipm/pmx032.

Frearson, D. J. T., Ferguson, A. W., Campbell, J. M. and Williams, I. H. 2005. The spatial dynamics of pollen beetles in relation to inflorescence growth stage of oilseed rape: implications for trap crop strategies. *Entomologia Experimentalis et Applicata* 116(1), 21–9. doi:10.1111/j.1570-7458.2005.00299.x.

Glare, T. R. 2015. Types of biopesticides. In: Nollet, L. M. L. and Rathore, H. S. (Eds), *Handbook of Biopesticides*. CRC Press, pp. 7–24. doi:10.1201/b18014-4.

Gupta, S. and Dikshit, A. K. 2010. Biopesticides: an ecofriendly approach for pest control. *Journal of Biopesticides* 3(1), 186–8.

Holland, J. M., Bianchi, F. J. J. A., Entling, M. H., Moonen, A. C., Smith, B. M. and Jeannerete, Ph. 2016. Structure, function and management of semi-natural habitats for conservation biological control: a review of European studies. *Pest Management Science* 72(9), 1638–51. doi:10.1002/ps.4318.

Holland, J. M., Douma, J. C., Crowley, L., James, L., Kor, L., Stevenson, D. R. W. and Smith, B. M. 2017. Semi-natural habitats support biological control, pollination and soil conservation in Europe. A review. *Agronomy for Sustainable Development* 37(4), 31. doi:10.1007/s13593-017-0434-x.

ISPM 5: Glossary of Phytosanitary Terms. FAO 2017, http://www.fao.org/fileadmin/user_upload/faoterm/PDF/ISPM_05_2016_En_2017-05-25_PostCPM12_InkAm.pdf.

Jones, E. E. and Stewart, A. 2011. *Coniothyrium minitans* survival in soil and ability to infect sclerotia of *Sclerotinia sclerotiorum*. *New Zealand Plant Protection* 64, 168–74. doi:10.30843/nzpp.2011.64.5977.

Khan, Z. R., Pittchar, J. O., Midega, C. A. O. and Pickett, J. A. 2018. Push-pull farming system controls fall armyworm: lessons from Africa. *Outlooks on Pest Management* 29(5), 220–4. doi:10.1564/v29_oct_09.

Kiss, J. 2008. Peaceful co-existence with WCR? A case of integrated production/farming from Hungary. In: Governatori, G. (Ed.), *Integrated Pest Management for Corn: a Necessity or an Opportunity? The Case of the Western Corn Rootworm*. ERSA – Regional Agency for Rural Development, Autonomous Region of Friuli Venezia Giulia, Gorizia Italy, pp. 78–84.

Kiss, J., Komáromi, J., Bayar, J. K., Edwards, C. R. and Hatala-Zsellér, I. 2005. Western corn rootworm (*Diabrotica virgifera virgifera* LeConte) and the crop rotation systems in Europe. In: Vidal, S., Kuhlmann, U. and Edwards, C. R. (Eds), *Western Corn Rootworm: Ecology and Management*. CABI Publishing, Wallingford, Oxon, UK, pp. 189–220.

Lamichhane, J. R., Akbaş, B., Andreasen, C., Arendse, W., Bluemel, S., Dachbrodt-Saaydeh, S., Fuchs, A., Jansen, J. P., Kiss, J., Kudsk, P., Larena, I., Malet, J. C., Marzetti, A., Zweifel, J. and Messean, A. 2016a. *Strategic Research Agenda for IPM in Europe*. INRA.

Lamichhane, J. R., Aubertot, J. N., Begg, G., Birch, A. N. E., Boonekamp, P., Dachbrodt-Saaydeh, S., Hansen, J. G., Hovmøller, M. S., Jensen, J. E., Nistrup Jørgensen, L. N., Kiss, J., Kudsk, P., Moonen, A. C., Rasplus, J. Y., Sattin, M., Streito, J. C. and Messéan, A. 2016b. Networking of Integrated Pest Management: a powerful approach to address common challenges in agriculture. *Crop Protection* 89, 139–51. doi:10.1016/j.cropro.2016.07.011.

Lamichhane, J. R., Bischoff-Schaefer, M., Bluemel, S., Dachbrodt-Saaydeh, S., Dreux, L., Jansen, J. P., Kiss, J., Köhl, J., Kudsk, P., Malausa, T., Messéan, A., Nicot, P. C., Ricci, P., Thibierge, J. and Villeneuve, F. 2017. Identifying obstacles and ranking common biological control research priorities for Europe to manage most economically important pests in arable, vegetable and perennial crops. *Pest Management Science* 73(1), 14–21. doi:10.1002/ps.4423.

Lele, S., Springate-Baginski, O., Lakerveld, R., Deb, D. and Dash, P. 2013. Ecosystem services: origins, contributions, pitfalls, and alternatives. *Conservation and Society* 11(4), 343–58. doi:10.4103/0972-4923.125752.

Martin, E. A., Dainese, M., Clough, Y., Báldi, A., Bommarco, R., Gagic, V., Garratt, M. P. D., Holzschuh, A., Kleijn, D., Kovács-Hostyánszki, A., Marini, L., Potts, S. G., Smith, H. G., Al Hassan, D., Albrecht, M., Andersson, G. K. S., Asís, J. D., Aviron, S., Balzan, M. V., Baños-Picón, L., Bartomeus, I., Batáry, P., Burel, F., Caballero-López, B., Concepción, E. D., Coudrain, V., Dänhardt, J., Diaz, M., Diekötter, T., Dormann, C. F., Duflot, R., Entling, M. H., Farwig, N., Fischer, C., Frank, T., Garibaldi, L. A., Hermann, J., Herzog, F., Inclán, D., Jacot, K., Jauker, F., Jeanneret, P., Kaiser, M., Krauss, J., Le Féon, V., Marshall, J., Moonen, A. C., Moreno, G., Riedinger, V., Rundlöf, M., Rusch, A., Scheper, J., Schneider, G., Schüepp, C., Stutz, S., Sutter, L., Tamburini, G., Thies, C., Tormos, J., Tscharntke, T., Tschumi, M., Uzman, D., Wagner, C., Zubair-Anjum, M. and Steffan-Dewenter, I. 2019. The interplay of landscape composition and configuration: new pathways to manage functional biodiversity and agroecosystem services across Europe. *Ecology Letters* 22(7), 1083–94. doi:10.1111/ele.13265.

Matyjaszczyk, E. 2018. 'Biorationals' in integrated pest management strategies. *Journal of Plant Diseases and Protection* 125(6), 523–7. doi:10.1007/s41348-018-0180-6.

Maumené, C. 2018. Diminution de l'IFT : remplacer le T1 par du biocontrôle. Perspectives Agricoles n°461. Available at: https://www.perspectives-agricoles.com/view-1015 category.html?lasturl=1012&lasturlobjecttype=category&numpa=461.

Mészáros, Z., Ádám, L., Balázs, K., Benedek, I., Csikai, C., Draskovits, Á. D., Kozár, F., Lövei, G., Mahunka, S., Meszleny, A., Mihályi, F., Mihályi, K., Nagy, L., Oláh, J., Papp, J., Papp, I., Papp, L., Polgár, L., Radwan, Z., Rácz, V., Ronkay, L., Solymosi, P., Soós, Á., Szabó, S., Szabóky, C., Szalay-Marzsó, L., Szarukán, I., Szelényi, G., Szentkirályi, F., Sziráki, C., Szőke, L. and Török, J. 1984a. Results of faunistical and floristical studies in Hungarian apple orchards. *Acta Phytopathologica Academiae Scientiarum Hungaricae* 18, 91–176.

Mészáros, Z., Ádám, L., Balázs, K., Benedek, I., Draskovits, Á. D., Kozár, F., Lövei, G., Mahunka, S., Meszleny, A., Mihályi, K., Nagy, L., Oláh, J., Papp, J., Papp, L., Polgár, L., Radwan, Z., Rácz, V., Ronkay, L., Soós, Á., Szabó, S., Szabóky, C., Szalay-Marzsó, L., Szarukán, I., Szelényi, G., Szentkirályi, F. 1984b. Results of faunistical studies in Hungarian maize stands. *Acta Phytopathologica Academiae Scientiarum Hungaricae* 18, 65–90.

Midega, C. A. O., Pittchar, J. O., Pickett, J. A., Hailu, G. W. and Khan, Z. R. 2018. A climate-adapted push-pull system effectively controls fall armyworm, *Spodoptera frugiperda* (J.E. Smith), in maize in East Africa. *Crop Protection* 105, 10–5. doi:10.1016/j.cropro.2017.11.003.

Millennium Ecosystem Assessment (MAE). 2005. http://www.millenniumassessment.org /en/About.html.

Note de suivi Écophyto. 2017. http://www.consultation-ecophyto2plus.gouv.fr/presentat ion-du-projet-de-plan-ecophyto-ii-a1.html.

Ooi, P. A. C. 1996. Experiences in educating rice farmers to understand biological control. *Entomophaga* 41(3–4), 375–85. doi:10.1007/BF02765791.

Papp-Komáromi, J., Terpó, I. and Tokaji, M. 2007. Working together – farmer field schools in Hungary. *Pesticides News* 78, 8–9.

Pelzer, E., Fortino, G., Bockstaller, C., Angevin, F., Lamine, C., Moonen, C., Vasileiadis, V., Guérin, D., Guichard, L., Reau, R. and Messéan, A. 2012. Assessing innovative cropping systems with DEXiPM, a qualitative multi-criteria assessment tool derived from DEXi. *Ecological Indicators* 18, 171–82. doi:10.1016/j.ecolind.2011.11.019.

Pesticides Sold National Database. 2017. Pesticide sold national database-banque nationale de données des ventes des distributeurs de produits phytosanitaires (BNV-D). Available at: https://www.data.gouv.fr/fr/datasets/donnees-de-vente-de-pesticides-par-departement/.

Pilz, Ch., Keller, S., Kuhlmann, U. and Toepfer, S. 2009. Comparative efficacy assessment of fungi, nematodes and insecticides to control western corn rootworm larvae in maize. *BioControl* 54(5), 671–84. doi:10.1007/s10526-009-9209-x.

Pontius, J., Diltz, R. and Bartlett, A. 2002. *From Farmer Field School to Community IPM: Ten Years of IPM Training in Asia*. Food and Agriculture Organization of the United Nations, Regional Office for Asia and the Pacific, Bangkok.

Ruck, L., Cadoux, S. and Robert, C. 2017. Agronomic practices to control cabbage stem flea beetle and rape winter stem weevil Integrated Control in Oilseed Crops. *IOBC-WPRS Bulletin* 136, 65–7.

Szalai, M., Kiss, J., Kövér, Sz. and Toepfer, S. 2014. Simulating crop rotation strategies with a spatiotemporal lattice model to improve legislation for the management of the maize pest *Diabrotica virgifera virgifera*. *Agricultural Systems* 124, 39–50. doi:10.1016/j.agsy.2013.10.009.

Thakore, Y. 2006. The biopesticide market for global agricultural use. *Industrial Biotechnology* 2(3), 194–208. doi:10.1089/ind.2006.2.194.

Toepfer, S., Kurtz, B. and Kuhlmann, U. 2010. Influence of soil on the efficacof entomopathogenic nematodes in reducing *Diabrotica virgifera virgifera* in maize. *Journal of Pest Science* 83(3), 257–64. doi:10.1007/s10340-010-0293-6.

Toepfer, S., Kuhlmann, U., Kansiime, M., Onyango, D. O., Davis, T., Cameron, K. and Day, R. 2019. Communication, information sharing, and advisory services to raise awareness for fall armyworm detection and area-wide management by farmers. *Journal of Plant Diseases and Protection* 126, 103–6. doi:10.1007/s41348-018-0202-4.

van Lenteren, J. C., Bolckmans, K., Köhl, K., Ravensberg, W. J. and Urbaneja, A. 2018. Biological control using invertebrates and microorganisms: plenty of new opportunities. *BioControl* 63, 39–59. doi:10.1007/s10526-017-9801-4.

Vasileiadis, V. P., Sattin, M., Otto, S., Veres, A., Pálinkás, Z., Ban, R., Pons, X., Kudsk, P., van der Weide, R., Czembor, E., Moonen, A. C. and Kiss, J. 2011. Crop protection in European maize-based cropping systems: current practices and recommendations for innovative Integrated Pest Management. *Agricultural Systems* 104(7), 533–40. doi:10.1016/j.agsy.2011.04.002.

Vasileiadis, V. P., Moonen, A. C., Sattin, M., Otto, S., Pons, X., Kudsk, P., Veres, A., Dorner, Z., van der Weide, R., Marraccini, E., Pelzer, E., Angevin, F. and Kiss, J. 2013. Sustainability of European maize-based cropping systems: economic, environmental and social assessment of current and proposed innovative IPM-based systems. *European Journal of Agronomy* 48, 1–11. doi:10.1016/j.eja.2013.02.001.

Veres, A., Tóth, F., Kiss, J., Fetykó, K., Orosz, Sz., Lavigne, C., Otto, S. and Bohan, D. A. 2012. Spatio-temporal dynamics of *Orius* spp. (Heteroptera: Anthocoridae) abundance in the Hungarian landscape. *Agriculture, Ecosystems and Environment* 162, 45–51. doi:10.1016/j.agee.2012.08.009.

Veres, A., Petit, S., Conord, C. and Lavigne, C. 2013. Does landscape composition affect pest abundance and their control by natural enemies? A review. *Agriculture, Ecosystems and Environment* 166, 110–7. doi:10.1016/j.agee.2011.05.027.

Chapter 3

Improving regulatory approval processes for biopesticides and other new biological technologies in agriculture

Wyn Grant, University of Warwick, UK; and Roma Gwynn, Biorationale, UK

1 Introduction

The emergence of a regulatory state is a phenomenon that has attracted extensive commentary by analysts. It has been driven by demands for more enhanced forms of social and environmental protection, supported by campaigns by a range of non-governmental organisations. The limited fiscal powers of the European Union (EU) have encouraged it to rely on regulation to expand and deepen its sphere of competence. The innovative nature of biological technologies has encouraged regulatory interventions by governments across the world and these will be reviewed here. These regulatory interventions take the form of schemes of regulation specifically designed to cater to the particular requirements of biopesticides or reduced-fee schedules such as the Biopesticides Scheme launched in the United Kingdom in 2006.

Pesticides have been subjected to a particularly rigorous regulation. Conventional chemical plant protection products are synthesised: they are

http://dx.doi.org/10.19103/AS.2020.0073.04

not generally found in nature (though some synthetic pesticides are derived from molecules occurring in nature). Regulation therefore has to be strict to anticipate and prevent potential harm to human health or the environment. All players in the system have been trained and embedded into this precautionary approach. There is an unavoidable tension between regulation and the agrochemical industry yet the linkages between them are strong in terms of exchanges of expertise and personnel so they are co-reliant. What is changing is that some regulators are choosing to break these linkages, giving less priority to the views of the agrochemical industry, and moving to support substances that intrinsically cause less harm than some long-established pesticides.

For a product and its producers, regulation is a double-edged sword. It provides a means of protection from producers who make exaggerated or scientifically unsound claims for their products, evident in relation to 'grey area' products like plant enhancers or strengtheners. In that sense it provides reputational protection. It also ensures that there are no harmful side effects from new products, for humans (operators, consumers, bystanders and residents), the environment (air, soil and water) and non-target organisms (aquatic organisms, birds, soil organisms, invertebrates including bees). In addition, compatibility of new products with existing methods and product that may be used in crop protection is evaluated. However, it can also provide a barrier to new entrants and in that sense can serve as a dampener on valuable innovation.

2 Establishing a regulatory framework for biopesticides

The challenge is to design a regulatory system for biological technologies that is robust and scientifically rigorous, but not totally inflexible as biological technologies are a diverse group of substances that have their origins in nature and consequently can often be accommodated in the agroecosystem with no or limited perturbations or unintended consequences. The unintended consequences of some conventional pesticides such as the well-documented detrimental human and environmental effects resulting from (over)use of organophosphates have rightly created a cautious approach to their regulation (Hertz-Picciotto et al., 2018).

This overall precautionary approach is carried over to biological technologies mainly because of the regulator's unfamiliarity with a new field of research, for example, with microbial ecology rather than synthetic chemistry. It means less-experienced regulators of biological technologies revert to a very precautionary approach and demand more data and effectively raise standards for a set of substances for which this is usually not needed. However, it should be noted that some biopesticides can be toxic to non-target organisms, for example, rotenone to fish, amphibians and macro-invertebrates. Thus there are specific cases where precaution is warranted for natural product biopesticides. Moreover, biopesticides need to be used with appropriate concentrations, timings and frequencies of application to avoid adverse effects on non-target organisms and consumers, for example, residue issues relating to human toxicity or flavour taint.

It is not a question of non-regulation, but ensuring that the regulations are well-designed and appropriate. The regulatory system must not impose undue transaction costs on the regulator and the regulated, while ensuring that all appropriate checks relating to toxicity and other issues are made. It must not degenerate into a bureaucratic box-ticking exercise that inhibits scientific progress or unduly delays the availability of new products that may be of real benefit. It must be open to dialogue with those it seeks to regulate. Responding effectively to these challenges is not easy as bureaucrats are typically risk averse and successful regulatory innovation requires a fortuitous combination of circumstances including both exogenous and endogenous pressures (Greaves, 2009). This is in contrast to developers of biological technologies who tend to be innovative and less risk averse. This leads to a tension for regulatory development, not least in how the different parties communicate.

For a new regulatory approach, data requirements themselves are not necessarily the primary issue, but deciding which data are relevant, how they are interpreted and the experience and expertise of the person asking the questions. For example, the European Food Safety Agency (EFSA) is meant to perform this role, but it still does not employ or retain the specialist services of microbial ecologists. It still believes that general skills in food microbiology are sufficient even, for example, in assessing environmental effects. Examination of the peer review reports and dialogue between applicants, regulatory agencies, the Commission and EFSA demonstrates a frequent mismatch between the parties in understanding where the relevant risks might lie. For example, regulators often state that the dossiers submitted by applicants are poor and propose to resolve this by offering training to applicants. However, in some cases the dossiers may be sufficient but, to correctly understand them, the evaluators also need training in the new technologies. To be effective, all parties are best served if there is trust and a good partnership. There can also be poor levels of trust between national authorities.

3 Pesticide regulation in the European Union (EU)

As noted in the previous section, in addition to the need for mutual understanding between regulators and product manufacturers, effective regional and international regulation relies on trust and cooperation between national regulatory agencies. After the implementation of directive 91/414/EEC it quickly became evident that mutual recognition between different EU Member States was not working (Chandler et al., 2011). Implementation of EC/1007/2009 had a specific aim to improve mutual recognition but, in the experience of the authors, the lack of trust between regulatory authorities and bodies continues eight years after its coming into force, inhibiting the system from working well. Trust is eroded because some regulatory bodies consider that they apply better standards than others. For example, when becoming a member of the European Economic Area Agreement in 1994 Norway decided not to be a part of the EU pesticide regulations, because the Norwegian regulations were assumed to provide higher protection to human health and the environment. In June 2015 Norway finally implemented Regulation EC 1107/2009 and a new Norwegian regulation came into force. In the Norwegian Parliament it was argued that it would be wise to implement the EU regulation, because the criteria for pesticide assessment would be very similar to current practice in Norway (Kvakkestad et al., 2020). The consequence of this precautionary approach and supposed raising of standards is that Norway has four biological technologies approved. This is in contrast with neighbouring Sweden which has engaged more in building capacity in biological technologies, and which has 20 biological products approved.

Trust is eroded between applicants and regulatory bodies when applicants are communicating to evaluators who are clearly not sufficiently skilled in the art. This probably explains why applicants tend to use the same evaluators for microbials, based on countries which are more supportive of biopesticide development (Fig. 1), while some EU Member States are never or rarely used (e.g. Bulgaria, Czech Republic, Greece, Finland, Hungary, Ireland, Lithuania, Latvia, Romania and Slovakia).

The EU approval system for plant protection products was originally designed to cater for synthetic or conventional chemical pesticides. The effect

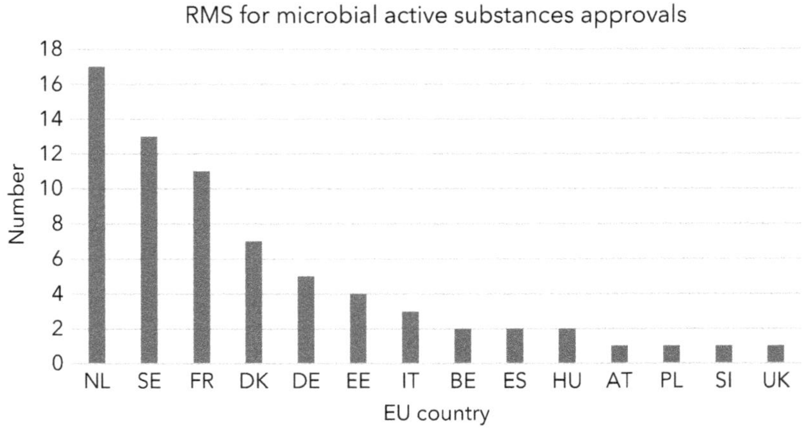

Figure 1 EU Rapporteur Member States responsible for approving microbial active substances (new, renewals and pending).

of the influence of conventional pesticides is found at all layers of the system because it was developed by and for them. Many of its procedures, or the questions that could be asked of applicants, were irrelevant for biocontrol products, for example, asking questions about the physical-chemical properties such as the boiling point of a microorganism but not asking about their biological properties. The majority of the companies researching and developing these products are micro companies, typically a spin-off from a university on a science park or innovation start-up. Staffed by scientists, they have limited understanding of the political process and few spare resources to devote for lobbying. They often find the checks and balances of regulation baffling and restrictive. They have lacked the political influence of the big agrochemical companies. Developing an effective representational structure for the emerging biopesticides industry took some time and continues to do so. Hence, a regulatory barrier was created to the entry of more of these sustainable products on to the market.

4 The development of the biopesticide sector and new regulatory requirements

The market share for biological technologies has grown from virtually nothing in 1993 to 4 billion US$ in 2018, in a span of 15 years, and is growing exponentially; forecast reveals the share to double and reach 8 billion US$ within the next 5 years. In terms of market share in 2018, Europe and the United States each hold 32%, Latin America 18% and Asia-Pacific 16% (Dunham, 2018). Of the different technologies globally available in 2018, microorganisms continue to dominate with 53% of the market, biochemicals 39% and macrobials 7.4% (Dunham, 2018). There has been some consolidation in the industry with multinational agrochemical companies such as Bayer, BASF and Syngenta and smaller multinational companies such as Certis having been brought into the sector by acquiring smaller companies. Individual spin-outs and start-ups have also been acquired by larger companies or by venture capital funders. The involvement of the larger pesticide companies in industry representation has strengthened the International Biocontrol Manufacturers Association (IBMA).

However, it can also be argued that the involvement of larger agrochemical companies is shifting the agenda, potentially increasing regulatory hurdles by making guidance documents more precautionary. This approach is informed by its existing experience from the conventional chemical regulatory framework. Well-resourced companies, embedded in using familiar conventional pesticide regulations, are used to a 'tick-box' approach to regulation where they, more or less, address every data requirement with a study, following established protocols. There is little consideration in presenting a reasoned-case approach for non-provision of study data. This is even when the nature of the substance being registered may make this applicable. This often ignores the explicit information in relevant guidance documents that a reasoned-case approach is accepted. This increases the information provided in a dossier of sometimes extraneous data and thereby increases the workload of assessors who are obliged to evaluate all information presented, relevant or not. The cost of providing this extra information is not a burden to a large company so there are fewer consequences to them from taking this approach.

The traditional approach of established pesticide manufacturers can be contrasted with the newer generation of biopesticide companies. They could develop products that are toxic or pathogenic (because these products work best against target pests) but they actively choose not to; they focus on substances which cause minimal harm. Those regulators who have developed expertise in relation to biological technologies have chosen a similar path: they are motivated to approve products that cause less harm. This approach is reinforced where there is a strong government policy to facilitate the availability of biological technologies such as the Netherlands and French governments' 'Green' agendas. One should add a note of caution. At present biological technologies are, by definition, seen to pose minimal harm to people or the environment. However, this may change if there are unforeseen adverse environmental impacts or harm to insects, animals or humans.

Nevertheless, biopesticide manufacturers have to overcome a framing of the crop protection dialogue around an approach that a pest controlled (> 80% efficacy) by a conventional chemical, leads to a good crop. This way of thinking has been embedded into the mental maps of farmers as well as support industries and regulators. Advocates for biopesticides are asked why products do not provide 80% efficacy, even though they do not need to. A 50% pest reduction twice equals 100%, so two applications are able to achieve an acceptable reduction in the target pest population. The aim is to reduce the pest population below a damage threshold, rather than eliminate the entire population. Expensive plant breeding programmes focus on overcoming resistance to conventional chemicals when they should be asking a different and more appropriate question about alternatives to conventional chemicals. Breeding programmes could focus more usefully on crop plants that have better natural defences and retain their natural ecology and relationship with microorganisms and so develop plants that have more resilience that many present varieties currently have.

5 Challenges in improving the regulatory framework for biopesticides

The biggest challenge that biopesticide manufacturers have faced in moving the debate forward has been the lack of a clear political constituency of support and hence of cohesive advocacy for biological technologies. Relatively few broadacre farmers have used them so there has been relatively little pressure from farming organisations for their wider availability (although the importance of the greenhouse sector creates a somewhat different political context in countries such as the Netherlands). In the United Kingdom, for example, the National Farmers Union (NFU) policy for integrated pest management (IPM) appears to be strongly influenced by and embedded in the use of agrochemicals. Farmers are subtly persuaded that they will fail if they do not have conventional chemistry and this is used to influence the conversation about what crop protection solutions will look like. The potentially influential organic sector is not really a source of support as they are suspicious of biological technologies or 'biopesticides' on the grounds that they are 'still pesticides'.

Retailers are a potential source of support because of their interest in food safety and quality issues. There has been a trend towards the privatisation of the regulation of such issues with big retail chains having their own supplementary systems operated by approved growing protocols which may prohibit their suppliers from using plant protection products approved by the state system. Many retailers certainly present themselves as 'greener' than their rivals as a useful means of market differentiation. However, they find it easier to give a message by promoting organics than by saying that their suppliers are users of IPM and biological technologies which does not convey a clear meaning to consumers.

In the United Kingdom at least, the biological technologies policy network has been weakly developed and relatively immature, particularly when compared with the agrochemicals network. An incomplete and loosely integrated policy network 'results in an underperforming network ... As one may expect, a very loose and incomplete network has little impact on observable policy outcomes' (Greaves and Grant, 2010: p. 35).

Nevertheless, some progress has been made at both the Member State level and the EU level. How has this been achieved? Analysis suggests that five factors have been important:

1 Interventions by global organisations such as OECD, IOBC, FAO and WHO which encourage international cooperation to develop harmonised, pragmatic and proportional regulatory requirements that are adapted to biological technologies. Certain EU agencies (EPPO[1] and MUCF[2]) and Member States (the Netherlands, France, Sweden, Germany and formerly the United Kingdom) are regular and active participants. Such activities support communication and shared experiences and

1 European Plant Protection Organisation: www.eppo.int.
2 Minor Use Coordination Facility: www.minoruses.eu.

engender trust between countries and develop guidelines that can be used by applicants and evaluators in the EU providing support where currently none exists.

2 Interventions by national governments who wish to enhance the contribution made by biological technologies to sustainability goals. The Biopesticides Scheme in the United Kingdom made a positive contribution in facilitating the successful completion of biopesticides registrations.
3 A greater commitment by EU regulatory agencies to biological technologies, in part stimulated by national governments policies to reduce chemical pesticide use and increase biological technology alternatives, but also in pursuit of 'first mover' advantage, although this commitment has its fragilities. Unfortunately, such commitments are often reliant on particular champions within agencies who may leave and they may also be affected by cutbacks in resources. However, this can work both ways. In some EU organisations there have been significant changes of personnel which bring in a new approach. There are early signs of a willingness to consider new ways of regulating biological technologies because of a desire to support food production that has less environmental impact.
4 The role of regulatory consultants both as a resource for the industry and as intermediaries between the industry and regulatory agencies. Their particular advantage is that they can look across the whole industry and have insight into its diversity whereas individual companies look inwards to seek solutions to show that their idea or technology can be proved to work. Consultants have also argued for greater political sophistication and the mobilisation of external expertise to assist this process.
5 Improvements in the capacities and sophistication of the European trade association, the IBMA which has had some impact on the policy-making process at the EU level. It has influenced the debate about regulatory changes and the guidance documents which inform their implementation. It has also raised the profile of biopesticides among decision-makers although this is an area in which more work needs to be undertaken.

The extent of this progress can be seen, for example, in current EU regulation in this sector.

6 Current EU regulation of biopesticides

In the EU, microbial, biochemical and semiochemical biological technologies are registered under the same regulatory framework as chemical pesticides although some special provisions have been developed. The legislation provides for measures that favour the registration of products defined as 'low risk' (most of the biological technologies) but the process has not worked well in practice. Although 'biological technologies or biopesticides' do not exist as a regulatory category, the pesticide categories 'basic substances' and 'low risk substances' were introduced in August 2017, as defined in Regulation 2017/1432, amending Regulation 1107/2009.

Before plant protection products (PPP) could be introduced to the European market, the active substance must be approved for the EU, after which authorisation of the product can be sought in individual Member States. For the approval of an active substance, the applicant submits a dossier to an individual Member State, the Rapporteur Member State. The Rapporteur Member State produces a Draft Assessment Report, which is submitted to the European Commission and the EFSA which is responsible for peer reviewing the risk assessments of active substances. Finally, the European Commission presents a review report to the Standing Committee for Food Chain and Animal Health (SCFCAH), who votes on approval or non-approval of the active substance.

It should be noted that the resources available to regulatory bodies in the Member States vary considerably and this in turn affects their competence and reputation. As a broad generalisation, those in Northern Europe and one or two Member States in Central Europe are the most innovative, while those in the rest of Central and Southern Europe have more limited capacity. This is reflected in the Member States that have invested capacity for evaluating microbial active substances (Fig. 1).

Under the current regulatory system (in Reg. (EC) 1107/2009), and depending on the complexity and completeness of the dossier, it takes 2.5–3.5 years to approve a new active substance. The subsequent authorisation of a product is carried out by the Member States, with the applicant specifying the Member State carrying out the evaluation which takes a further 1.5–2 years. Following authorisation, a company may apply for mutual recognition within one of the three geographic zones (North, Central and South) for products with the same use(s) under similar agricultural conditions. The approval process takes the same time regardless of the type of active substance or the dossier size (e.g. the majority of biopesticide dossiers contain about 10% of the paperwork of a conventional pesticide). Biopesticide dossiers are smaller because they usually do not trigger the need for higher level studies, for example, for many biopesticides a reasoned case can be provided to justify the lack of relevant residues whereas extensive studies are needed for most conventional pesticides.

The IBMA White Paper comments, 'At the latest count, 60% of new submissions to the EU are bioprotection technologies but evidence indicates that, the existing approach is not working well in Europe, despite guidance produced for use in Reg. (EC) 1107/2009 for these types of technologies' (IBMA, 2018). It suggests a single EU body for regulating bioprotection technologies. This would coordinate registration of active substance and products integrally (together) and would ensure evaluations are provided by experts with the proper field of expertise and relevant scientific knowhow. However, the chances of securing such a body are not good at this time with the EU distracted by so many issues.

7 Global national initiatives in biopesticide regulation

Within the developed world, the United States has attracted praise for its proactive approach to the encouragement of biological technologies or biopesticides. In the United States, in 1994, the Environmental Protection Agency (EPA) established the Biopesticides and Pollution Prevention Division (BPPD) in the Office of Pesticide Programs to facilitate the registration of biopesticides. The BPPD aims were to promote the use of safer pesticides, including biopesticides, as components of IPM programmes. The success of a biopesticide-specific registration programme is demonstrated by the United States having the highest number of active substances available worldwide. In early 2019 there were 366 registered biopesticide active ingredients compared to 135 approved and 16 pending in the EU. A 'white paper' from the Europe-based IBMA states:

> The USA EPA process for regulating biopesticides is the best example of a system which has been set up solely to regulate bioprotection products. It has a track record of delivering new product approvals through robust rigorous review within a year of submission since it was established in 1994. Importantly it follows a tiered approach and reviews are conducted by dedicated evaluators with relevant expertise (IBMA, 2018).

More recently, national initiatives on biological technologies registration have been assisted by the production of a joint Food and Agriculture Organization (FAO) and World Health Organisation (WHO) document 'Guidelines for the registration of microbial, botanical and semiochemical pest control agents for plant protection and public health uses' (FAO and WHO, 2017) which provides a guidance framework for national policy and replaces an earlier 1988 document. The FAO and WHO documents emphasise that, in relation to the registration of biological pest control agents, 'Harmonization of the different approaches is important for streamlining and facilitating the research, development, commercialization and use of biopesticides for plant protection and public health. Using similar data requirements and evaluations should make it easier for applicants to submit applications to different countries and for regulatory agencies to benefit from each other's reviews' (FAO and WHO, 2017).

A number of countries outside Europe have been taking initiatives to boost the use of biological technologies, foremost among them China, Kenya and Cuba. Over the past five years, the number of biological technologies registered in China has enjoyed a 16% increase while growth of conventional chemical products grew just 3%. They have adopted the FAO and WHO guidelines. The Chinese government is 'eager to find safer alternatives to the conventional chemicals currently used domestically. Food safety is a massive concern in China and the government has tried to drive substitution of hazardous conventional chemicals domestically by subsidizing biopesticide users and offering favorable registration policies'. The authorities started to draw up a supplementary document for data requirement on biological technology registration in 2010 and a revised draft of the Data Requirement on Pesticide Registration was released for public consultation the following year. However, the consultation never progressed any further, partly because of delays in approving the overarching regulation being developed for the sector, the 'Regulation on Pesticide Administration'. The biggest obstacle facing China's biopesticide registration policy remains the delay in the legislative process for approving this overarching regulation.

Horticulture is a vital industry for Kenya with its products accounting for 92% of exports. The adoption of maximum residue levels (MRLs) by the EU in 1994 provided a stimulus for the extensive use of biological control agents. About 30 products derived from plants or microbes are registered in Kenya. 'In 2003, in response to horticulture industry needs, Kenya's pesticide regulatory authority, the Pest Control Products Board (PCPB), supported by a range of stakeholders such as the UK Department for International Development (DFID) Crop Protection Programme, developed biological technology-specific registration pathways, including one for microbial pest control products' (Gwynn and Maniania, 2010). The total number of biological technology products available in Kenya in 2008 was 25. By 2018 this had doubled to 51. Thirty-four products were active against insects: 12 microorganisms based on *Bacillus thuringiensis* (Bt), 12 other microorganism non-Bt and 10 botanical. There were 13 products active against diseases and 4 against nematodes and 3 with other activity.

Cuba is characterised by intensively cropped shareholdings, often located in residential areas, which use biomanagement strategies for crop protection, although there is a shortage of supply of the required biological technologies. However, it has achieved a substantial reduction in imports and made suitable products available for small-scale farmers. 'The regulation system has to be flexible and not too costly or too lengthy but it must have the full and justified trust of the public. As current legislation stands, there are certain categories of microbial pesticides that have an easier and quicker passage to registration than others' (Diaz, 2004).

8 Developing good regulatory practice

In the EU over 30% of approved active substances are now biological technologies. Yet, with one or two Member States that are good exceptions, the problems within the EU regulatory systems for 'alternative' active substances and products are similar to those in many other regions or countries:

- Lack of specific data requirements for this new class of products;
- Limited knowledge and experience within government regulatory agencies;
- Lack of government financial support to regulatory agencies to build their capacity; and
- No mechanisms for a fast track system for products that are, in the main, inherently less harmful to humans and the environment.

Organisations such as the IBMA and its sister organisations globally have, for at least the last 15 years, publicised these shortcomings and barriers and proposed well thought-out solutions. The continuing inertia, to reduce the regulatory barriers to allow better crop protection solutions to get onto the market, is a disservice in achieving a more sustainable future.

However, the activities of international bodies such as OECD, FAO and WHO, and good national programmes, show that there are good examples of initiatives pushing for good regulatory practices that support disciplined and robust regulation but that remain proportional to the risks that biological technologies may present to consumers and the environment. As outlined in this chapter, there are some countries which have been proactive to reduce the barriers and consequently, they have more biological technologies available for farmers.

It is all too easy for any activity on biological technologies regulation to remain embedded in the detail of how to prepare guidance. This creates the illusory sense of progress but, without a framework for implementing this guidance, it does not substantially change timelines for regulatory approvals. What would good regulatory practice be? Certainly, it would be one that balances properly the need to be safe with the need for resilient crop protection tools that fit good quality IPM and supports farmers to grow a good quality crop with good yields at economically viable prices.

To better regulate biological technologies and emerging technologies is to ensure all participants in the process are well skilled in the relevant technologies. This needs investment because of the diversity of technology types. These skills are needed by the applicants, the contract research organisations running the testing (with appropriate testing methods), the evaluators, the peer review organisations and finally the decision-makers (risk assessment and risk management). Any break in this chain can lead to difficulties and delays, for example, if the lack of expertise acts as a barrier and leads to delays and usually additional costs through requests for extra studies and delays in getting products onto the market.

Technology transfer and knowledge exchange for biological technologies need to be actively pursued by governments to resource their representatives adequately so they can acquire the skills to fulfil their regulatory roles. Governments and public policies are requiring that food is grown following IPM practices and that alternative technologies are the mainstay of this more environmentally friendly approach. Yet they are failing to resource their regulatory agencies sufficiently to allow them to fulfil this obligation so combined with the inherent inertia progress to adopt new technologies progress is slow. There need to be additional efforts and resources for regulators to build their capacity in new technologies. Many regulatory agencies work on a full-cost recovery basis within which there is no mechanism to fill the funding gap for capacity building in new technologies.

More precisely 'Good Regulatory Practice Guide for biological technologies (alternatives)' is one that upholds the intent of government policy for IPM which promotes alternatives and indicates that good practice involves:

1 Dedicated regulators (at all levels) that have the necessary skills for biological solutions (microbial, natural substances-botanical, semiochemical).
2 Having a strong partnership using a clear, structured pre-submission process that engages equally both evaluators and applicants for example:
   a initial enquiry with evaluators, ideally evaluators visiting the applicants research facility, to discuss broad principles while the applicant is still developing new active substances
   b early-stage meeting once the active substance specification is established
   c mid-stage meeting during initial dossier development
   d late-stage meeting once studies have been done and the dossier has been prepared
   e completeness check

   During this process there should be interim exchanges of emails to confirm studies proposed and so on.
3 Providing a high-level framework for the principle that data are excluded 'except when ...' and the exceptions are explained: for example, are

ecotoxicology studies for microbials needed when it is known from their ecology they are specifically plant pathogen hyperparasites? In principle it could be stated that microbials are of concern only for arthropod non-targets when they are insect pathogenic.

4 Trusted and respectful partnership between evaluators for quick resolution of any issues and for promoting harmonised outcomes.
5 Harmonisation of evaluations that the same standards are applied across evaluations.
6 Reciprocity of evaluations, if a dossier is accepted in one country at a suitable standard then it is automatically accepted by another, without re-evaluation. There are already agreements on acceptable standards so this can be implemented quickly if there is a political will.
7 Reciprocity for efficacy: data from one region should be accepted in another area, provided there are no major differences in uses and agronomic practices.

The unrealised advantage of regulating biological technologies better is that the focus can be on where the real risks are and it frees up resources to pay correct attention to these. This in turn should speed up the evaluation process supporting innovative biological technology companies and provide farmers with more much-needed IPM tools. The consequent benefits also accrue to society.

9 Conclusions

Biological technologies have a key but specific role to play in crop protection. Synthetic chemical pesticides are a precious resource for crop protection and should be treated as such. However, the occurrence of human and environmental harm because of misuse or overuse of pesticides, problems of resistance to pesticides and withdrawal of some products means that fewer chemical pesticides are available in the market. Alternative crop protection tools are needed and biological technologies have an increasing and important role to play in crop protection, especially as part of IPM programmes.

Biological technologies should be regulated. Just because something is natural does not mean it is safe. However, the regulatory system has tended to follow a chemical pesticide model that does not facilitate the efficient registration of biological technologies. An international 'epistemic community' has been formed among experts, facilitated by organisations such as FAO, WHO and OECD, which has been able to identify the key unresolved issues, recommend solutions to technical challenges, promote harmonisation of regulation and suggest ways forward.

However, this agenda for more effective forms of regulation will only be fulfilled with sufficient political support. There is greater interest in biological technologies than in the past and an increased recognition of the contribution they can make to environmentally sustainable solutions to plant protection. This reflects a number of factors, including greater effectiveness on the part of the industry representative body, the IBMA, including a willingness to build coalitions with environmental groups. It also reflects the greater involvement of major agrochemical companies in the sector as market opportunities improve. However, public knowledge and understanding of biological technologies remains limited and this means that there is no substantial constituency of public support for their greater use. This in turn influences the priority that political decision-makers give towards improving the responsiveness and speed of regulation that allows new products to enter the market.

10 Where to look for further information

Each country maintains information and resources (usually internet-based) for the countries' regulations for plant protection produces, depending on the countries expertise in biopesticides, these resources can provide good guidance, for example, the Netherlands regulatory agency has guidance on how to prepare a dossier for biopesticides. Details of any guidance or other provisions for biopesticide registration are provided by organisations such as the EU department for the environment (DG SANTE), by the OECD Pesticides facility and by the UN agency, FAO on their IPM webpages.

Organisations such as IBMA and its sister organisations worldwide, actively participate to get provisions made for biopesticide regulation and for regulations to be adapted for them.

Research that relates to the regulatory risk areas for biopesticides supports the efforts of industry and provides good-quality scientific evidence to underpin the development of appropriate and proportional regulation. There is a need to consider the impact of Brexit on regulation in the UK and how this interacts with EU regulation. More generally, the responsiveness of regulation to a variety of stakeholders requires analysis and drawing an appropriate balance between risk avoidance and facilitating innovation.

11 References

Chandler, D., Bailey, A. S., Tatchell, G. M., Davidson, G., Greaves, J. and Grant, W. P. 2011. The development, regulation and use of biopesticides for integrated pest management. *Philos. Trans. R. Soc. Lond., B, Biol. Sci.* 366(1573), 1987–98. doi:10.1098/rstb.2010.0390.

Diaz, L. H. 2004. Registration of biological pesticides in Cuba. In: Wabule, M. N., Ngaruiya, P. N., Kimmins, F. K. and Silverside, P. J. (eds), Registration for Biocontrol Agents in Kenya. Kenya Agricultural Research Institute, Nairobi, pp. 117–34.

Dunham, W. 2018. Biocontrol's position in the global marketplace. Presentation to *ABIM Conference*, Basel.

FAO and WHO. 2017. International Code of Conduct on Pesticide Management: guidelines for the registration of microbial, botanical and semiochemical pest control agents for plant protection and public health uses.

Greaves, J. 2009. Biopesticides, regulatory innovation and the regulatory state. *Public Policy Admin.* 24(3), 245–64. doi:10.1177/0952076709103810.

Greaves, J. and Grant, W. 2010. Underperforming policy networks: the biopesticides policy network in the United Kingdom. *Br. Pol.* 5(1), 14–40. doi:10.1057/bp.2009.15.

Gwynn, R. L. and Maniania, J. N. K. 2010. Africa with special reference to Kenya. In: Kabaluk, J. T., Svircev, A. M., Goettel, M. S. and Woo, S. G. (Eds), *The Use and Regulation of Microbial Pesticides in Representative Jurisdictions Worldwide*. IOBC Global, 99p.

Hertz-Picciotto, I., Sass, J. B., Engel, S., Bennett, D. H., Bradman, A., Eskenazi, B., Lanphear, B. and Whyatt, R. 2018. Organophosphate exposures during pregnancy and child neurodevelopment: recommendations for essential policy reforms. *PLoS Med.* 15(10), e1002671. doi:10.1371/journal.pmed.1002671.

IBMA. 2018. A new EU regulatory framework for bioprotection agents: IBMA vision on how to improve regulation in the European Union. IBMA White Paper.

Kvakkestad, V., Gwynn, R., Sundbye, A. and Klingen, I. 2020. Authorization of microbial plant protection products in the Scandinavian countries: a comparative analysis. *Environmental Science and Policy* 106, 115–24. doi:10.1016/j.envsci.2020.01.017.

Part 2

Microbial biopesticides, entomopathogenic nematodes and mites

Chapter 4

Advances in the use of entomopathogenic fungi as biopesticides in suppressing crop pests

Enrique Quesada-Moraga, Meelad Yousef-Naef and Inmaculada Garrido-Jurado, University of Córdoba, Spain

1 Introduction

Today's agriculture faces multiple challenges; the two foremost are ensuring food security for a world population expected to grow to nearly 9 billion people by 2050 through highly efficient and sustainable plant production and food safety as the toxic materials that are routinely used for crop protection may persist in the food we eat (Hillocks, 2012). Despite the use of pest control measures, global figures for crop losses still show that pests, diseases, and weeds are reducing around 30–35% of the total agricultural production (Oerke, 2006).

Most of the current pest management practices are mainly based on the use of synthetic pesticides, whereas alternatives to broaden the suite of applicable tools beyond these chemicals have been sought due to increasing awareness concerning the possible negative impacts of pesticides on human health, as well as impacts on non-target organisms, and soil and aquatic ecosystems. Indeed, increased public awareness and societal demands about

http://dx.doi.org/10.19103/AS.2020.0073.05

the potential risks of pesticides have resulted not only in the withdrawal of many active ingredients of pesticides (i.e. more than 70% of them in the EU in the last 10 years) but also in continuous bans of other key conventional pesticides and heavy restrictions of new active compounds being marketed (Lamichhane et al., 2016). Therefore, the demand for alternative and feasible safe plant protection approaches and solutions, as offered by integrated pest management (IPM) methods, is increasingly required.

In IPM, the integration of the various cultural, chemical, plant-resistance, and biological control measures employed in a combined way where pesticides are used only as a last resort ensures that pests remain below the economic threshold, thus supporting food security and safety and international market access. Hence, one of the major current challenges of the agriculture is that effective and economically feasible alternatives to synthetic pesticides are not yet sufficiently available and even some of them are still under development (Furlan et al., 2018).

Microbial control of insect pests is considered to be among the most viable alternatives to synthetic chemical pesticides (Eilengberg and Hokkanen, 2006). However, not all entomopathogenic microorganisms invade susceptible hosts in the same manner. For instance, viruses, bacteria, and protozoa must be ingested, while entomopathogenic fungi (EPF), with emphasis on hypocrealean ascomycetes, may enter their hosts by direct penetration through the cuticle, which, together with their natural incidence on the pest populations, put them at the forefront of the global development, innovation, and discovery of alternative control strategies and as a significant component of any IPM (Quesada-Moraga and Santiago-Álvarez, 2008).

2 Natural occurrence and biodiversity

Traditionally, the soil, together with their insect hosts, has been considered the main reservoir of EPF (Klingen and Haukeland, 2006). However, recently, important ecological niches also occupied by EPF have been highlighted: endophytic relationships of EPF with the plants (Quesada-Moraga et al., 2006b; Vega, 2018) and presence in the plant phylloplane (Garrido-Jurado et al., 2015), in the rhizosphere (Hu and St Leger, 2002), and even in the air (Aira et al., 2007). Nevertheless, these newly discovered ecological niches, where EPF can be found, do not exclude their traditional role as insect pathogens, but they also offer a bifunctional lifestyle in association with plants (Wang and St Leger, 2007). The unexpected roles of EPF provide new potential applications in agriculture in response to abiotic stresses and phytopathogen organisms (Ownley et al., 2008, 2010; Sánchez-Rodríguez et al., 2015, 2016).

The genetic evolution of EPF for the wide range of lifestyles is poorly understood. Recent phylogenetic studies on the genus *Metarhizium*

(Ascomycota: Hypocreales) has shed light on host-pathogen interactions. The 12 species of *Metarhizium* show differences on host specificity probably due to their adaptation to the immune system of the insect host (Bischoff et al., 2009; Wang et al., 2016). In addition, the speciation of *Metarhizium* suggests that it occurs via intermediate host range from specialists to generalists, and it has been speculated that the ancestor of this EPF was a plant-related fungus, which later acquired the entomopathogenic feature (Fig. 1) (Hu et al., 2014). *Metarhizium* spp. are usually found colonizing roots and infecting insects principally in the soil, leading to the hypothesis that the dual lifestyle is due to the nutrient transfer from insects to plants (Meyling et al., 2011; Behie et al., 2013). In contrast, *Beauveria bassiana* Bals. (Vuill) (Ascomycota: Hypocreales) is usually found endophytically colonizing plants and infecting insects above ground (Meyling et al., 2011; Behie et al., 2015). In fact, phylogenetic studies using the EF1-α gene and the intergenic nuclear region BLOC suggest that *B. bassiana* in the phylloplane could have resulted from the dispersal of fungal propagules from the soil, whereas some isolates

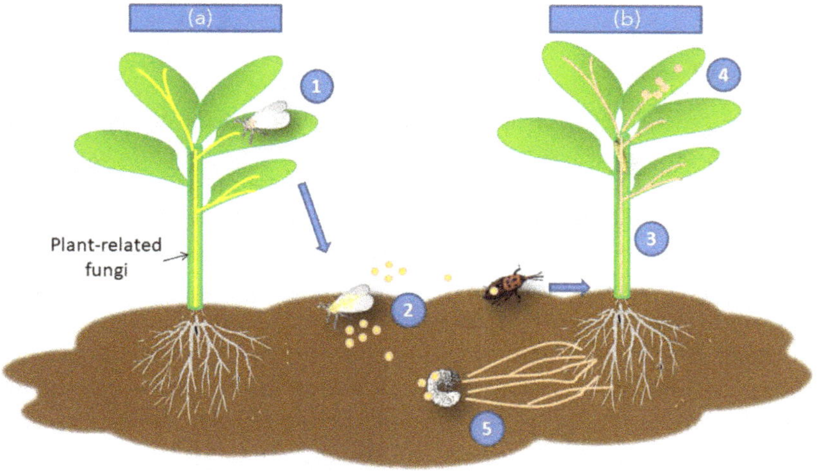

Figure 1 Diagram illustrating the hypothetical scenario in which the insects have acquired the entomopathogenicity feature and the development of new ecological roles. Situation (a) represents a past time and situation (b) the present time. The yellow color represents a non-entompathogenic fungus, while the orange one shows the acquisition of the entomopathogenicity. (1) A plant-related fungus (yellow) is able to infect a sucking insect for first time. (2) The infected insect disperses the fungal inoculum (yellow and orange) and it can infect other insects above and below the ground. (3) After a long time of surviving in the soil and insects, the fungus (orange) can colonize plant endophytically because of the fungal infection of boring insects principally and the consequent adaption to the new conditions inside the plants. (4) Some fungi can survive in or on the plant without the presence of insects. (5) The fungus may also transfer nutrients from the insects to the plant becoming rhizosphere competent.

seem to be true epiphytes (Garrido-Jurado et al., 2015; Fernández-Bravo et al., 2016; Fig. 1).

The new discovery of these diverse ecological roles supports the importance of diverse EPF communities in crop systems. Usually, natural habitats with low human disturbance have more diversity of EPF communities, and it is not related to the seasonal factors (Meyling and Eilenberg, 2007; Garrido-Jurado et al., 2015). This diversity is influenced by soil parameters, since some physico-chemical factors such as texture or pH determine the presence and availability of the EPF (Quesada-Moraga et al., 2007; Garrido-Jurado et al., 2011a,b). *Metarhizium* spp. are more abundant in cultivated habitats, while *Beauveria* spp. are usually isolated from undisturbed natural areas, probably because of the abundance and diversity of insect hosts (Quesada-Moraga et al., 2007; Meyling et al., 2009). Phylogenetic studies on fungal communities performed with microsatellites have shown multispecies communities, with one species being dominant, and also mating-type loci have revealed that natural habitats enable the sexual recombination of *B. bassiana*, although local populations keep the asexual structure (Meyling et al., 2009).

3 Mode of action

The adhesion of conidia or specialized fungal structures to the cuticle is only the beginning of the invasion of the host. Germination, penetration into the host, modulation of cellular and humoral defenses, and fungal growth inside the hemocoel conclude with the death of the host, which is caused by, among other factors, nutrient depletion, the invasion of tissues and organs, and asphyxia due to the development of the fungus in the respiratory system and/or the production of toxic fungal compounds (Fig. 2). The life cycle of the fungus is completed with sporulation, when hyphae emerge from the cadaver and produce conidiophores and conidia, which allow for horizontal transmission. Recently, the possibility of 'accidental' death during the pathogenesis cycle has also been discovered. In this case, both conidia and fungal compounds may produce stress in the insects and consequently their death (Butt et al., 2013; Garrido-Jurado et al., 2016; Alkhaibari et al., 2018).

4 Delivery methods

Mycoinsecticides can often be applied with existing spray equipment, in spray treatments with liquid formulations targeting the soil or the plant or granules usually targeting the soil. A more recently described strategy to expose insects to EPF named 'attract-and-infect' is to lure them to baits in which a high concentration of fungal spores are present and thereby picked up by the

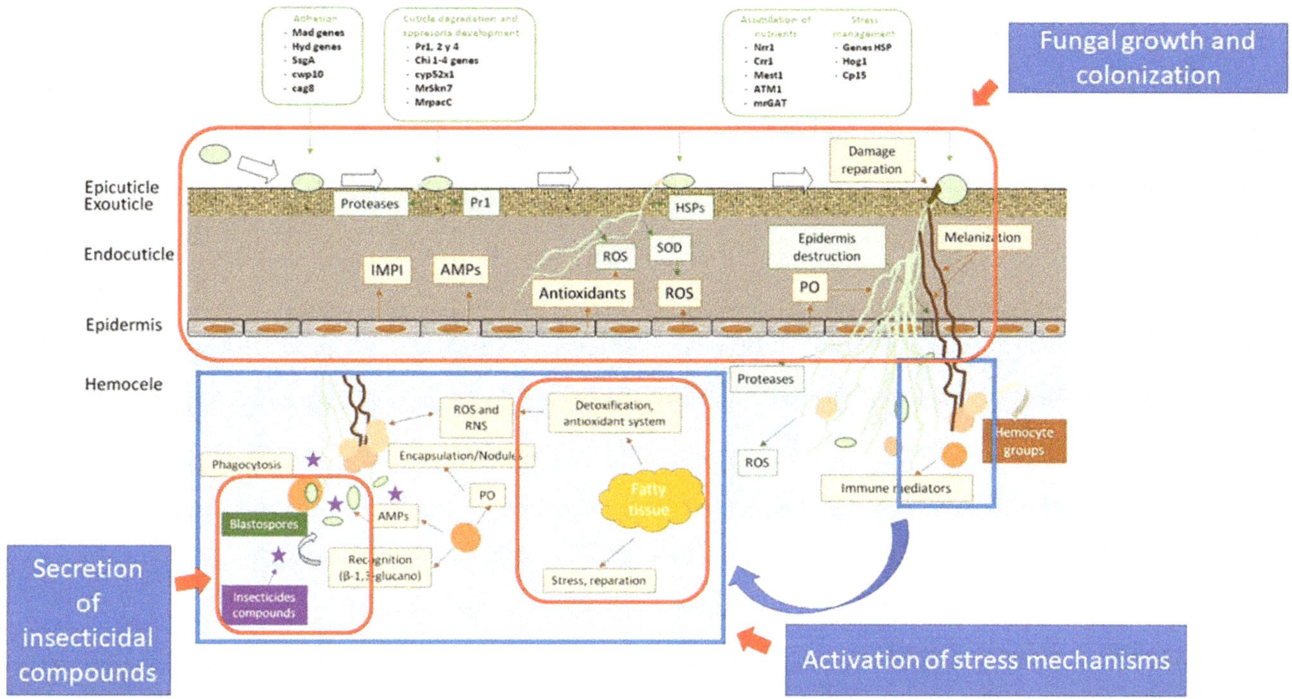

Figure 2 Pathogenesis cycle of an entomopathogenic mitosporic ascomycete and possible interactions fungus-insect in each stage of the infection process. Own preparation based on Butt et al. (2016) and Wang and Wang (2017). Red cycles indicate the three mechanisms that can cause the insect death.

attracted insects after landing on the bait. Hence, the chance of a successful infection of the insect is increased because it will pick up higher doses of the spores via the baits than possible via crop spraying, especially for dipteran adults such as spotted wing drosophila *Drosophila suzukii* Matsumura (Diptera: Drosophilidae) and Mediterranean fruit fly (medfly), *Ceratitis capitata* (Wied.) (Diptera: Tephritidae) (Yousef et al., 2018a). Moreover, these 'attract-and-infect' devices are provided with holes that allow the free entry and exit of the adults, which improve the autodissemination of the fungal inoculum in the population and therefore its efficacy by horizontal transmission from males to females and vice versa (Quesada-Moraga et al., 2006a). Additional advantages of this technique are the reduced volume of spores that have to be used under field conditions compared with the traditional inundation method and, therefore, the reduction of fungal residues on the crop. Furthermore, this technique lowers inoculum at the field level, minimizing the possible effect on non-target organisms.

The additional ecological roles previously described may provide additional delivery strategies of EPF. Some EPF have been identified as natural endophytes in more than one host plant species [i.e. *B. bassiana*, *Lecanicillium lecanii* (Zimm.) Zare & W. Gams (Ascomycota: Hypocreales)] (Quesada-Moraga et al., 2014; Vega, 2018), which indicates that they are multi-host endophytes capable of colonizing several plant species. Artificial inoculation of other EPF has been achieved in numerous plant hosts (Quesada-Moraga et al., 2014; Vega, 2018). The endophytic behavior of EPF has altered the rationale behind the use of EPF as biocontrol agents in agriculture, providing new approaches to pest management. Indeed, there have been several studies aimed at using these endophytic strains as biological control agents against specific pests, whose life cycle (e.g. they feed endophytically producing extensive tunneling in stems, pseudostems, rhizomes, roots, and seeds) seriously limits the effectiveness of chemical insecticides and other control methods (Quesada-Moraga et al., 2006b, 2013).

Compared to conventional mycoinsecticides, the use of EPF as artificially introduced endophytes has the advantage of targeting the pest within the plant at reduced application costs because little inoculum is required in cases where colonization is systemic. Furthermore, the endophytic fungus is protected inside the plant from detrimental abiotic and biotic factors that would limit its use as an epiphyte. Most of these studies have completed only the first stage, that is, inoculation into the plant, although some of them have gone further, stating that the endophytic colonization of the plant by an EPF affects the survivorship and development of the cryptic insects, while reducing plant damage (Quesada-Moraga et al., 2009).

Meanwhile, it has been recently demonstrated in our laboratory that spray application with conidial suspensions targeting chewing and sucking insect pests can be accompanied by a transient endophytic colonization of plant tissues, leading to a mortality rate that is additional to that of the direct spray (Garrido-Jurado et al., 2016; Resquín-Romero et al., 2016). In addition, it has been revealed that *B. bassiana* and *M. brunneum* Petch (Ascomycota: Hypocreales) may initiate infection cycles in chewing and sucking insects feeding on endophytically colonized tissues (Garrido-Jurado et al., 2016; Resquín-Romero et al., 2016), whereas killing strategies were different for both fungal species and for both feeding habits (Garrido-Jurado et al., 2016; Resquín-Romero et al., 2016).

5 Use of entomopathogenic fungi as biopesticides

Over 700 different fungal species are known to be pathogenic to insects. However, only a few genera are well recognized as entomopathogens, including *Beauveria*, *Metarhizium*, *Isaria*, *Lecanicillium*, and *Hirsutella*. Various fungal-based products containing *B. bassiana*, *Metarhizium anisopliae*, *Isaria* spp., and *Lecanicillium* spp. have been developed for use against a wide variety of pests in forest, field, and greenhouse environments, and against structural, household, and storage pests. Commercial formulations of a range of EPF are available to farmers in most parts of the world, and in the twenty-first century, due to greater commercialization, EPF use has expanded.

As described before, not all entomopathogenic microorganisms invade susceptible hosts in the same way. While viruses and bacteria have to be ingested (usually with food), entomopathogenic ascomycetes enter via the exoskeleton, a mode of action by contact that makes them an attractive component of IPM programs for the biological control of several groups of pests of agricultural, forestry, veterinary, and medical importance by using classical biological control, augmentation, or conservation (Fig. 3). The emphasis for presenting the selected examples in the following subsections was to focus on ongoing programs mainly employing ascomycetes in the genera *Metarhizium* and *Beauveria*. A recent review by Vega and Kaya (2012) provides further information and case studies.

5.1 Locust and grasshoppers

Acridid control is necessary to prevent locust damage to crops, and alternatives to synthetic pesticide spray treatments are needed. EPF act as key natural enemies of locust and grasshoppers. Several mycoinsecticides such as the *Metarhizium*-based Green Muscle® and Green Guard® have been successfully applied in Africa and Australia for locust control. Insects such as locusts and grasshoppers can reduce the effectiveness of pathogens and parasites by adopting different defense strategies. Apart from the cost

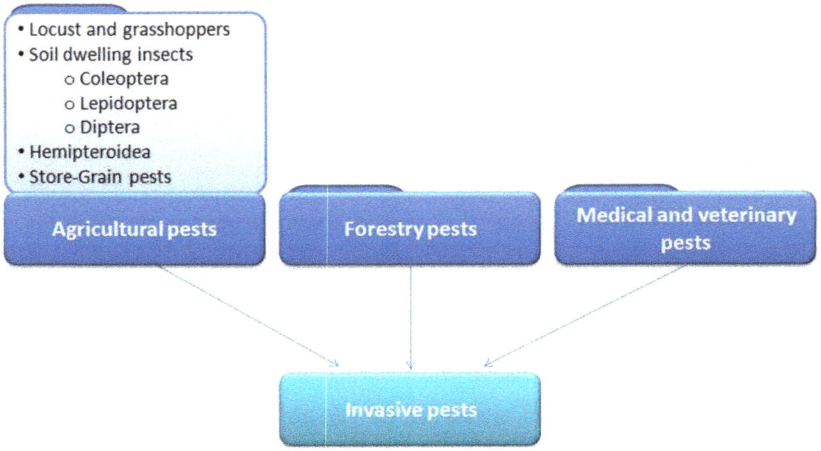

Figure 3 Groups of pests where entomopathogenic fungi are used as components of IPM program.

and environmental-effectiveness analysis of these fungal treatments, the most recent research has focused on the impact of locust thermoregulation on strain selection and efficacy. The behavioral thermopreference of *Locusta migratoria manilensis* (Meyen) (Orthoptera: Acrididae) induced by the fungus *B. bassiana*, and the impact of this behavior on the fungal mycosis, has been studied under laboratory conditions (Sangbaramou et al., 2018). By basking in higher-temperature locations, infected nymphs elevated their thoracic temperature to 30.0–32.6°C, which is higher than the optimum temperature (25°C) for *B. bassiana* conidial germination and hyphal development. A minimum thermoregulation period of 3 h/day increased the survival of infected locusts by 43%. The therapeutic effect decreased when thermoregulation was delayed after the initial infection. The fungus grew and overcame the locusts as soon as the thermoregulation was interrupted, indicating that thermoregulation helped the insects to cope with infection but did not completely rid them of the fungus, which was apparently accomplished through an increase in the levels of hemocytes, leading to greater phagocytic activity under certain environmental conditions (Sangbaramou et al., 2018).

Strains IMI 330189 of *Metarhizium acridum* (Driver & Milner) Bisch., Rehner & Humber (Ascomycota: Hypocreales) and EABb 90/2-Dm of *B. bassiana* are a promising biocontrol tool of the Moroccan locust *Dociostaurus maroccanus* (Thunberg) (Orthoptera: Acrididae). Strain IMI 330189 had greater fitness at temperatures above 27°C, and EABb 90/2-Dm strain in the temperature range of 10–25°C. These effects were mirrored in vivo, where locust thermoregulation caused a marked reduction in the virulence of EABb 90/2-Dm in comparison to no thermoregulation conditions but resulted in only a moderate reduction in the virulence of IMI 330189 (Valverde-García et al., 2018a,b). Thermal gradient experiments revealed that strain IMI 330189 induced behavioral fever in *D. maroccanus*, but strain EABb 90/2-Dm did not induce such a clear behavioral response. Under the temperature conditions of the main breeding areas of the Moroccan locust in the Iberian Peninsula, strain IMI 330189 is likely a better candidate for use in biocontrol, although strain EABb 90/2-Dm could also be a good alternative in more temperate environments either as a stand-alone control product or in mixed combinations of the two fungal strains, potentially providing more effective control over a broader range of temperatures. Hence, adult locusts (2–3 days post fledging) treated with low dosages of both fungal pathogens alone and in mixture suffered a reduction in per capita feeding and female fecundity, whereas only IMI 330189 isolate caused such impact on the locust biotic potential under thermoregulatory regimes (Valverde-García et al., 2018a,b).

Locust treatments with EPF in the field represent the lowest hazard to the sprayed ecosystems as shown in *M. acridum* (Green Guard®) aerial treatments in either a barrier or blanket ULV at replicated sites, which mimicked the techniques employed for locust control operations in Australia that represented a relatively low and transient hazard to most non-target arthropod taxa (Maute et al., 2017).

5.2 Soil-dwelling insect pests

Approximately 80% of insect pests spend part of their life cycle in the soil (Tremblay, 1993). Soil pests are more difficult to control partly due to insufficient tools for monitoring them, and, very often, damage is done before control measures can be implemented. Few pesticides are available to protect the plant against subterranean pests, even less after the exclusion of neonicotinoids (i.e. in the EU), which results in highly increased risk of resistance development through the extensive use of the remaining ones. This is even worse if it is considered that usually higher dosages have to be applied to the soil to control subterranean pests than that used to control canopy pests due to the buffering capacity of soils.

Most of the soil-dwelling pests are polyphagous, widespread and highly damaging, and hereby it is provided information on the main groups and how EPF are being investigated as a key component of IPM programs. Indeed, several works have shown that exploiting synergies of EPF with low dosages of other natural insecticides could greatly reduce or, in some instances, eliminate chemical pesticide inputs.

5.2.1 Soil-inhabiting coleopterans

5.2.1.1 White grubs

White grub is the name for the larvae of scarab beetles (Coleoptera: Scarabaeidae) and includes species like *Melolontha melolontha* L. (European chafer), *M. hippocastani* F. (Forest chafer), *Phyllopertha horticola* L. (Garden chafer), *Amphimallon* spp, and *Hoplia philanthus* Füessly (Welsh chafer). In general, the larvae are polyphagous and feed on the roots of a wide range of crops worldwide, causing damage of billions of euros per annum. BIPESCO 2 strain of *B. brongniartii* (Sacc.) Petch (Ascomycota: Hypocreales) was registered in Austria, France, and Switzerland for the control of white grubs (Sierpińska et al., 2015). The most recent work focuses on the occurrence and pathogenicity of *Beauveria* spp. in forest soils in Poland, in outbreak areas of *M. melolontha* (Niemczyk et al., 2019). *M. pingshaense* Q.T. Chen & H.L. Guo (Ascomycota: Hypocreales) has been evaluated as a control agent of the white grub *Anomala cincta* Say applied as a seed treatment or directly to the compost around the plant roots (Peña-Peña et al., 2015). Although the greatest infection (93%) occurred in the direct inoculation treatment, 50% of larvae still became infected by *M. pingshaense* applied as a seed treatment. The fungus persisted in the compost and also colonized the roots of the developing maize plants endophytically (Peña-Peña et al., 2015).

Experimental and commercial preparations of *M. brunneum* (Petch) strain F52 were evaluated for the control of Japanese beetle, *Popillia japonica* Newman larvae, in the laboratory and under field conditions (Behle et al., 2015). Experimental preparations consisted of granule and liquid formulations made using microsclerotia produced in vitro, which are intended to produce infective spores after the application. These formulations were compared against commercial insecticides (imidacloprid and trichlorfon) and commercial formulations of *M. brunneum* F52 (Met 52) containing only conidia. Grub densities were reduced most consistently by autumn applications targeting later instars (second- and third-instar grubs), with significantly lower grub densities compared with untreated plots, providing 38.6–69.2% control, which sometimes equaled the levels of control with chemical insecticides (Behle et al., 2015).

5.2.1.2 Wireworms

Wireworm is the larval stage of the click beetles (Coleoptera: Elateridae), several species of polyphagous species with larvae feeding on the roots of worldwide-distributed key crops such as potatoes, cereals, pasture, root vegetables, and sunflower, causing damage of billions of euros per annum worldwide. Innovative wireworm EPF control strategies are required to implement IPM on the basis of different regulations in the world. Applications of *M. brunneum* isolate LRC112 conidia caused high mortality to *Agriotes obscurus* L. in field trials by using three strategies: banded conidiated rice, conidia dust, and an aqueous conidia suspension spray, outperforming spinosad in all cases and revealing opportunities for controlling click beetles using fungal entomopathogens, including the potential for the horizontal transmission of the inoculum (Kabaluk, 2014). In the search for the best fungus delivery strategy, three EPF (*M. brunneum* F52, *B. bassiana* GHA, and *M. robertsii* DWR 346) were tested as seed-coating, in-furrow granular, and soil band-over-row drench applications, in addition to a chemical control (Reddy et al., 2014). Formulated granules or soil drenches resulted in significantly higher plant stand counts and yields than the fungus-coated seed treatments and provided an efficacy comparable or superior to imidacloprid, supporting their value in the management of this pest (Reddy et al., 2014).

In a two-year study, wireworm control with a novel attract-and-kill strategy aimed at enhancing *M. brunneum* efficacies in organic potato production systems in Lower Saxony, Germany, was investigated (Brandl et al., 2017). The approach was based on the attraction of wireworms (*Agriotes* spp.) toward an artificial carbon dioxide–emitting source, using baker's yeast (*Saccharomyces cerevisiae* Meyen ex Hansen) (to produce CO_2), in combination with *M. brunneum* conidia for wireworm infection. Both components were encapsulated in alginate as a carrier material and applied in a mixture with two types of beads (one for encapsulated yeast and one for *M. brunneum* conidia) (Brandl et al., 2017). An application of these beads within the potato rows during potato planting reduced wireworm tuber damage by 37–75% relative to the untreated control and helped enhance the efficacy of *M. brunneum* by up to 35% through an attract-and-kill approach compared to beads without a carbon dioxide source. This indicated not only the potential of the strategy to promote biological wireworm control as an alternative to insecticide but also the impact of this strategy on reducing the inoculum compared to an inundate *M. brunneum* conidia release strategy (Brandl et al., 2017).

The preventive application of *M. brunneum* conidia to reduce wireworm populations in cover crops, preceding a damage-sensitive crop like potatoes, may also be a promising biocontrol strategy as revealed in semi-field pot experiments for the control of wireworms *Agriotes obscurus* with *M. brunneum* strain ART2825 (Rogge et al., 2017). Pots were treated in the late summer during sowing of spring oat as a cover crop. The survival of wireworms was assessed 4 weeks after their release in October 2013 and 30 weeks after their release in April 2014. The concentration of the fungus in the soil in October 2013 increased with increasing concentrations of applied conidia, and no significant reduction was observed at the second evaluation date in April

2014. Increasing conidia application rates significantly increased mycosis and reduced wireworm survival, to a level comparable to that of the treatment using insecticide-coated oat seeds (Brandl et al., 2017).

5.2.1.3 Black wine weevil

The black vine weevil, *Otiorhynchus sulcatus* F. (Coleoptera: Curculionidae), is a polyphagous pest and a major problem on soft fruit (strawberries, raspberries, blackcurrants) and vines. The larvae feed on the roots, whereas the adults feed on foliage. This pest is found throughout most European countries and is also a major pest in the USA and Canada. The global damage to soft fruit and vines caused by this pest is several hundred million euros per annum but is much higher if other commodities are included such as ornamentals, hardy nursery stock, and herbaceous perennials. Fungal isolates from Norway have been shown to kill *O. sulcatus* larvae at low temperatures, with a Norwegian *M. brunneum* isolate proliferating more in the rhizosphere than an exotic isolate. Besides, this work is the first report of the rhizosphere competence for *B. pseudobassiana* Rehner & Humber (Ascomycota: Hypocreales), since up to now this species was frequently found in the aerial parts of the crops (Klingen et al., 2015; Meyling et al., 2011).

Laboratory and polytunnel experiments have been performed to investigate a novel control system based on the autodissemination of spores of an EPF to populations of adult vine weevils (Pope et al., 2018). Three isolates of EPF were applied as conidial powders in artificial refuges placed among populations of adult weevils held in experimental boxes in the laboratory at 20°C, with 70–90% mortality of adult weevils over 28 days (Pope et al., 2018). The polytunnel experiment tested the efficacy of conidial powders of *M. brunneum* placed in artificial refuges to increase vine weevil mortality. Overall weevil mortality was relatively low (26–41%) but was significantly higher in cages in which the conidial powders were placed in refuge traps than in cages with control traps (Pope et al., 2018). It was concluded that the lower weevil mortality recorded in the polytunnel experiment compared to the laboratory test was most likely a consequence of the greater amounts of inoculum required to kill adult weevils when conditions fluctuate between favorable and unfavorable temperatures (Pope et al., 2018).

5.2.1.4 Western corn rootworm

The western corn rootworm, *Diabrotica virgifera virgifera* LeConte (Coleoptera: Chrysomelidae), probably originates from Central America and became a major pest of maize in North America in the twentieth century, with an expanding range associated with continuous maize growing. Indeed, it was accidentally introduced into Europe, where *D. virginifera* is now an invasive pest. EPF also play a key role on the development of IPM strategies against this pest.

An attract-and-kill strategy for *D. virginifera* control with *B. bassiana* based on CO_2-emitting formulations has been investigated (Vemmer et al., 2016). Formulations containing *S. cerevisiae*, starch, and *B. bassiana* were attractive for western corn rootworm larvae within the first 4 h following the exposure; however, when considering the whole testing period, the maize root systems remained more attractive for the larvae than beads (Vemmer et al., 2016). This is still a very promising strategy even if the attractiveness needs to be increased by phagostimuli to extend contact between larvae and the EPF growing out of these formulations (Vemmer et al., 2016).

Recently, a two-year field study using a granular formulation of *M. brunneum* in conjunction with entomopathogenic nematodes and chemical insecticides revealed the potential of the fungus against western corn rootworm. However, one of the major outcomes of the investigation was that the economic threshold of this pest is considerably higher than the stipulated economic threshold under favorable growing conditions (Rauch et al., 2017).

5.2.1.5 Other soil-inhabiting coleopterans

The Mediterranean flatheaded peachborer, *Capnodis tenebrionis* (L.) (Coleoptera: Buprestidae), poses an enormous threat to stone-fruit crops of the family Rosaceae in the Mediterranean basin due to the feeding behavior of both adults and larvae. Adults cause defoliation by feeding on twigs and young branches throughout the warm season while neonate boring larvae, which hatch from eggs deposited on the ground close to the base of the tree stem, reach the host by crawling through the soil and may cause tree death or severe plant weakening in consequence of root/collar tunneling under the bark. The efficacy of *M. brunneum* EAMA 01/58-Su strain has been ascertained, targeting adults through fiber band application and eggs and neonate larvae in a potted plant bioassay with soil application, respectively, demonstrating key strategies for new stone-fruit crop plantations (Marannino et al., 2006, 2008).

5.2.2 Cutworms

Cutworm is the general name for lepidopteran larvae, mostly in the Noctuidae family [i.e. *Agrotis segetum* (Schiff.) and *A. ipsilon* (Hufn.)] that hide under litter or soil during the day, coming out in the dark to feed on the collar and stem of the plant, often at the seedling stages, and consequently cutting it down.

There are a very few studies using EPF for the control of cutworms under field conditions. Several isolates of *B. bassiana* have been evaluated under laboratory conditions against second-instar larvae of diamondback moth,

Plutella xylostella (L.); European corn borer, *Ostrinia nubilalis* (Hübner); corn earworm, *Helicoverpa zea* (Boddie); fall armyworm, *Spodoptera frugiperda* (Smith); beet armyworm, *Spodoptera exigua* (Hübner); black cutworm, *Agrotis ipsilon* (Hufnagel); cabbage worm, *Pieris rapae* L.; and cabbage looper *Trichoplusia ni* (Iliibner), with different virulence levels (Wraight et al., 2010). Similarly, laboratory studies revealed the potential of *M. anisopliae* strains toward *Spodoptera litura* (F.) and safety for its predator *Rhynocoris kumarii* Ambrose and Livingstone (Hemiptera: Reduviidae) (Sahayaraj et al., 2018).

More recently, the persistence of *Metarhizium* spp. and *B. bassiana* in tobacco soils and their potential as biocontrol agents of *S. litura* have been investigated under glasshouse pot and field experiments (Lepidoptera: Noctuidae) in the Guangdong Province in China (Yang et al., 2018). The persistence of *Metarhizium* spp. in the soil was longer (close to 3 months) than that of *B. bassiana*, which resulted in the highest rates of mortality of *S. litura* (Yang et al., 2018). Similarly, the stability and insecticidal activity of a Colombian isolate of *Metarhizium rileyi* (Farl.) Kepler, Rehner & Humber biopesticide to control *S. frugiperda* has been evaluated under glasshouse conditions (Grijalba et al., 2018), demonstrating its potential to control this pest in maize even if the formulation is still under development (Grijalba et al., 2018). To this end, transient endophytic colonization of plants improved the outcome of foliar applications of mycoinsecticides against chewing insects such as *S. littoralis* (Resquín-Romero et al., 2016) and even the safety of the treatments with *M. brunneum* BIPESCO5 and EAMa 01/58-Su strains for the generalist predator *Chrysoperla carnea* (Stephens) (Neuroptera: Chrysopidae) (Ríos-Moreno et al., 2018).

5.2.3 Soil-inhabiting dipterans

5.2.3.1 Cabbage, carrot, and onion root flies

Larvae of the cabbage root fly *Delia radicum* (L.) (Diptera: Anthomyiidae) negatively affect the yield and quality of cruciferous plants, with emphasis on brassicas (i.e. cauliflowers, broccoli, Brussels sprouts, cabbages) and oil crops (i.e. oilseed rape, mustard). This is also the case for the onion fly [*D. antique* (Meigen) (Diptera: Anthomyiidae)], the wheat bulb fly [*D. coarctata* (Fallen) (Diptera: Anthomyiidae)], the turnip root fly [*D. floralis* (Fallen) (Diptera: Anthomyiidae)], and the bean seed fly [*D. platura* Meigen (Diptera: Anthomyiidae)]. The interest in the use of soil treatments with EPF for the control of these pests has increased due to the progressive reduction in the availability (if any) of registered chemical insecticides. To this end, conidial suspensions of several strains of *Beauveria* spp. and *Metarhizium* spp. have been evaluated for virulence against cabbage maggot eggs, mimicking pathways of natural exposure of various insect life stages to the fungal strains (Razinger et al., 2014), whereas ovicidal activities of these fungal species are known to be very low. Chandler and Davidson (2005) also evaluated *M. anisopliae* (Metschn.) Sorokīn (Ascomycota: Hypocreales) against soil-dwelling stages of cabbage maggot in glasshouse and field experiments. In glasshouse studies, the effect of drenching a suspension of conidia onto the base of cabbage plants infested with *D. radicum* eggs was compared with mixing conidial suspension into compost modules used to raise seedlings, with drench application reducing the mean number of larvae and pupae recovered per plant by up to 90%, and both strategies reducing the emergence of adult flies from pupae by up to 92% (Chandler and Davidson, 2005).

In a field experiment, drench applications of *M. anisopliae* to the base of cauliflower plants did not control *D. radicum* populations, although up to 30% of larval cadavers recovered supported sporulating mycelium. Drench applications often exhibited considerable lateral movement on the soil surface before penetrating the ground, which may have reduced the amount of inoculum in contact with *D. radicum* larvae (Chandler and Davidson, 2005). Similarly, direct plantlet inoculation with *Beauveria* spp. and *Metarhizium* spp. for cabbage root fly maggots in vitro and glasshouse experiments caused significant pathogenicity and evidence of rhizosphere competence and endophytic capacity (Razinger et al., 2014). These fungal species have also been evaluated against *D. radicum* by pre-transplantation inoculation, in which case drench of cauliflower plantlets with spinosad outperformed a commercial *B. bassiana* mycoinsecticide and a non-formulated fungal isolate (Razinger et al., 2017).

5.2.3.2 *Tipulids*

Larvae of the crane fly *Tipula paludosa* Meigen (Diptera: Nematocera) are economically important pests of horticultural crops, with prominence in nurseries and those grown in organic farming and grasslands. The use of EPF targeting soil-inhabiting stages of crane fly has also been explored, with *M. robertsii* Bisch., Rehner & Humber (Ascomycota: Hypocreales) V1005 showing considerable potential for the control of *T. paludosa* under laboratory and greenhouse conditions (Ansari and Butt, 2012).

5.2.3.3 *Tephritids*

During their life cycle, prepupariating third-instar larvae of all fruit flies (Diptera: Tephritidae) drop from fruits to the ground and burrow in the soil to locate pupation sites, and they pupate within the first soil centimeters. Thus, the soil-dwelling stage of tephritid flies offers an excellent opportunity for an effective management using EPF as revealed by recent investigations.

The medfly, *C. capitata*, is arguably one of the most serious world tephritid pests, and the above-mentioned strategy has been explored, with soil treatments being highly effective for preimaginal control independently of the soil texture and ionic strength of the soil (Garrido-Jurado et al., 2011a). The use of a multiple logistic regression model has allowed evaluating the effects of soil moisture and temperature on the virulence of EPF medfly preimaginals, with lower mycosis values observed under extreme moisture conditions (1% and 17% wt.:wt.), and higher virulence at moderate humidity and temperature conditions, even if each fungal species and strain is an independent biological entity with different responses to environmental conditions (Garrido-Jurado et al., 2011b). It is extremely important that fungal strains for pest control should have a range of temperatures and RH conditions that can be matched to the desired environment (Garrido-Jurado et al., 2011b). Soil treatments with EPF have been shown to be safe for soil-dwelling non-target arthropods at a commercial olive orchard (Garrido-Jurado et al., 2011c).

The monophagous olive fruit fly *Bactrocera oleae* (Rossi) (Diptera: Tephritidae) is the most destructive pest of olive crops worldwide. As in most tephritid species, in the last generation, at the end of the third larval stage, the prepupariating larva jumps from the olive fruit to pupate on the ground beneath the tree canopy. The effectiveness of soil applications of the *M. brunneum* EAMa 01/58-Su strain beneath the tree canopy targeting *B. oleae* preimaginals to reduce the adult fly population has been investigated in a 4-year field experiment in the Cordoba Province in Andalusia (Spain) (Yousef et al., 2017). During the four seasons, a 50–70% reduction of adult olive fruit fly spring populations was observed in the treated plots compared with the untreated ones, which could help in minimizing the number of treatments, if any, within an olive fruit fly IPM strategy (Yousef et al., 2017). These treatments have also shown to be effective in the Tarragona Province in Catalonia in Northern Spain (Yousef et al., 2018b). The strategy was safe for the *B. oleae* cosmopolitan parasitoid *Psyttalia concolor* (Szépligeti) (Hymenoptera: Braconidae) (Yousef et al., 2018b), and the strain was even demonstrated to be compatible with the herbicide oxyfluorfen commonly used in olive orchards. Soil application of the fungus reduced the *B. oleae* population density up to 50% even when it was mixed with the herbicide in the same tank (Yousef et al., 2013, 2015, 2018b).

Field trials were also conducted for two seasons in mango orchards at Nthagaiya, Kenya, to evaluate the efficacy of soil inoculation with *M. anisopliae* for the suppression of the invasive fruit fly *Bactrocera invadens* Drew, Tsuruta & White (Diptera: Tephritidae). During the 2006–2007 season, post-treatment samples showed that *B. invadens* catches and fruit infestations from the control orchards were lower than those observed in the plots receiving *M. anisopliae*

(Ekesi et al., 2011). In the 2007–2008 season, average *B. invadens* post-treatment samples in the control orchards were seven times higher than the treatment with *M. anisopliae*, with fruit infestation of 38% in the orchards treated with *M. anisopliae* and 52% in the controls (Ekesi et al., 2011).

It has to be highlighted that EPF have been also evaluated toward tephritid adults following different strategies such as autoinoculation devices and sterile fruit fly males for dissemination of conidia of *B. bassiana* and *M. anisopliae* for the control of *C. capitata* (Navarro-Llopis et al., 2015; Toledo et al., 2017). These devices have shown efficacy at about 3 months and can reduce the population in field over 90%.

5.2.3.4 Fungus gnats

The control of fungus gnats with EPF has received increasing attention. *Lycoriella ingenua* Dufour (Diptera: Sciaridae) is a major pest species in commercial mushroom [*Agaricus bisporus* (Lange) Imbach] production throughout the world, with very few alternatives available for its control. A commercial mycoinsecticide has been evaluated against *L. ingenua* when incorporated at spawning, with the fungus not being detrimental to mushroom yield but was also ineffective in controlling the development of *L. ingenua* larvae in artificially infested compost (Andreadis et al., 2016).

5.3 Hemipteroidea (piercing and sucking insect and mite pests)

Due to their contact mode of action, one of the main targets of EPF have been the piercing and sucking insects and mite pests, with emphasis on the hemipterans and thysanopteran (Vega, 2018; Wu et al., 2018). There are several ongoing research projects on new formulations, combinations with chemical insecticides, and delivery strategies of EPF against these pests.

Many thrip species (Thysanoptera: Thripidae) spend part of their life cycles below ground (mainly as prepupae and pupae) in which moment they can be targeted by EPF soil treatments. *B. bassiana* granules have been evaluated toward the soil-dwelling stage of western flower thrips, *Frankliniella occidentalis* Pergande (Lee et al., 2017). In a glasshouse trial, granules of BbERL836 and BbERL1578 isolates were applied to the soil surface of tomatoes growing in pots and female *F. occidentalis* were released on tomato plants 3 days after the fungal application, with population reductions between 75% and 90% at 20 and 40 days after the application (chemical control: 85% reduction). In a field trial with cucumbers, these soil treatments showed ca. 90% control efficacy, similar to that of spinetorm, a semi-synthetic insecticide (85% efficacy), revealing the effectivity of this strategy to manage *F. occidentalis* populations in tomato and cucumber (Lee et al., 2017).

Other strategies have also been evaluated for thrips control with EPF such as spraying onto the plant as reported by Aristizabal et al. (2017) against chilli thrips, *Scirtothrips dorsalis* Hood (Thysanoptera: Thripidae), infesting roses under nursery conditions, with variable efficacy percentages, and the 'attract-and-infect' strategy for the application of *M. anisopliae* for the control of bean flower thrips in cowpea (Mfuti et al., 2017). In this last case it is concluded that spot spraying provided advantage when compared with cover spray application in terms of reduction in labor and the quantity of inoculum used (Mfuti et al., 2017).

Enhanced mortality of *Bemisia tabaci* (Gennadius) (Hemiptera: Aleyrodidae) nymphs by the application of *Isaria javanica* (Bally) Samson & Hywel-Jones (Ascomycota: Hypocreales) has been obtained by combining the fungus with sublethal doses of the chemical insecticides buprofezin and spiromesifen (Dos Santos et al., 2018) and combining *Isaria fumosorosea* Wize (Ascomycota: Hypocreales) and Buprofezin for the mitigation of *Aleurodicus rugioperculatus* Martin (Hemiptera: Aleyrodidae) (Kumar et al., 2018). The toxicological and biochemical basis of synergism between *Lecanicillium muscarium* (Petch) Zare & Gams (Ascomycota: Hypocreales) and the insecticide matrine against *B. tabaci* has been elucidated (Ali et al., 2017).

Compatibility of *B. bassiana* spores with a sprayable bioplastic formulation for application in the biocontrol of tarnished plant bug in cotton has been investigated (Portilla et al., 2019). Plots sprayed with *B. bassiana* showed at least a twofold decrease in tarnished plant bug adults 3 days after treatment compared with control plots, whereas little to no variation was observed in tarnished plant bug nymph populations between treated and untreated plots (Portilla et al., 2019).

EPF are also a major tool for regulating populations of other hemipteran vectors such as the Asian citrus psyllid (ACP), *Diaphorina citri* Kuwayama (Hemiptera: Liviidae), the vector of the pathogen *Candidatus Liberibacter asiaticus* (Las) that causes Huanglongbing (HLB) or citrus greening disease, which is a serious threat to citrus industry worldwide. Field trials with *B. bassiana* in commercial Tahiti lime orchards reduced immature ACP individuals between 50% and 75% under field conditions, demonstrating the potential of the fungus (Ramirez-Godoy et al., 2018). The efficacy of *B. bassiana* and *I. fumosorosea* against adult *Diaphorina citri* Kuwayama (Homoptera: Psyllidae) has been assessed through monthly applications during 1 year in a commercial citrus grove located in Itapetininga, Sao Paulo, Brazil. The two fungal strains were as effective as the chemical insecticides in most trials, supporting their use as an eco-friendly biopesticide for the integrated management of *D. citri*.

The potential of EPF for the control of mites has also been explored, with recent innovations to deliver fungus-treated foam on the upper or lower parts against the third-instar *Tetranychus evansi* Baker & Pritchard

(Acari: Tetranychidae)-infested tomato leaf, with this strategy shown as an effective method to optimize the use of this fungus in screenhouse conditions (Hounmalon et al., 2018).

5.4 Store-grain pests

EPF are also among the best biocontrol tools for the management of store-grain pests, as recently reviewed by Batta and Kavallieratos (2018). The potential for using entomopathogenic nematodes and fungi in the management of the maize weevil, *Sitophilus zeamais* (Motschulsky) (Coleoptera: Curculionidae), has been demonstrated (Mbata et al., 2018). In a recent innovative approach, storage bags for maize (jute bags) were treated with a wettable powder of *B. bassiana*, and this was effective to control adult weevils that walked for 30 min on the treated jute bags at 14 days post-inoculation (Mbata et al., 2018).

5.5 Forestry pests

One of the most successful historical uses of EPF for the control of forest pest was for the Asian longhorned beetle *Anoplophora glabripennis* Motschulsky (Coleoptera: Cerambycidae), a pest native to China and Korea, that was found for the first time outside its native habitat in the USA in 1996, with subsequent detections being made in Canada and several European countries (Hu et al., 2009). *M. brunneum* has been developed as a control method using the 'lure-and-infect' strategy, and it has been demonstrated that the fungal infection impairs the female reproduction before death. Indeed, fewer surviving larvae were produced by fungal-infected females compared with uninfected, with the fungal infection affecting female fitness by decreasing female longevity, by decreasing female oviposition before death, and through horizontal transmission of the fungus to offspring. Consequently, overall, this strategy is highly diminishing of the next generation (Hajek et al., 2008).

B. bassiana was tested for its ability to reduce the survival and reproduction of spruce beetle, *Dendroctonus rufipennis* (Kirby) (Coleoptera: Scolytinae) as applied to the surface of felled and pheromone-baited Engelmann spruce (*Picea engelmannii*) trees using a backpack sprayer. Total F-1 reproduction was reduced by 17%, and the emergence from logs was reduced by 13% in treated logs, but considerable variability in reproduction and emergence was observed (Davis et al., 2018). It was concluded that matching environmental tolerances of biocontrol fungi to field conditions can likely improve their usefulness for the control of spruce beetle in windthrown trees (Davis et al., 2018). *B. bassiana* has also demonstrated high efficacy against the small southern pine engraver *Ips avulsus* Eichhoff (Coleoptera: Scolytidae) in a loblolly pine bolt assay in the Kisatchie National Forest (Rapides Parish,

Louisiana, USA) (Olatinwo et al., 2018). Results showed that the *I. avulsus* adults successfully colonized both the *B. bassiana*-treated bolts and the control bolts within the first week of field exposure. Although the formulation did not protect bolts from attack, it significantly resulted in the higher mortality of *I. avulsus* adults in *B. bassiana*-treated bolts than in the control and significantly reduced the number of adults that reemerged. Approximately 84% of *I. avulsus* adults found in the *B. bassiana*-treated bolts died from *B. bassiana* infection and never reemerged, compared to 14% in the control bolts (Olatinwo et al., 2018). Although effects from *B. bassiana* were too late to stop *I. avulsus* from entering host bolts, the high rates of beetle mortality observed under the bark suggested potential utility of the formulation for managing pine bark beetles (Olatinwo et al., 2018).

The efficacy of *M. brunneum, B. bassiana*, and *B. caledonica* Bissett & Widden (Ascomycota: Hypocreales) against large pine weevil, *Hylobius abietis* (L.) (Coleoptera: Curculionidae), an important pest of reforestation in northern Europe, is assessed by weevil infection in stumps and by numbers of adult weevils emerging in traps above the stumps. *M. brunneum* could persist for at least 2 years in the forest setting (Mc Namara et al., 2018). Brabbs et al. (2015) reviewed the prospects for the use of EPF against *Anoplophora* spp. (Coleoptera: Cerambycidae) in Europe, which highlighted the potential for use in Europe of *B. brongniartii,* and there is even the development of a commercial product in Japan (Brabbs et al., 2015).

5.6 Invasive pests

EPF have become also a main tool for the IPM of invasive pests such as the red palm weevil *Rhynchophorus ferrugineus* (Olivier) (Coleoptera: Curculionidae) (Dembilio et al., 2010; Ortega-García et al., 2017), and an attract-and-infect system to control this curculionid with *B. bassiana* has been developed (Dembilio et al., 2010). Furthermore, our recent studies showed the effectiveness of the attract-and-infect strategy using an autoinoculation device impregnated with conidia of *M. brunneum* EAMa 01/58-Su strain for the control of the invasive *D. suzukii*, with high adult mortality, further sublethal effects on fecundity of females, and horizontal transmission of the fungus from male to female and vice versa (Yousef et al., 2018a). Reduced fecundity of the fly due to fungal infection has also been reported by Cossentine et al. (2016).

5.7 Medical and veterinary pests

Current strategies of Chagas disease vector control, based on chemical insecticide spraying, are increasingly threatened by the emergence of pyrethroid-resistant *Triatoma infestans* (Klug) (Hemiptera: Reduviidae) populations (Forlani et al., 2015). A pheromone-containing infective *Beauveria* box has been shown as a promising new tool against indoor populations of this Chagas disease vector, with the number of boxes per house being the main driver of the reduction of the total domestic bug population (Forlani et al., 2015). This ecologically safe approach is the first-proven alternative to chemical insecticides in the control of *T. infestans* (Forlani et al., 2015). EPF have been also investigated for the control of triatomine vectors of the protozoan *Trypanosoma cruzi*, the etiological agent of Chagas disease (García et al., 2016). While some isolates have been found to be highly virulent against the vector [i.e. *Rodnius prolixus* Stål (Hemiptera: Reduviidae)], a critical problem for this biological tool could be that the parasite *T. cruzi* appears to confer a measure of resistance to the insect against the potential biopesticide agent (García et al., 2016).

Many mosquito (Diptera: Culicidae) species are zoonotic vectors responsible for numerous infectious diseases of medical and veterinary importance. Recent reviews on the biocontrol approaches for mosquito (Diptera: Culicidae) control have been published (Bawin et al., 2015; Evans et al., 2018). Furthermore, the search for mosquito control methods based on the use of EPF has been extensively studied.

The house fly *Musca domestica* L. (Diptera: Muscidae) and stable fly *Stomoxys calcitrans* (L.) (Diptera: Muscidae) are major pests of livestock, and several commercial EPF formulations have been shown to have potential for IPM of both species (Weeks et al., 2017).

The increasing problem of bed bugs requires the development of new control strategies, and insect pathogenic fungi can contribute toward their management (Aak et al., 2018). EPF have been evaluated in the laboratory against *Cimex lecticularius* (Hemiptera: Cimicidae). In a further experiment, an oil formulation of a selected *B. bassiana* strain induced horizontal transfer, mainly adult to nymphs, to elevate mortality in a 10-day arena bioassay. It is proposed using EPF as a part of an IPM strategy, with *B. bassiana* carried to harborages to target hidden individuals (Aak et al., 2018). It has also been revealed that a newly developed *B. bassiana* biopesticide is equally effective against insecticide-susceptible and insecticide-resistant bed bugs and could provide pest control operators with a promising new tool for the control of bed bugs and insecticide-resistance management (Barbarin et al., 2017).

A field microcosm experiment was conducted to assess the combined use of *M. brunneum* Met52 strain and the predator spider *Schizocosa ocreata* against the tick *Ixodes scapularis* Say (Acari: Ixodidae), vector of Lyme disease (Fischhoff et al., 2018). Evidence that the biopesticide Met52 and spider *S. ocreata* each reduced the survival of *I. scapularis* was found; however, Met52 also reduced spider survival (Fischhoff et al., 2018).

6 Advantages and limitations of entomopathogenic fungi as biopesticides

Potential mycoinsecticides are often selected based on the factors such as speed of kill, economic cost of production, and compatibility with existing spray equipment. Despite the potential of EPF for development as inundative/inoculative agents, often it is still underestimated mainly due to inconsistent field results as a consequence of sensitivity to abiotic stresses and naturally low virulence (i.e. slow kill and high amount of inoculum) (Lovett and St Leger, 2018).

The idea that pathogenic microbes are endowed with certain components that give them the capacity for virulence is the central theme of the virulence factor concept. Although what constitutes a virulence factor is varied and controversial, any property that enables a microorganism to establish itself within the host of a particular species may be defined as the virulence factor. Genetic engineering to improve virulence has focused on reducing both the time to kill and the required lethal conidial dosage (Lovett and St Leger, 2018). With this aim, different strategies have been investigated and developed (Lovett and St Leger, 2018), such as exploiting the insect pathogenic fungi themselves as a resource of genes (virulence factors) for strain improvement, either using enzymes to penetrate the insect cuticle, designing synthetic genes from EPF (i.e. secondary metabolite cyclosporin), enhancing hemolymph nutrition, or modifying the host range (Lovett and St Leger, 2018). The expression of host insect proteins has been also investigated by EPF by engineering host to pathogen gene transfer or exploiting host homeostasis (Lovett and St Leger, 2018). Of particular scientific relevance has been the expression of genes from insect predators and other insect pathogens, with a focus on engineering arthropod-derived toxins into EPF and enhancing virulence with bacterial toxins toward insects (Lovett and St Leger, 2018).

Together with the speed of kill, one of the greatest challenges for improving fungi as mycoinsecticides is to identify and establish a hierarchy of pertinent environmental constraints and to develop ways to overcome them. Therefore, control strategies that entail insect pathogenic fungi required a biologically rational approach for the selection of strains based not only on intrinsic virulence to the target host but also on their ability to operate over the range of environmental conditions that they would encounter in the field, while limiting impact on non-target natural enemies. In general, the best strategy is evaluating individual isolates for their tolerance for environmental constrains and match these to the environment in which they are to be used as biocontrol agents on the basis of their tolerance to the major climatic stresses, solar radiation, temperature, and moisture (Quesada-Moraga and Santiago-Álvarez, 2008).

A major problem in the use of pathogens as bioinsecticides is the relative persistence of fungal propagules on leaf surfaces mainly due to the detrimental effect of the solar radiation. UV light protectants in conidial formulations are used to counteract sunlight effects, but in many cases, only limited protection is achieved. Thus, a second strategy to counteract the detrimental effect of solar radiation may consist of selecting fungal isolates with increased resistance to irradiation. However, little is known regarding both inter- and intraspecific ranges of tolerance to solar radiation of entomopathogenic hypocrealean ascomycetes. In general, germination rates are below 30% after the exposition to 1200 mW m^{-2} for 6 h with no relationship between the strain of fungus, the isolation habitat, or fungal species in both *Beauveria* and *Metarhizium* (Fernández-Bravo et al., 2016, 2017). Some isolates have shown an exceptional response to this condition, which is 30% of ozone depletion, with *Beauveria* isolates being more resistant than *Metarhizium* (Fernández-Bravo et al., 2016, 2017).

Thermal conditions strongly influence germination and sporulation, as well as the growth of EPF. Thus, the knowledge of fungal growth in response to temperature could be considered as a logical starting point for strain selection based on the tolerance to temperature constraints. The thermotolerance of the new isolates originating from the phylloplane and soil of the aforementioned ecosystems have been examined, as these isolates may have diverse thermal requirements as they have been isolated at different orientations, heights (soil, leaves, etc.), seasons, and ecosystems. Molecular characterization conducted by using the EF-1α gene, the intergenic nuclear region Bloc, and 15 microsatellites yielded no relationship between the effect of temperature and isolation habitats of the multilocus genotypes, but it showed the exceptional response of some isolates to suboptimal conditions (Fernández-Bravo et al., 2016).

Germination behavior is dependent not only on temperature but also on the presence of water and is influenced by cuticle components. Despite the established importance of moisture on fungal pathogenesis, moisture influence on conidial germination has seldom been investigated with entomopathogenic hypocrealean ascomycetes. Moisture requirements are more striking than those for temperature; for instance, optimal development of most entomopathogenic hypocrealean ascomycetes occurs if the relative humidity is above 97%, and the critical water activity for conidial germination is about 0.92 aw (Vidal and Fargues, 2007). The evaluation of tolerance to low water activity and humidity of the soil and phylloplane isolates has shown, and as with the thermal conditions, no relationship between the effect of moisture and isolation habitats of the multilocus genotypes (Fernández-Bravo et al., 2016).

7 Market overview

Sales of biopesticides products have been increasing in the last 20 years, with a forecast reaching of around $6.6 billion for 2022, that is, the 7% of the total crop protection market (Marketsandmarketc.com, 2016; Dunham and Trimmer, 2017). This sector has a consolidated annual growth rate of 15% in the last 10 years (Fig. 4), which is double than the annual growth rate of the conventional pesticides (Dunham and Trimmer, 2017). By 2050 the market size of biopesticides is expected to equalize with synthetics (Orson, 2015). However, the global market of biopesticides shows unequal distribution, since the EU regulations make more difficult for biopesticide registration, resulting in a lower number of active substances registered in the EU than in the USA, China, Brazil, or India (Balog et al., 2017).

Biocontrol products are based on microorganisms, macroorganisms, plant extracts, and pheromones. Of them all, the microorganisms constitute the major segment of the biocontrol market with the 58% of the products, in particular, the formulations of *Bacillus thuringiensis*. This segment will continue to grow nearly 60% of market through 2025, reaching $5 and $1 billion for the sales of bacteria and fungi formulations, respectively (Trimmer, 2017). The fungal-based biopesticide market sold $401.6 million in 2016, with an annual growth rate of 19.1% (Marketreportsworld.com, 2019). Today, there are around 30 formulations of EPF commercialized in the world, with 13 different fungal species as active agents, but more than 80% are from the genera *Beauveria* and *Metarhizium* (Butt et al., 2016). Currently, *B. bassiana* sales have reached $48 million in 2019, and it is expected to reach $130 million in 2024, with North America being the largest consumer of this product followed by Brazil, which has a 23% world market share (Mascarin et al., 2018; Reportsmonitor.com, 2019).

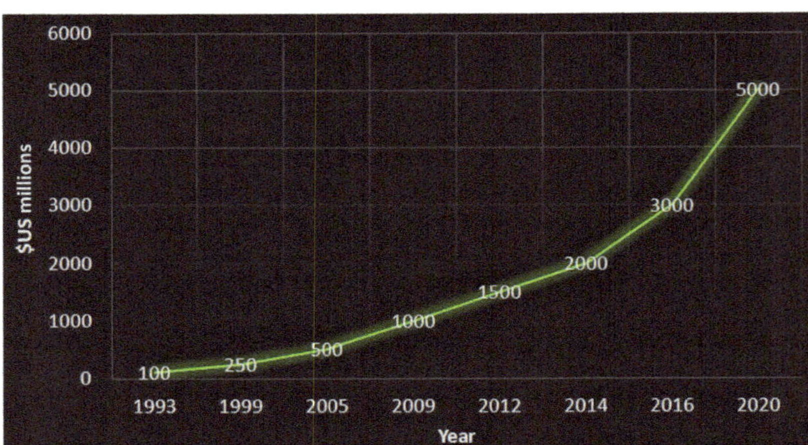

Figure 4 Evolution of the global biopesticide market and estimation until 2020. Source: adapted from Dunham and Trimmer (2017).

8 Conclusion and future trends

IPM, which is currently implemented in the European Union Member States (Directive 2009/128/EC) as well as the USA (U.S. Code § 136r-1), promotes the use of environmentally friendly methods contributing to reduce the input of conventional chemical pesticides. In this sense, EPF have burst on to IPM programs because their use fits perfectly with these requirements. These microorganisms may be found causing natural infection in many insect and mite hosts, since they are present in almost all terrestrial ecosystems and habitats, playing many different roles. Although the soil is their main reservoir, they have been found also in the phylloplane of several crops and forest plants, but their role is still unknown. EPF have been recently discovered to have multiple ecological niches such as plant endophytes and rhizosphere competency that can be exploited for the control of insects in cryptic habits such as borers with endophagous behavior, which can be difficult to access. Likewise, the efficacy of different species of EPF has been proved against several insect pests under laboratory and field conditions. Because of their unique mode of action between the microorganisms, including the ability to penetrate through the insect cuticle, they can also control sucking insect pests. Indeed, natural presence and intraspecific variability in their adaptation to different environmental conditions has encouraged the search of well-adapted strains to the environment for the development of biological control products. Horizontal transmission from infected insects or cadavers to untreated insects is also expected. This autodissemination potential may be used in management strategies with 'lure-and-infect' devices, which could be very important in forest areas. Another aspect to consider is the sublethal effects caused by these EPF, which further contribute to the reduction of pest populations by these fungi. In conclusion, EPF are a strong alternative to be taken into account in IPM programs.

9 Where to look for further information

The European policies and directives drive towards a reduction in synthetic chemical pesticides, but the requisite regulatory framework varies by country. European regulations of obliged fulfillment are as listed below:

- Directive 2009/128/EC of the European Parliament and Council of 21 October 2009 establishing a framework for Community action to achieve the sustainable use of pesticides.
- EC regulation 1107/2009 of the European Parliament and of the Council of 21 October 2009 concerning the placing of plant protection products on the market and repealing Council Directives 79/117/EEC and 91/414/EEC.

The European Food Safety Authority (EFSA) conducts consultations to the Member states about the public health risks associated with the use of these microorganisms:

- https://www.efsa.europa.eu/en/consultations/call/public-consultation-active-substance-metarhizium.
- https://www.efsa.europa.eu/en/efsajournal/pub/2498.

The OECD has published working documents regarding the risk assessment of the use of mycoinsecticides:

- http://www.oecd.org/officialdocuments/publicdisplaydocumentpdf/?cote=env/jm/mono(2018)33/ann1&doclanguage=en.
- http://www.oecd.org/officialdocuments/publicdisplaydocumentpdf/?cote=env/jm/mono(2018)33&doclanguage=en.

The International Organization for Biological and Integrated Control (IOBC), West Palearctic Regional Section (WPRS) working group Microbial and Nematode Control of Invertebrate Pests offers a forum for exchange of information and to stimulate contacts and cooperation between scientists in the field of insect pathology, with a key section focusing on the use of entomopathogenic fungi for pest control:

- https://www.iobc-wprs.org/expert_groups/03_wg_control_invertebrate_pests.html.

Finally, the European Cooperation in Science and Technology (COST) funding organizations for the creation of research networks. Theses COST actions are multi-stakeholder, often involving the private sector, policymakers as well as civil society. Examples of COST Actions past:

- Biological control of pest insects and mites with special reference to Entomophthorales: https://www.cost.eu/actions/842/#tabs|Name:overview.
- Using three-way interactions between plants, microbes and arthropods to enhance crop protection and production: https://www.cost.eu/actions/FA1405/#tabs|Name:overview.
- Towards a sustainable and productive EU organic greenhouse horticulture: https://www.cost.eu/actions/FA1105/#tabs|Name:overview.
- Endophytes in Biotechnology and Agriculture: https://www.cost.eu/actions/FA1103/#tabs|Name:overview.

On the other hand, further information on EPF can be found at the homepage of Society for Invertebrate Pathology at: www.sipweb.org.

There are a number of journals in which readers can find valuable and novel scientific studies about the EPF including:

- *Journal of Invertebrate Pathology* https://www.journals.elsevier.com/journal-of-invertebrate-pathology.
- *Journal of Pest Science* https://link.springer.com/journal/10340.
- *Journal of Pest Management Science* https://onlinelibrary.wiley.com/journal/15264998.

10 References

Aak, A., Hage, M., Rukke, B. A. and Rukke, B. A. 2018. Insect pathogenic fungi and bed bugs: behaviour, horizontal transfer and the potential contribution to IPM solutions. *Journal of Pest Science* 91, 823–35. doi:10.1007/s10340-017-0943-z.

Aira, M. J., Jato, V., Stchigel, A. M., Rodríguez-Rajo, F. J. and Piontelli, E. 2007. Aeromycological study in the Cathedral of Santiago de Compostela (Spain). *International Biodeterioration and Biodegradation* 60(4), 231–7. doi:10.1016/j.ibiod.2007.02.007.

Ali, S., Zhang, C., Wang, Z. Q., Wang, X. M., Wu, J. H., Cuthbertson, A. G. S., Shao, Z. F. and Qiu, B. L. 2017. Toxicological and biochemical basis of synergism between the entomopathogenic fungus *Lecanicillium muscarium* and the insecticide matrine against *Bemisia tabaci* (Gennadius). *Scientific Reports* 7, 46558. doi:10.1038/srep46558.

Alkhaibari, A. M., Lord, A. M., Maffeis, T., Bull, J. C., Olivares, F. L., Samuels, R. I. and Butt, T. M. 2018. Highly specific host-pathogen interactions influence *Metarhizium brunneum* blastospore virulence against *Culex quinquefasciatus* larvae. *Virulence* 9(1), 1449–67. doi:10.1080/21505594.2018.1509665.

Andreadis, S. S., Cloonan, K. R., Bellicanta, G. S., Paley, K., Pecchia, J. and Jenkins, N. E. 2016. Efficacy of *Beauveria bassiana* formulations against the fungus gnat *Lycoriella ingenua*. *Biological Control* 103, 165–71. doi:10.1016/j.biocontrol.2016.09.003.

Ansari, M. A. and Butt, T. M. 2012. Evaluation of entomopathogenic fungi and a nematode against the soil-dwelling stages of the crane fly *Tipula paludosa*. *Pest Management Science* 68(10), 1337–44. doi:10.1002/ps.3338.

Aristizabal, L. F., Chen, Y., Cherry, R. H., Cave, R. D. and Arthurs, S. P. 2017. Efficacy of biorational insecticides against chilli thrips, *Scirtothrips dorsalis* (Thysanoptera: Thripidae), infesting roses under nursery conditions. *Journal of Applied Entomology* 141(4), 274–84. doi:10.1111/jen.12340.

Balog, A., Hartel, T., Loxdale, H. D. and Wilson, K. 2017. Differences in the progress of the biopesticide revolution between the EU and other major crop-growing regions. *Pest Management Science* 73(11), 2203–8. doi:10.1002/ps.4596.

Barbarin, A. M., Bellicanta, G. S., Osborne, J. A., Schal, C. and Jenkins, N. E. 2017. Susceptibility of insecticide-resistant bed bugs (*Cimex lectularius*) to infection by fungal biopesticide. *Pest Management Science* 73(8), 1568–73. doi:10.1002/ps.4576.

Batta, Y. A. and Kavallieratos, N. G. 2018. The use of entomopathogenic fungi for the control of stored-grain insects. *International Journal of Pest Management* 64(1), 77–87. doi:10.1080/09670874.2017.1329565.

Bawin, T., Seye, F., Boukraa, S., Zimmer, J.-Y., Delvigne, F. and Francis, F. 2015. La lutte contre les moustiques (Diptera: Culicidae): diversité des approches et application du contrôle biologique. *The Canadian Entomologist* 147(4), 476-500. doi:10.4039/tce.2014.56.

Behie, S. W., Padilla-Guerrero, I. E. and Bidochka, M. J. 2013. Nutrient transfer to plants by phylogenetically diverse fungi suggests convergent evolutionary strategies in rhizospheric symbionts. *Communicative and Integrative Biology* 6(1), e22321. doi:10.4161/cib.22321.

Behie, S. W., Jones, S. J. and Bidochka, M. J. 2015. Plant tissue localization of the endophytic insect pathogenic fungi *Metarhizium* and *Beauveria*. *Fungal Ecology* 13, 112-9. doi:10.1016/j.funeco.2014.08.001.

Behle, R. W., Richmond, D. S., Jackson, M. A. and Dunlap, C. A. 2015. Evaluation of *Metarhizium brunneum* F52 (Hypocreales: Clavicipitaceae) for control of Japanese beetle larvae in turfgrass. *Journal of Economic Entomology* 108(4), 1587-95. doi:10.1093/jee/tov176.

Bischoff, J. F., Rehner, S. A. and Humber, R. A. 2009. A multilocus phylogeny of the *Metarhizium anisopliae* lineage. *Mycologia* 101(4), 512-30. doi:10.3852/07-202.

Brabbs, T., Collins, D., Herard, F., Maspero, M. and Eyre, D. 2015. Prospects for the use of biological control agents against *Anoplophora* in Europe. *Pest Management Science* 71(1), 7-14. doi:10.1002/ps.3907.

Brandl, M. A., Schumann, M., Przyklenk, M., Patel, A. and Vidal, S. 2017. Wireworm damage reduction in potatoes with an attract-and-kill strategy using *Metarhizium brunneum*. *Journal of Pest Science* 90(2), 479-93. doi:10.1007/s10340-016-0824-x.

Butt, T. M., Greenfield, B. P., Greig, C., Maffeis, T. G., Taylor, J. W., Piasecka, J., Dudley, E., Abdulla, A., Dubovskiy, I. M., Garrido-Jurado, I., Quesada-Moraga, E., Penny, M. W. and Eastwood, D. C. 2013. *Metarhizium anisopliae* pathogenesis of mosquito larvae: a verdict of accidental death. *PLoS ONE* 8(12), e81686. doi:10.1371/journal.pone.0081686.

Butt, T. M., Coates, C. J., Dubovskiy, I. M. and Ratcliffe, N. A. 2016. Entomopathogenic fungi: new insights into host-pathogen interactions. *Advances in Genetics* 94, 307-64. doi:10.1016/bs.adgen.2016.01.006.

Chandler, D. and Davidson, G. 2005. Evaluation of entomopathogenic fungus *Metarhizium anisopliae* against soil-dwelling stages of cabbage maggot (Diptera: Anthomyiidae) in glasshouse and field experiments and effect of fungicides on fungal activity. *Journal of Economic Entomology* 98(6), 1856-62. doi:10.1093/jee/98.6.1856.

Cossentine, J., Robertson, M. and Buitenhuis, R. 2016. Impact of acquired entomopathogenic fungi on adult *Drosophila suzukii* survival and fecundity. *Biological Control* 103, 129-37. doi:10.1016/j.biocontrol.2016.09.002.

Davis, T. S., Mann, A. J., Malesky, D., Jankowski, E. and Bradley, C. 2018. Laboratory and Field evaluation of the entomopathogenic fungus *Beauveria bassiana* (Deuteromycotina: Hyphomycetes) for population management of spruce beetle, *Dendroctonus rufipennis* (Coleoptera: Scolytinae), in felled trees and factors limiting pathogen success. *Environmental Entomology* 47(3), 594-602. doi:10.1093/ee/nvy036.

Dembilio, O., Quesada-Moraga, E., Santiago-Álvarez, C. and Jacas, J. A. 2010. Potential of an indigenous strain of the entomopathogenic fungus *Beauveria bassiana* as a biological control agent against the Red Palm Weevil, *Rhynchophorus ferrugineus*. *Journal of Invertebrate Pathology* 104(3), 214-21. doi:10.1016/j.jip.2010.04.006.

Dos Santos, T. T. M., Quintela, E. D., Mascarin, G. M. and Santana, M. V. 2018. Enhanced mortality of *Bemisia tabaci* nymphs by *Isaria javanica* combined with sublethal doses of chemical insecticides. *Journal of Applied Entomology* 142(6), 598–609. doi:10.1111/jen.12504.

Dunham, B. and Trimmer, M. 2017. A sustained pace of mergers and acquisitions. *New AG International*, March/April, pp. 34–42.

Eilengberg, J. and Hokkanen, H. M. T. 2006. *An Ecological and Societal Approach to Biological Control*. Springer, Dordrecht.

Ekesi, S., Maniania, N. K. and Mohamed, S. A. 2011. Efficacy of soil application of *Metarhizium anisopliae* and the use of GF-120 Spinosad bait spray for suppression of *Bactrocera invadens* (Diptera: Tephritidae) in mango orchards. *Biocontrol Science and Technology* 21(3), 299–316. doi:10.1080/09583157.2010.545871.

Evans, H. C., Elliot, S. L. and Barreto, R. W. 2018. Entomopathogenic fungi and their potential for the management of *Aedes aegypti* (Diptera: Culicidae) in the Americas. *Memorias do Instituto Oswaldo Cruz* 113(3), 206–14. doi:10.1590/0074-02760170369.

Fernández-Bravo, M., Garrido-Jurado, I., Valverde-García, P., Enkerli, J. and Quesada-Moraga, E. 2016. Responses to abiotic environmental stresses among phylloplane and soil isolates of *Beauveria bassiana* from two holm oak ecosystems. *Journal of Invertebrate Pathology* 141, 6–17. doi:10.1016/j.jip.2016.09.007.

Fernández-Bravo, M., Flores-León, A., Calero-Lopez, S., Gutierrez-Sanchez, F., Valverde-Garcia, P. and Quesada-Moraga, E. 2017. UV-B radiation-related effects on conidial inactivation and virulence against *Ceratitis capitata* (Wiedemann) (Diptera; Tephritidae) of phylloplane and soil *Metarhizium* sp. strains. *Journal of Invertebrate Pathology* 148, 142–51. doi:10.1016/j.jip.2017.06.012.

Fischhoff, I. R., Burtis, J. C., Keesing, F. and Ostfeld, R. S. 2018. Tritrophic interactions between a fungal pathogen, a spider predator, and the blacklegged tick. *Ecology and Evolution* 8(16), 7824–34. doi:10.1002/ece3.4271.

Forlani, L., Pedrini, N., Girotti, J. R., Mijailovsky, S. J., Cardozo, R. M., Gentile, A. G., Hernández-Suarez, C. M., Rabinovich, J. E. and Juarez, M. P. 2015. Biological control of the Chagas disease vector *Triatoma infestans* with the entomopathogenic fungus *Beauveria bassiana* combined with an aggregation cue: field, laboratory and mathematical modeling assessment. *PLoS Neglected Tropical Diseases* 9(5), e0003778. doi:10.1371/journal.pntd.0003778.

Furlan, L., Pozzebon, A., Duso, C., Simon-Delso, N., Sánchez-Bayo, F., Marchand, P. A., Codato, F., Bijleveld van Lexmond, M. and Bonmatin, J. M. 2018. An update of the Worldwide Integrated Assessment (WIA) on systemic insecticides. Part 3: alternatives to systemic insecticides. *Environmental Science and Pollution Research International*, 1–23. doi:10.1007/s11356-017-1052-5.

García, A. R. M., Rocha, A. P., Moreira, C. C., Rocha, S. L., Guarneri, A. A. and Elliot, S. L. 2016. Screening of fungi for biological control of a triatomine vector of Chagas disease: temperature and trypanosome infection as factors. *PLoS Neglected Tropical Diseases* 10(11), e0005128. doi:10.1371/journal.pntd.0005128.

Garrido-Jurado, I., Valverde-García, P. and Quesada-Moraga, E. 2011a. Use of a multiple logistic regression model to determine the effects of soil moisture and temperature on the virulence of entomopathogenic fungi against pre-imaginal Mediterranean fruit fly *Ceratitis capitata*. *Biological Control* 59(3), 366–72. doi:10.1016/j.biocontrol.2011.09.011.

Garrido-Jurado, I., Torrent, J., Barrón, V., Corpas, A. and Quesada-Moraga, E. 2011b. Soil properties affect the availability, movement, and virulence of entomopathogenic fungi conidia against puparia of *Ceratitis capitata* (Diptera: Tephritidae). *Biological Control* 58(3), 277–85. doi:10.1016/j.biocontrol.2011.05.017.

Garrido-Jurado, I., Ruano, F., Campos, M. and Quesada-Moraga, E. 2011c. Effects of soil treatments with entomopthogenic fungi on soil dwelling non-target arthropods at a commercial olive orchard. *Biological Control* 59(2), 239–44. doi:10.1016/j.biocontrol.2011.07.001.

Garrido-Jurado, I., Fernández-Bravo, M., Campos, C. and Quesada-Moraga, E. 2015. Diversity of entomopathogenic Hypocreales in soil and phylloplanes of five Mediterranean cropping systems. *Journal of Invertebrate Pathology* 130, 97–106. doi:10.1016/j.jip.2015.06.001.

Garrido-Jurado, I., Alkhaibari, A., Williams, S. R., Oatley-Radcliffe, D. L., Quesada-Moraga, E. and Butt, T. M. 2016. Toxicity testing of *Metarhizium* conidia and toxins against aquatic invertebrates. *Journal of Pest Science* 89(2), 557–64. doi:10.1007/s10340-015-0700-0.

Grijalba, E. P., Espinel, C., Cuartas, P. E., Chaparro, M. L. and Villamizar, L. F. 2018. *Metarhizium rileyi* biopesticide to control *Spodoptera frugiperda*: stability and insecticidal activity under glasshouse conditions. *Fungal Biology* 122(11), 1069–76. doi:10.1016/j.funbio.2018.08.010.

Hajek, A. E., Lund, J. and Smith, M. T. 2008. Reduction in fitness of female Asian longhorned beetle (*Anoplophora glabripennis*) infected with *Metarhizium anisopliae*. *Journal of Invertebrate Pathology* 98(2), 198–205. doi:10.1016/j.jip.2007.12.003.

Hillocks, R. J. 2012. Farming with fewer pesticides: EU pesticide review and resulting challenges for UK agriculture. *Crop Protection* 31(1), 85–93. doi:10.1016/j.cropro.2011.08.008.

Hounmalon, G. Y. A., Maniania, N. K., Niassy, S., Fellous, S., Kreiter, S., Deletre, E., Fiaboe, K. K. M. and Martin, T. 2018. Performance of *Metarhizium anisopliae*-treated foam in combination with *Phytoseiulus longipes* Evans against *Tetranychus evansi* Baker & Pritchard (Acari: Tetranychidae). *Pest Management Science* 74(12), 2835–41. doi:10.1002/ps.5073.

Hu, G. and St Leger, R. J. 2002. Field studies using a recombinant mycoinsecticide (*Metarhizium anisopliae*) reveal that it is rhizosphere competent. *Applied and Environmental Microbiology* 68(12), 6383–7. doi:10.1128/aem.68.12.6383-6387.2002.

Hu, J., Angeli, S., Schuetz, S., Luo, Y. and Hajek, A. E. 2009. Ecology and management of exotic and endemic Asian longhorned beetle *Anoplophora glabripennis*. *Agricultural and Forest Entomology* 11(4), 359–75. doi:10.1111/j.1461-9563.2009.00443.x.

Hu, X., Xiao, G., Zheng, P., Shang, Y., Su, Y., Zhang, X., Liu, X., Zhan, S., St Leger, R. J. and Wang, C. 2014. Trajectory and genomic determinants of fungal-pathogen speciation and host adaptation. *PNAS* 111(47), 16796–801. doi:10.1073/pnas.1412662111.

Kabaluk, T. 2014. Targeting the click beetle *Agriotes obscurus* with entomopathogens as a concept for wireworm biocontrol. *BioControl* 59(5), 607–16. doi:10.1007/s10526-014-9603-x.

Klingen, I. and Haukeland, S. 2006. The soil as a reservoir for natural enemies of pest insects and mites with emphasis on fungi and nematodes. In: Eilenberg, J. and Hokkanen, H. M. T. (Eds), *An Ecological and Societal Approach to Biological Control*. Springer, Dordrecht, pp. 145–211.

Klingen, I., Westruma, K. and Meyling, N. V. 2015. Effect of Norwegian entomopathogenic fungal isolates against *Otiorhynchus sulcatus* larvae at low temperatures and persistence in strawberry rhizospheres. *Biological Control* 81, 1-7. doi:10.1016/j.biocontrol.2014.10.006.

Kumar, V., Francis, A., Avery, P. B., McKenzie, C. L. and Osborne, L. S. 2018. Assessing compatibility of *Isaria fumosorosea* and buprofezin for mitigation of *Aleurodicus rugioperculatus* (Hemiptera: Aleyrodidae): an invasive pest in the Florida landscape. *Journal of Economic Entomology* 111(3), 1069–79. doi:10.1093/jee/toy056.

Lamichhane, J. R., Dachbrodt-Saaydeh, S., Kudsk, P. and Messéan, A. 2016. Toward a reduced reliance on conventional pesticides in European agriculture. *Plant Disease* 100(1), 10–24. doi:10.1094/PDIS-05-15-0574-FE.

Lee, S. J., Kim, S., Kim, J. C., Lee, M. R., Hossain, M. S., Shin, T. S., Kim, T. H. and Kim, J. S. 2017. Entomopathogenic *Beauveria bassiana* granules to control soil-dwelling stage of western flower thrips, *Frankliniella occidentalis* (Thysanoptera: Thripidae). *BioControl* 62(5), 639–48. doi:10.1007/s10526-017-9818-8.

Lovett, B. and St Leger, R. J. 2018. Genetically engineering better fungal biopesticides. *Pest Management Science* 74(4), 781-9. doi:10.1002/ps.4734.

Marannino, P., Santiago-Álvarez, C., de Lillo, E. and Quesada-Moraga, E. 2006. A new bioassay method reveals pathogenicity of *Metarhizium anisopliae* and *Beauveria bassiana* against early stages of *Capnodis tenebrionis* (Coleoptera; Buprestidae). *Journal of Invertebrate Pathology* 93(3), 210–3. doi:10.1016/j.jip.2006.08.002.

Marannino, P., Santiago-Álvarez, C., de Lillo, E. and Quesada-Moraga, E. 2008. Evaluation of *Metarhizium anisopliae* Metch (Sorok.) to target larvae and adults of *Capnodis tenebrionis* (L.) (Coleoptera: Buprestidae) in soil and fiber band applications. *Journal of Invertebrate Pathology* 97(3), 237–44. doi:10.1016/j.jip.2007.09.006.

Marketreportsworld.com. 2019. Global agricultural microbial market analysis (edition:2019) by growth rate, opportunity, key industry peers and forecast to 2023. Available at: https://honestbusinessman24.com/2019/02/global-agricultural-microbial-market-analysis-edition2019-by-growth-rate-opportunity-key-industry-peers-and-forecast-to-2023/ (accessed on 01 March 2019).

Marketsandmarketc.com. 2016. Biopesticides market by type, origin, mode of application, formulation, & Crop Type – Global Forecast to 2022. Available at: https://www.prnewswire.com/news-releases/biopesticides-market-by-type-origin-mode-of-application-formulation--crop-type---global-forecast-to-2022-300415432.html (accessed on 15 January 2019).

Mascarin, G. M., Lopes, R. B., Delalibera, Í Jr., Fernandes, É. K. K., Luz, C. and Faria, M. 2018. Current status and perspectives of fungal entomopathogens used for microbial control of arthropod pests in Brazil. *Journal of Invertebrate Pathology* 165, 46–53. doi:10.1016/j.jip.2018.01.001.

Maute, K., French, K., Story, P., Bull, C. M. and Hose, G. C. 2017. Short and long-term impacts of ultra-low-volume pesticide and biopesticide applications for locust control on non-target arid zone arthropods. *Agriculture, Ecosystems and Environment* 240, 233–43. doi:10.1016/j.agee.2017.02.024.

Mbata, G. N., Ivey, C. and Shapiro-Ilan, D. 2018. The potential for using entomopathogenic nematodes and fungi in the management of the maize weevil, *Sitophilus zeamais* (Motschulsky) (Coleoptera: Curculionidae). *Biological Control* 125, 39–43. doi:10.1016/j.biocontrol.2018.06.008.

McNamara, L., Kapranas, A., Williams, C. D., O'Tuama, P., Kavanagh, K. and Griffin, C. T. 2018. Efficacy of entomopathogenic fungi against large pine weevil, *Hylobius abietis*, and their additive effects when combined with entomopathogenic nematodes. *Journal of Pest Science* 91(4), 1407–19. doi:10.1007/s10340-018-0994-9.

Meyling, N. V. and Eilenberg, J. 2007. Ecology of the entomopathogenic fungi *Beauveria bassiana* and *Metarhizium anisopliae* in temperate agroecosystems: potential for conservation biological control. *Biological Control* 43(2), 145–55. doi:10.1016/j.biocontrol.2007.07.007.

Meyling, N. V., Lübeck, M., Buckley, E. P., Eilenberg, J. and Rehner, S. A. 2009. Community composition, host range and genetic structure of the fungal entomopathogen *Beauveria* in adjoining agricultural and seminatural habitats. *Molecular Ecology* 18(6), 1282–93. doi:10.1111/j.1365-294X.2009.04095.x.

Meyling, N. V., Thorup-Kristensen, K. and Eilenberg, J. 2011. Below-and aboveground abundance and distribution of fungal entomopathogens in experimental conventional and organic cropping systems. *Biological Control* 59(2), 180–6. doi:10.1016/j.biocontrol.2011.07.017.

Mfuti, D. K., Niassy, S., Subramanian, S., du Plessis, H., Ekesi, S. and Maniania, N. K. 2017. Lure and infect strategy for application of entomopathogenic fungus for the control of bean flower thrips in cowpea. *Biological Control* 107, 70–6. doi:10.1016/j.biocontrol.2017.01.011.

Navarro-Llopis, V., Ayala, I., Sanchis, J., Primo, J. and Moya, P. 2015. Field efficacy of a *Metarhizium anisopliae*-based attractant-contaminant device to control *Ceratitis capitata* (Diptera: Tephritidae). *Journal of Economic Entomology* 108(4), 1570–8. doi:10.1093/jee/tov157.

Niemczyk, M., Sierpinska, A., Tereba, A., Sokołowski, K. and Przybylski, P. 2019. Natural occurrence of *Beauveria* spp. in outbreak areas of cockchafers (*Melolontha* spp.) in forest soils from Poland. *BioControl* 64(2), 159–72. doi:10.1007/s10526-019-09927-3.

Oerke, E.-C. 2006. Crop losses to pests. *The Journal of Agricultural Science* 144(1), 31–43. doi:10.1017/S0021859605005708.

Olatinwo, R., Walters, S. and Strom, B. 2018. Impact of *Beauveria bassiana* (Ascomycota: Hypocreales) on the small southern pine engraver (Coleoptera: Scolytidae) in a loblolly pine bolt assay. *Journal of Entomological Science* 53(2), 180–91. doi:10.18474/JES17-58.1.

Orson, S. 2015. An analysis of the biopesticide market now and where it is going. *Outlooks on Pest Management* 26(5), 203–6. doi:10.1564/v26_oct_04.

Ortega-García, L., Tabone, E., Beaudoin-Ollivier, L., Ment, D., Buradino, M., Jaques, J. A., Garrido-Jurado, I., Dembilio, O. and Quesada Moraga, E. 2017. Natural enemies of *Rhynchophorus ferrugineus* and *Paysandisia archon*. In: *Handbook of Major Palm Pests: Biology and Management*. John Wiley & Sons Ltd, United Kingdom, pp. 171–86.

Ownley, B. H., Griffin, M. R., Klingeman, W. E., Gwinn, K. D., Moulton, J. K. and Pereira, R. M. 2008. *Beauveria bassiana*: endophytic colonization and plant disease control. *Journal of Invertebrate Pathology* 98(3), 267–70. doi:10.1016/j.jip.2008.01.010.

Ownley, B. H., Gwinn, K. D. and Vega, F. E. 2010. Endophytic fungal entomopathogens with activity against plant pathogens: ecology and evolution. *BioControl* 55(1), 113–28. doi:10.1007/s10526-009-9241-x.

Peña-Peña, A. J., Santillan-Galicia, M. T., Hernandez-Lopez, J. and Guzman-Franco, A. W. 2015. *Metarhizium pingshaense* applied as a seed treatment induces fungal infection in larvae of the white grub *Anomala cincta*. *Journal of Invertebrate Pathology* 130, 9–12. doi:10.1016/j.jip.2015.06.010.

Pope, T. W., Hough, G., Arbona, C., Roberts, H., Bennison, J., Buxton, J., Prince, G. and Chandler, D. 2018. Investigating the potential of an autodissemination system for managing populations of vine weevil, *Otiorhynchus sulcatus* (Coleoptera: Curculionidae) with entomopathogenic fungi. *Journal of Invertebrate Pathology* 154, 79–84. doi:10.1016/j.jip.2018.04.002.

Portilla, M., Abbas, H. K., Accinelli, C. and Luttrell, R. 2019. Laboratory and field investigations on compatibility of *Beauveria bassiana* (Hypocreales: Clavicipitaceae) spores with a sprayable bioplastic formulation for application in the biocontrol of tarnished plant bug in cotton. *Journal of Economic Entomology* 112(2), 549–57. doi:10.1093/jee/toy382.

Quesada-Moraga, E. and Santiago-Álvarez, C. 2008. Hongos entomopatógenos. In: Jacas, J. A. and Urbaneja, A. (Eds), *Control biológico de plagas agrícolas*. S.P. Phytoma, Madrid, pp. 98–120.

Quesada-Moraga, E., Ruiz-García, A. and Santiago-Álvarez, C. 2006a. Laboratory evaluation of entomopathogenic fungi *Beauveria bassiana* and *Metarhizium anisopliae* against puparia and adults of *Ceratitis capitata* (Diptera: Tephritidae). *Journal of Economic Entomology* 99(6), 1955–66. doi:10.1603/0022-0493-99.6.1955.

Quesada-Moraga, E., Landa, B. B., Muñoz-Ledesma, J., Jiménez-Díaz, R. M. and Santiago-Álvarez, C. 2006b. Endophytic colonisation of opium poppy, *Papaver somniferum*, by an entomopathogenic *Beauveria bassiana* strain. *Mycopathologia* 161(5), 323–9. doi:10.1007/s11046-006-0014-0.

Quesada-Moraga, E., Navas-Cortes, J. A., Maranhao, E. A. A., Ortiz-Urquiza, A. and Santiago-Alvarez, C. 2007. Factors affecting the occurrence and distribution of entomopathogenic fungi in natural and cultivated soils. *Mycological Research* 111(8), 947–66. doi:10.1016/j.mycres.2007.06.006.

Quesada-Moraga, E., Muñoz-Ledesma, F. J. and Santiago-Álvarez, C. 2009. Systemic protection of *Papaver somniferum* L. against *Iraella luteipes* (Hymenoptera: Cynipidae) by an endophytic strain of *Beauveria bassiana* (Ascomycota: Hypocreales). *Environmental Entomology* 38(3), 723–30. doi:10.1603/022.038.0324.

Quesada-Moraga, E., Yousef, M., Ortiz, A., Ruíz-Torres, M., Garrido-Jurado, I. and Estevez, A. 2013. *Beauveria bassiana* (Ascomycota: Hypocreales) wound dressing for the control of *Euzophera pinguis* (Lepidoptera: Pyralidae). *Journal of Economic Entomology* 106(4), 1602–7. doi:10.1603/ec13203.

Quesada-Moraga, E., López-Díaz, C. and Landa, B. B. 2014. The hidden habit of the entomopathogenic fungus *Beauveria bassiana*: first demonstration of vertical plant transmission. *PLoS ONE* 9(2), e89278. doi:10.1371/journal.pone.0089278.

Ramírez-Godoy, A., Vera-Hoyos, M. D., del Pilar Vera-Hoyos, M., Jimenez-Beltran, N. and Restrepo-Diaz, H. 2018. Evaluation of three commercial biopesticides for the reduction of *Diaphorina citri* (Hemiptera: Liviidae) populations. *HortScience* 53, 1453–60. doi:10.21273/HORTSCI13349-18.

Rauch, H., Steinwender, B. M., Mayerhofer, J., Sigsgaard, L., Eilenberg, J., Enkerli, J., Zelger, R. and Strasser, H. 2017. Field efficacy of *Heterorhabditis bacteriophora* (Nematoda: Heterorhabditidae), *Metarhizium brunneum* (Hypocreales: Clavicipitaceae), and chemical insecticide combinations for *Diabrotica virgifera virgifera* larval management. *Biological Control* 107, 1–10. doi:10.1016/j.biocontrol.2017.01.007.

Razinger, J., Lutz, M., Schroers, H. J., Palmisano, M., Wohler, C., Urek, G. and Grunder, J. 2014. Direct plantlet inoculation with soil or insect-associated fungi may control cabbage root fly maggots. *Journal of Invertebrate Pathology* 120, 59–66. doi:10.1016/j.jip.2014.05.006.

Razinger, J., Zerjav, M., Zemljic-Urbancic, M., Modic, Š, Lutz, M., Schroers, H. J., Grunder, J., Fellous, S. and Urek, G. 2017. Comparison of cauliflower-insect-fungus interactions and pesticides for cabbage root fly control. *Insect Science* 24(6), 1057–64. doi:10.1111/1744-7917.12534.

Reddy, G. V. P., Tangtrakulwanich, K., Wu, S. H., Miller, J. H., Ophus, V. L., Prewett, J. and Jaronski, S. T. 2014. Evaluation of the effectiveness of entomopathogens for the management of wireworms (Coleoptera: Elateridae) on spring wheat. *Journal of Invertebrate Pathology* 120, 43-9. doi:10.1016/j.jip.2014.05.005.

Reportsmonitor.com. 2019. Global *Beauveria bassiana* market growth (status and outlook) 2019–2024. Available at: https://www.reportsmonitor.com/report/351930/Beauveria-Bassiana-Market (accessed on 01 March 2019).

Resquín-Romero, G., Garrido-Jurado, I., Delso, C., Ríos-Moreno, A. and Quesada-Moraga, E. 2016. Transient endophytic colonizations of plants improve the outcome of foliar applications of mycoinsecticides against chewing insects. *Journal of Invertebrate Pathology* 136, 23–31. doi:10.1016/j.jip.2016.03.003.

Ríos-Moreno, A., Quesada-Moraga, E. and Garrido-Jurado, I. 2018. Treatments with *Metarhizium brunneum* BIPESCO5 and EAMa 01/58-Su strains (Ascomycota: Hypocreales) are low risk for the generalist predator *Chrysoperla carnea*. *Journal of Pest Science* 91(1), 385–94. doi:10.1007/s10340-017-0905-5.

Rogge, S. A., Mayerhofer, J., Enkerli, J., Bacher, S. and Grabenweger, G. 2017. Preventive application of an entomopathogenic fungus in cover crops for wireworm control. *BioControl* 62(5), 613–23. doi:10.1007/s10526-017-9816-x.

Sahayaraj, K., Subash, N., Allingham, R. W., Kumar, V., Avery, P. B., Mehra, L. K., McKenzie, C. L. and Osborne, L. S. 2018. Lethal and sublethal effects of three microbial biocontrol agents on *Spodoptera litura* and its natural predator *Rhynocoris kumarii*. *Insects* 9(3), 101. doi:10.3390/insects9030101.

Sánchez-Rodríguez, A. R., Del Campillo, M. C. and Quesada-Moraga, E. 2015. *Beauveria bassiana*: an entomopathogenic fungus alleviates Fe chlorosis symptoms in plants grown on calcareous substrates. *Scientia Horticulturae* 197, 193–202. doi:10.1016/j.scienta.2015.09.029.

Sánchez-Rodríguez, A. R., Barrón, V., del Campillo, M. C. and Quesada-Moraga, E. 2016. The entomopathogenic fungus *Metarhizium brunneum*: a tool for alleviating Fe chlorosis. *Plant and Soil* 406(1–2), 295–310. doi:10.1007/s11104-016-2887-0.

Sangbaramou, R., Camara, I., Huang, X. Z., Shen, J., Tan, S. Q. and Shi, W. P. 2018. Behavioral thermoregulation in *Locusta migratoria manilensis* (Orthoptera: Acrididae) in response to the entomopathogenic fungus, *Beauveria bassiana*. *PLoS ONE* 13(11), e0206816 doi:10.1371/journal.pone.0206816.

Sierpińska, A., Popowska-Nowak, E. and Bednarek, A. 2015. *Beauveria brongniartii* Sacc. (Petch) against *Melolontha* spp. white grubs in forest nurseries with different soil pH. *Folia Forestalia Polonica* 57(4), 210–7. doi:10.1515/ffp-2015-0021.

Toledo, J., Flores, S., Campos, S., Villasenor, A., Enkerlin, W., Liedo, P., Valle, Á and Montoya, P. 2017. Pathogenicity of three formulations of *Beauveria bassiana* and efficacy of autoinoculation devices and sterile fruit fly males for dissemination of conidia for the control of *Ceratitis capitata*. *Entomologia Experimentalis et Applicata* 164(3), 340–9. doi:10.1111/eea.12608.

Tremblay, E. 1993. *Entomologia Applicata*. Tercera edición. Liguori Editore (vols, I–IV).

Trimmer, M. 2017. Biological control. Global market overview. Available at: http://wrir4.ucdavis.edu/events/2017_SLR_Meeting/Presentations/GeneralPresentations/1%20Trimmer%20-%20Global%20Biocontrol%20Market%202017.pdf (accessed on 02 March 2019).

Valverde-García, P., Santiago-Alvarez, C., Thomas, M. B., Garrido-Jurado, I. and Quesada-Moraga, E. 2018a. Comparative effects of temperature and thermoregulation on candidate strains of entomopathogenic fungi for Moroccan locust *Dociostaurus maroccanus* control. *BioControl* 63(6), 819–31. doi:10.1007/s10526-018-9904-6.

Valverde-Garcia, P., Santiago-Alvarez, C., Thomas, M. B., Maranhao, E. A. A., Garrido-Jurado, I. and Quesada-Moraga, E. 2018b. Sublethal effects of mixed fungal infections on the Moroccan locust, *Dociostaurus maroccanus*. *Journal of Invertebrate Pathology* 161, 61–9. doi:10.1016/j.jip.2018.12.009.

Vega, F. E. 2018. The use of fungal entomopathogens as endophytes in biological control: a review. *Mycologia* 110(1), 4–30. doi:10.1080/00275514.2017.1418578.

Vega, F. E. and Kaya, H. K. 2012. *Insect Pathology*. Academic Press, San Diego, CA.

Vemmer, M., Schumann, M., Beitzen-Heineke, W., French, B. W., Vidal, S. and Patel, A. V. 2016. Development of a CO_2-releasing coformulation based on starch, *Saccharomyces cerevisiae* and *Beauveria bassiana* attractive towards western corn rootworm larvae. *Pest Management Science* 72(11), 2136–45. doi:10.1002/ps.4245.

Vidal, C. and Fargues, J. 2007. Climatic constraints for mycoinsecticides. In: Maniana, K. and Ekesi, S. (Eds), *Use of Entomopathogenic Fungi in Biological Pest Management*. Research Sign Posts, Trivandrum, India.

Wang, C. and St Leger, R. J. 2007. The *Metarhizium anisopliae* perilipin homolog MPL1 regulates lipid metabolism, appressorial turgor pressure, and virulence. *The Journal of Biological Chemistry* 282(29), 21110–5. doi:10.1074/jbc.M609592200.

Wang, C. and Wang, S. 2017. Insect pathogenic fungi: genomics, molecular interactions, and genetic improvements. *Annual Review of Entomology* 62, 73–90. doi:10.1146/annurev-ento-031616-035509.

Wang, J. B., St Leger, R. J. and Wang, C. 2016. Advances in genomics of entomopathogenic fungi. *Advances in Genetics* 94, 67–105. doi:10.1016/bs.adgen.2016.01.002.

Weeks, E. N. I., Machtinger, E. T., Gezan, S. A., Kaufman, P. E. and Geden, C. J. 2017. Effects of four commercial fungal formulations on mortality and sporulation in house flies (*Musca domestica*) and stable flies (*Stomoxys calcitrans*). *Medical and Veterinary Entomology* 31(1), 15–22. doi:10.1111/mve.12201.

Wraight, S. P., Ramos, M. E., Avery, P. B., Jaronski, S. T. and Vandenberg, J. D. 2010. Comparative virulence of *Beauveria bassiana* isolates against lepidopteran pests of vegetable crops. *Journal of Invertebrate Pathology* 103(3), 186–99. doi:10.1016/j.jip.2010.01.001.

Wu, S., Tang, L., Fang, F., Li, D., Yuan, X., Lei, Z. and Gao, Y. 2018. Screening, efficacy and mechanisms of microbial control agents against sucking pest insects as thrips. *Advances in Insect Physiology* 55, 199–217. doi:10.1016/bs.aiip.2018.07.005.

Yang, H., Qin, C., Chen, Y., Zhang, G., Dong, L. and Wan, S. 2018. Comparative virulence of *Beauveria bassiana* isolates against lepidopteran pests of vegetable crops. *Journal of Invertebrate Pathology* 103, 186–99..

Yousef, M., Lozano-Tovar, M. D., Garrido-Jurado, I. and Quesada-Moraga, E. 2013. Biocontrol of the olive fruit fly *Bactrocera oleae* Gmelin (Diptera: Tephritidae) with *Metarhizium brunneum* Petch. and its extracts. *Journal of Economic Entomology* 106(3), 1118–25. doi:10.1603/ec12489.

Yousef, M., Quesada-Moraga, E. and Garrido-Jurado, I. 2015. Compatibility of herbicides used in olive orchards with a *Metarhizium brunneum* strain used for the control of olive fly preimaginals in the soil. *Journal of Pest Science* 88, 605–12. doi:10.1007/s10340-014-0632-0.

Yousef, M., Garrido-Jurado, I., Ruíz-Torres, M. and Quesada-Moraga, E. 2017. Reduction of adult olive fruit fly populations by targeting preimaginals in the soil with the entomopathogenic fungus *Metarhizium brunneum*. *Journal of Pest Science* 90(1), 345–54. doi:10.1007/s10340-016-0779-y.

Yousef, M., Aranda-Valera, E. and Quesada-Moraga, E. 2018a. Lure-and-infect and lure-and-kill devices based on *Metarhizium brunneum* for spotted wing Drosophila control. *Journal of Pest Science* 91(1), 227–35. doi:10.1007/s10340-017-0874-8.

Yousef, M., Alba-Ramírez, C., Garrido-Jurado, I., Mateu, J., Raya-Díaz, S., Valverde-García, P. and Quesada-Moraga, E. 2018b. *Metarhizium brunneum* (Ascomycota; Hypocreales) treatments targeting olive fly in the soil for sustainable crop production. *Frontiers in Plant Science* 9, 1. doi:10.3389/fpls.2018.00001.

Chapter 5

Advances in the use of entomopathogenic bacteria/microbial control agents (MCAs) as biopesticides in suppressing crop insect pests

Tejas Rao and Juan Luis Jurat-Fuentes, University of Tennessee, USA

1 Introduction

Bacteria are single-cell prokaryotic microorganisms lacking a nuclear membrane and genetic material (genophore) localized to a cytoplasmic region generally called the nucleoid. Entomopathogenic bacteria, loosely defined as producing pathogenic factors allowing them to cause disease in insects, have been described from diverse bacterial taxonomic groups. For the purpose of this chapter, focus will be on commercialized entomopathogenic bacteria (Fig. 1) and their classification will follow the most current version of the 'List of Prokaryotic Names with Standing in Nomenclature' database (Parte, 2018).

For the purpose of this chapter, we define bacterial biopesticides as products manufactured from a living microbe (bacteria) and sold for insect

http://dx.doi.org/10.19103/AS.2020.0073.06

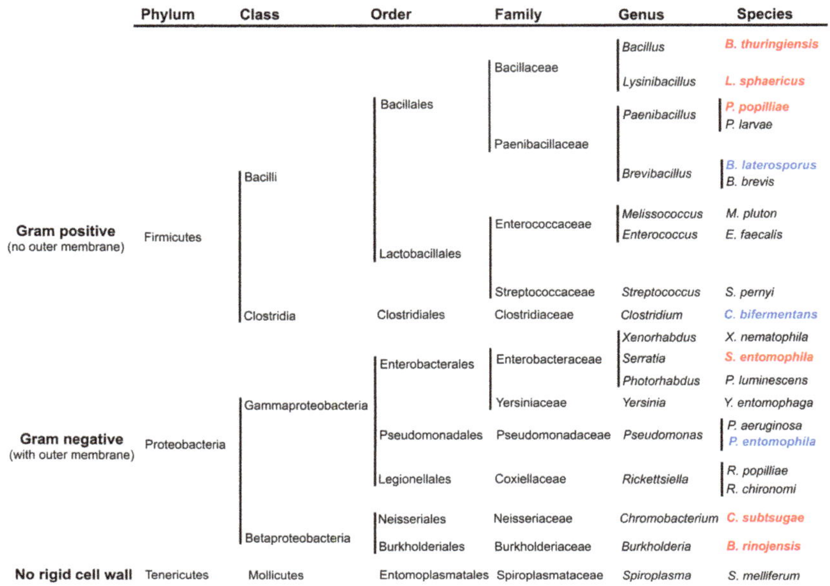

Figure 1 Classification of bacterial entomopathogens with species commercialized as biopesticides against insect pests and discussed in this chapter highlighted in red font. Species with potential for commercialization discussed in this chapter are highlighted in blue font. Source: modified from Jurat-Fuentes and Jackson (2012, p. 269), with permission from Elsevier.

control. Within bacterial biopesticides, we concentrate on entomopathogenic bacteria commercially used to control agricultural pests and vectors of disease. The use of insecticidal proteins from bacteria as plant-incorporated protectants (PIPs), in transgenic crops, and entomopathogenic bacteria associated with nematodes are discussed elsewhere in this book and will be mentioned only briefly.

Microbial pesticides based on entomopathogenic bacteria and their pathogenic factors dominate biopesticide sales, which amounts to ~6% of the global pesticide market (Marrone, 2019). Despite this low relative representation among all pesticides, bacterial biopesticides are the fastest growing segment among global pesticides, twice faster than synthetic chemicals (Ruiu, 2015). As an example, 75% of all biopesticide use involves products based on the bacterium *Bacillus thuringiensis* (Olson, 2015). Success of commercialized entomopathogenic bacteria is fueled by high efficacy and specificity, environmental and health safety, and usually easier registration process compared to synthetics. Major pesticide corporations globally are increasingly getting involved in licensing, acquiring and developing biopesticide products

(Marrone, 2019), which should contribute to increase commercial interest on entomopathogenic bacteria. Historical shortcomings of bacterial biopesticides include shorter field and storage life, and lower cost-effectiveness compared to synthetic alternatives.

The hallmark of bacterial biopesticides compared to other microbial products is their typical ability to quickly reduce feeding by infected insects, minimizing plant damage and maintaining populations below economic threshold levels. This fast action is due to the production of bacterial insecticidal proteins, which provide high specificity and convey greater environmental safety versus synthetic pesticides. Contrary to customary definitions in the literature, we refrain as much as possible from referring to these insecticidal proteins as 'toxins', as this definition is not exact given their focused activity against insects and implications on their safety when improperly defining them in a broader sense as 'toxins'.

Given their low risk to humans, bacterial biopesticides are exempted from residue tolerance requirements (i.e. limits to amount of pesticide allowed in food) and have zero or low re-entry and handling intervals (i.e. minimum amount of time after application considered safe for humans without protective clothing and equipment) in the USA (EPA, 2019; Seiber et al., 2014). In effect, bacterial pesticides may be the only available tool to control a pest close to harvest time, and the lack of residue requirements greatly benefits marketing by avoiding challenges related to maximum residue limits for exportation.

Biopesticides based on entomopathogenic bacteria are often used as standalone treatments in both organic and conventional production systems. Given their safety, they can be combined or rotated with other pesticides to increase efficacy and sustainability, being highly desirable components of robust integrated pest management (IPM) tactics (Ruiu, 2015; Chattopadhyay et al., 2017).

An additional benefit of bacterial versus synthetic pesticides is their reduced requirements and lower submission fees for registration dossiers in the USA (Marrone, 2019). However, this is not true for registrations in the European Union, which are comparatively more complex and lengthy, resulting in higher costs (Czaja et al., 2015; Frederiks and Wesseler, 2019).

Despite the above advantages, the use of bacterial pesticides is mostly limited to organic agriculture and IPM programs, mostly due to cost-competiveness versus synthetic pesticides. Areas needing further research and improvement include overcoming limitations from field and shelf life (especially with non-spore-forming bacteria) that result in unpredictable/insufficient efficacy, developing more cost-effective production application methods and the harmonization and simplification of registration procedures.

2 Overview of commercialized entomopathogenic bacteria

2.1 Commercialized bacterial entomopathogens

The first commercialized and most successful entomopathogenic bacterium is *Bacillus thuringiensis* (*Bt,* Bacillaceae), estimated to represent 95% of the microorganisms used for pest control (Mnif and Ghribi, 2015). The bacterium was originally described from diseased flour moths (*Ephestia kuehniella*) by Berliner (Berliner, 1915), and a similar isolate in 1938 was used for the development and production of the first *Bt* pesticide in France (Sporeine) to control flour moths (Steinhaus, 1975). Formation of a spore amenable to long-term storage without affecting virulence, ease of production and environmental safety have helped maintain the continued commercial success of *Bt* as a commercial biopesticide.

Pesticides containing other entomopathogenic spore-forming bacteria, namely *Paenibacillus popilliae* and *P. lentimorbus* (Paenibacillaceae, but formerly classified as *Bacillus*), were registered between 1948 and1961 in the USA, with registrations in Europe occurring shortly thereafter (Sundh and Goettel, 2013). Another commercially relevant spore-forming entomopathogen, *Lysinibacillus sphaericus* (Bacillaceae, formerly *Bacillus sphaericus*), was first registered in the USA for control of mosquito larvae in 1991. More recently, strains of *Brevibacillus laterosporus* (Paenibacillaceae) have raised interest for potential commercial development to control insects in the order of Coleoptera, Lepidoptera and Diptera as well as nematodes and molluscs (Ruiu, 2013).

Limitations mostly related to the preservation of activity during storage and application have contributed to a relatively slower development and commercialization of pesticides based on non-spore-forming entomopathogenic bacteria. A notable exception, the gram-negative pathogen, *Serratia entomophila* (Yersinaceae), was commercially developed as a soil-applied pesticide to control its host, the New Zealand grass grub (*Costelytra zealandica*) (Jackson, 2007). Optimization of *S. entomophila* formulation and application methods have greatly contributed to continued commercialization and efficacy of this bacterium, although geographic market expansion is challenging due to its restricted activity range.

Another example of commercial development of a non-spore-forming bacterium is *Chromobacterium subtsugae* (Neisseriaceae). Pesticides based on this bacterium contain live bacteria and fermentation residues, and display a wide range of insecticidal activity (Martin, 2007). Similarly, *Burkholderia rinojensis* (Burkholderiaceae) has been commercialized for control of a wide range of insect pests and mites (Cordova-Kreylos et al., 2013). Reported insecticidal activity for non-spore-forming entomopathogenic bacteria supports emerging interest in research towards their commercial development.

Promising candidates include *Yersinia entomophaga* (Yersiniaceae) active against Lepidoptera, Coleoptera and Orthoptera (Hurst et al., 2011a), and *Pseudomonas entomophila* (Pseudomonadaceae) with activity against species in Diptera (Vodovar et al., 2006).

2.2 General pathogenicity and resistance to commercialized bacterial entomopathogens

Despite their taxonomic diversity, commercialized bacterial entomopathogens share similar pathogenicity. A general description (Fig. 2), includes entering the host by ingestion, overcoming the gut epithelial barrier through insecticidal proteins, and then invading the hemocoel to proliferate, causing septicaemia and host death (de Bortoli and Jurat-Fuentes, 2019). The Insect Resistance Action Committee (IRAC) (www.irac-online.org) identifies insecticidal proteins produced by entomopathogenic bacteria as disruptors of midgut membranes, and are classified in mode of action group 11. After host death, the carcass serves as a nutrient source temporarily supporting growth until resource depletion, when the bacteria enter a dormant life cycle (spore), or are ingested by a new host.

Entomopathogenic bacteria produce a number of virulence factors (phospholipases, proteases, secondary metabolites, etc.), yet insecticidal proteins are the most critical factor in determining specificity and pesticide efficacy. These insecticidal proteins vary greatly between bacterial species, yet they target cells in the intestinal epithelium through recognition and binding to specific receptors on their surface.

The genes encoding these insecticidal proteins are commonly located in plasmids exchangeable through conjugation, although there are also examples of insecticidal protein genes located in the genome. Binding leads to subsequent processes that differ among insecticidal proteins, but they commonly result in enterocyte death and disruption of the intestinal epithelial barrier, facilitating

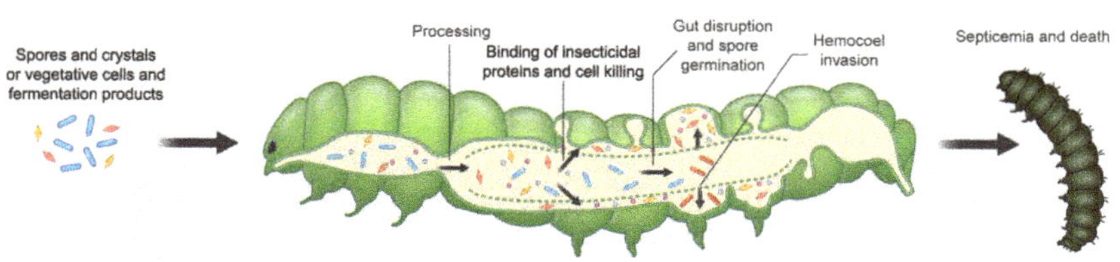

Figure 2 Schematic view of the mode of action of entomopathogenic bacteria. Presented are the main steps in pathogenesis resulting in the disruption of the epithelial gut barrier to allow vegetative cells accessing the main body cavity to cause lethal septicemia. This schematic model is applicable to pathogenesis of most commercialized bacterial entomopathogens; only the insecticidal proteins produced and their molecular mode of action may differ between species.

passage of bacterial cells and/or spores into the main body cavity (hemocoel) to cause lethal septicaemia (de Bortoli and Jurat-Fuentes, 2019).

Resistance to commercialized entomopathogenic bacteria is reported for field and greenhouse applications (de Bortoli and Jurat-Fuentes, 2019). The most commonly reported mechanisms in these cases involve mutations resulting in altered receptor function, preventing insecticidal protein binding. Cross-resistance commonly develops insecticidal proteins that recognize shared midgut receptors, but it does not involve insecticidal proteins with unique binding sites. Consequently, bacteria producing multiple insecticidal proteins recognizing distinct receptors are desirable in delaying the evolution of resistance. However, the effect of genetic regulators controlling expression of multiple genes or the alteration of a common step in the mode of action could also result in resistance against a bacterial pesticide even if it included multiple insecticidal proteins targeting distinct receptors.

Currently commercialized entomopathogenic bacteria with their main virulence factors and target insect hosts are shown in Table 1. Products containing *Bt* strains dominate and represent 75–95% of the biopesticide

Table 1 Commercialized entomopathogenic bacteria, main virulence factor/s and targeted pests. Also shown are commercial brand examples.

Bacteria	Target pest group	Virulence factor/s	Brand examples
Bacillus thuringiensis subsp. aizawai	Lepidopteran larvae	Cry1Aa, Cry1Ab, Cry1C, Cry1D	XenTari, FlorBac
Bacillus thuringiensis subsp. kurstaki	Lepidopteran larvae	Cry1Aa, Cry1Ab, Cry1Ac, Cry2Aa	Dipel, Lipel, Deliver, Foray
Bacillus thuringiensis subsp. israelensis	Mosquito, gnats and blackfly larvae	Cry4Aa, Cry4Ba, Cry10Aa, Cry11Aa, Cyt1Aa, Cyt2Ba	Teknar, VectoBac, Vectobar
Bacillus thuringiensis subsp. tenebrionis	Coleopteran larvae	Cry3Aa	Novodor, Trident
Lysinibacillus sphaericus	Mosquito larvae	Bin	VectoLex, VectoMax
Paenibacillus popilliae	Coleopteran larvae (grubs)	Multiple insecticidal factors	Doom, Grub Attack
Burkholderia spp.	Chewing and sucking insects and mites; nematodes	Multiple insecticidal factors	Majestene
Saccharopolyspora spinosa	Two-spotted spider mites	Spinosyns	Tracer 120, Conserve
Chromobacterium subtsugae	Chewing and sucking insects and mites	Multiple insecticidal factors	Grandevo
Serratia entomophila	New Zealand grass grub	SepTc and Afp	

market share (Olson, 2015; Vimala Devi et al., 2019). However, current interest in commercial development of entomopathogenic bacteria is driving research on alternative bacterial entomopathogens (Marrone, 2019).

3 Commercialized spore-forming bacterial entomopathogens

3.1 Bacillus thuringiensis *(*Bt*)*

The *Bt* bacterium was originally isolated by Ishiwata as a causative agent for the sudden ('sotto' in Japanese) disease in silkworms, before being formally described by Berliner (Berliner, 1915). Together with *Bacillus cereus* and *Bacillus anthracis*, *Bt* is part of the *B. cereus sensu lato* group of gram-positive spore-forming bacteria (Rasko et al., 2005).

The natural ecology of *Bt* is still a matter of debate, but the most generally accepted view is that the bacterium is a pathogen needing a host for reproduction, yet it can also be ubiquitously found in diverse habitats due to vectoring systems such as soil invertebrates (Ruan et al., 2015). Screening efforts have identified *Bt* isolates covering activities against a wide range of insects in diverse taxonomic groups (van Frankenhuyzen, 2009). Despite this diversity in specificity, commercialized *Bt* biopesticides target insect pests in the order of Lepidoptera, Diptera and Coleoptera. Production of diverse antifungal and antimicrobial compounds may increase the commercial interest of *Bt* strains (Salazar-Marroquín et al., 2016).

Classification of *Bt* isolates has been conducted by various morphological, biochemical and antigenic methods (Jurat-Fuentes and Jackson, 2012). Traditionally, classification of *Bt* isolates followed reactivity to the H flagellar antigen (Lecadet et al., 1999). Limitations in antisera availability and the fast development of next generation sequencing has opened the avenue for alternative *Bt* classification methods, as multi-locus sequence typing (MLST) (Wang et al., 2018).

While 85 *Bt* subspecies have been serotyped (Jurat-Fuentes and Jackson, 2012), strains of five *Bt* subspecies alone have been commercialized. These commercialized *Bt* subspecies are active ingredients in pesticides targeting Lepidoptera (subsp. *kurtstaki*, *aizawai* and *galleriae*), Coleoptera (subsp. *morrisoni* biovar *tenebrionis*, *thuringiensis* and *galleriae*) and Diptera (susbp. *israelensis*) (Kabaluk et al., 2010). Products based on recombinant *Bt* (Raven, Lepinox and Crymax), *Pseudomonas fluorescens* (MVP series and Mattch) or *Clavibacter xyli susbsp. Cynodontis* (InCide) producing *Bt* insecticidal proteins have also been commercialized (Glare and O'Callaghan, 2000).

In some cases, recombinant bacteria produce combinations of insecticidal *Bt* proteins targeting multiple insect orders, for example, Cry1Ac (Lepidoptera) and Cry3Aa (Coleoptera) in Raven, that are not found in natural commercialized *Bt* strains, broadening host range. In other cases, recombinant pesticides contain insecticidal *Bt* proteins that target the same group of insects but that represent different modes of action (i.e. insecticidal proteins binding to different receptors), such as in Crymax containing Cry1Ac, Cry2A and Cry1Ca proteins. This strategy increases efficacy and delays the evolution of resistance.

Biopesticides based on *Bt* contain fermentation solids together with *Bt* spores and insecticidal crystal (Cry) proteins (Fig. 3a and b), which are the most important determinants of specificity and activity of the product. Production of Cry proteins occurs during sporulation and they are stored in crystalline parasporal bodies. Genes encoding insecticidal proteins are usually located in plasmids such as the pBtoxis megaplasmid in *Bt* subsp. *israeliensis*, which contains four *cry* (Cry4Aa, Cry4Ba, Cry10Aa and Cry11Aa) and two cytolytic (Cyt1Aa and Cyt2Ba) genes that dictate pathogenicity (Berry et al., 2002). The bacterium also produces vegetative insecticidal proteins (Vip) which could also have a role in activity of a biopesticide once the *Bt* spores germinate.

There are currently 818 and 148 identified Cry and Vip proteins, respectively, which are classified according to their amino acid sequence identity into 78 Cry and 4 Vip primary families (Crickmore et al., 2019). Diverse protein structures are present among Cry proteins, although the majority follow a three-domain model (Adang et al., 2014). Current initiatives propose increased focus on using protein structural similarity to enhance the information already provided by the current sequence identity-based classification (Berry and Crickmore, 2017).

Upon ingestion by a susceptible host, the parasporal crystals are solubilized in the gut digestive fluids to release a pro-protein (protoxin) form of the host. This pro-protein form, usually around 120 kDa in size, is then processed by gut digestive proteases through removal of most of the C-terminus and ~43 amino acids from the N terminus to yield an activated insecticidal protein (toxin) form (normally 55–60 kDa in size). Some Cry proteins, such as Cry3Aa active against beetles or Cry2Aa active against caterpillars, have smaller pro-protein forms (approx. 75 kDa in size) and are mostly processed at their N terminus to render the activated insecticidal protein.

The activated Cry protein must cross the peritrophic matrix to bind receptors on the brush border membrane of midgut cells. In some cases, retention of the protein limits Cry activity by preventing interactions with receptors (Guo et al., 2019). The interactions between Cry proteins and their receptors and how they lead to gut cell death have been extensively investigated (Adang et al., 2014), yet a number of relevant questions remain (Vachon et al., 2012). Research efforts have attempted to develop a model for the Cry mode of action that is inclusive of all Cry proteins and insect hosts, yet discrepancies observed suggest that while some generalities may be conserved, the mode of action may be heavily influenced by the Cry protein and insect considered.

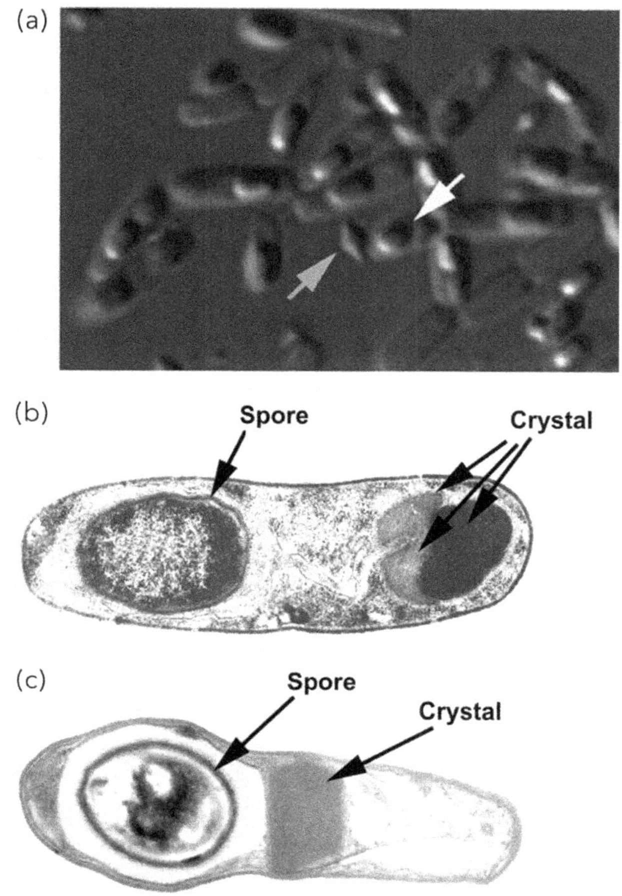

Figure 3 Morphology of commercialized spore-forming *Bacillus* and *Lysinibacillus* entomopathogenic bacteria. (a) Differential interference contrast (DIC) micrograph detailing the late sporulation phase of a *Bacillus thuringiensis* subsp. *kurstaki* strain HD73 culture. Spore (white arrow) and released rhombohedral Cry1Ac crystal (gray arrow) are visible. (b) Electron micrograph of a sporulated cell of *B. thuringiensis* var. *israelensis* displaying the spore and three different parasporal bodies produced (as indicated), which are composed of different insecticidal proteins [Courtesy of B. Federici and H-W. Park]. (c) Electron micrograph of a sporulated *Lysinibacillus sphaericus* cell (courtesy of Dr. Jean-François Charles, Institut Pasteur, Paris), note the characteristic spherical spore located in a terminal position within the swollen sporangium (as indicated). Source: (b) and (c) are reprinted from Jurat-Fuentes and Jackson (2012, p. 273), with permission from Elsevier.

Multiple proteins and glycolipids have been proposed as receptors for Cry proteins in insects (Adang et al., 2014), including cadherins, aminopeptidases, alkaline phosphatases and ATP binding cassette (ABC) transporters. Functional evidence for the role of these proteins in Cry intoxication has been generally provided by studies of resistant insect lines or heterologous expression of candidate receptors. Accumulated functional evidence supports a critical role for cadherins in Cry1 and Cry3 intoxication (Fabrick and Wu, 2015), and for members of distinct subfamilies of ABC transporters as functional receptors for Cry1A, Cry1F, Cry2A and Cry3A insecticidal proteins (Sato et al., 2019).

The formation of toxin oligomeric structures (before or after binding to receptors) and their insertion on the midgut cell membrane to form pores that lead to osmotic cell death are commonly accepted steps in the Cry intoxication process (Vachon et al., 2012). The participation of intracellular cell death pathways was also proposed, although evidence has been mostly from experiments with cultured insect cells (Zhang et al., 2005). It is worth noting that the pore formation and intracellular pathway models are not exclusive and it is plausible for both processes occurring *in vivo* depending on different parameters such as Cry protein and insect considered, or the concentration of Cry protein affecting the gut cell.

Eventually, in both of the proposed intoxication pathways, the collapsing midgut epithelial barrier allows passage of bacteria into the main body cavity where septicemia ensues and eventually kills the host (Raymond et al., 2010). The host cadaver serves as a nutrient-rich environment for cell division; once depleted, the vegetative cells sporulate (Verplaetse et al., 2015).

Spores and insecticidal proteins in *Bt* biopesticides persist in the environment for several years (depending on subspecies and environment), partially as a result of recycling in hosts (Vettori et al., 2003). There is also evidence for the transfer of insecticidal genes from *Bt* pesticides to indigenous *Bacillus* spp. in soil under field conditions (Donnarumma et al., 2010). Insecticidal proteins from *Bt* sprays persist and are protected in the environment by associating with soil particles, yet their uptake by plants has not been conclusively demonstrated (Icoz et al., 2009).

3.2 Lysinibacillus sphaericus

Originally named as *Bacillus sphaericus* but later re-classified as *Lysinibacillus sphaericus* (Ahmed et al., 2007), this bacterium is defined by the production of a spherical spore that locates to one of the ends in the sporangium (Fig. 3c). Advantages of *L. sphaericus* biopesticides include their safety to non-target organisms, stability in storage and handling (Su et al., 2018), and their good persistence in the field even in polluted waters that serve as breeding sites for target species (Silva Filha et al., 2014). The addition of UV protectants reduces degradation of *L. sphaericus* and improves stability of virulence in the field (Hadapad et al., 2009).

The *L. sphaericus* bacterium produces, among other minor virulence factors, binary (Bin) proteins that structurally relate to binary (Bin-like) Cry proteins, such as Cry34/Cry35 and Cry48/Cry49 (Adang et al., 2014). These Bin proteins are responsible for the mosquitocidal activity in commercialized strains (Silva Filha et al., 2014). Biopesticides based on *L. sphaericus* target *Culex* and *Anopheles* mosquito larvae of medical significance, as well as black flies and non-biting midges. While they do not effectively control *Aedes aegypti*, they are active against *Aedes vexans*, and *Ochlerotatus* spp. (Mulligan et al., 1978; Wraight et al., 1987).

Mosquitocidal activity of Bin proteins is dependent on the combined action of two protein components, BinA and BinB, which fulfill distinct roles in the intoxication process. Both components are co-transcribed from a single operon just before the sporulation phase (Ahmed et al., 1995) and are stored as parasporal bodies in the sporangium (Jones et al., 2007). Negligible activity is observed for BinA alone compared to the complexed Bin protein (Broadwell et al., 1990).

Upon ingestion, BinB (51.4 kDa) is processed at the N- and C-termini to a 43–46 kDa protein (Davidson et al., 1990) that binds to specific α-glucosidases (maltases) tethered to the membrane of gastric caecae and posterior midgut cells by a glycosylphosphatidylinositol (GPI) anchor. The first α-glucosidase binding BinB (Cpm1) was identified in *Culex pipiens* (Silva-Filha et al., 1999). Since then, orthologues have been identified in *Culex quinquefasciatus* (Cqm1) (Romao et al., 2006), *Anopheles gambiae* (Agm1) (Opota et al., 2008) and *Aedes aegypti* (Aam1) (Ferreira et al., 2010).

The importance of BinB binding to α-glucosidases for susceptibility is supported by the epitope required for BinB binding to Cqm1 missing in Aam1, explaining lack of Bin activity against *A. aegypti* (Ferreira et al., 2014). In addition, resistance to *L. sphaericus* is linked in the vast majority of cases to unavailability of α-glucosidase receptors (Silva-Filha, 2017). The numerous cases of field resistance to *L. sphaericus* sprays are expected when considering that commercialized strains rely on a single virulence factor (the Bin protein) for mosquitocidal activity (Silva Filha et al., 2014). Commercial combinations of *L. sphaericus* with *Bt* subsp. *israelensis* (VectoMax) expand activity range but also delay resistance evolution by providing insecticidal proteins with distinct mode of action.

Proteolysis of BinA (41.9 kDa) in digestive fluids of susceptible and non-susceptible insects produces a 40 kDa protein (Nicolas et al., 1990). Oligomerization of two BinA and BinB subunits in solution results in formation of a tetramer (Smith et al., 2005), leading to structural changes that seem to facilitate membrane insertion directed by the BinB subunits (Boonserm et al., 2006).

Intoxication of cultured *C. quinquefasciatus* cells with Bin protein results in rapid responses commonly observed with Cry proteins, such as dilation of the endoplasmic reticulum, condensation of mitochondria and gradual loss of membrane integrity (Davidson and Titus, 1987). However, while binding of

Bin to receptors also results in pore formation as in the Cry mode of action, cell lysis is not observed as part of Bin cytotoxicity (Pauchet et al., 2005). In addition, Bin intoxication includes unique features that correlate with cytotoxicity, such as internalization of the Bin protein (Oei et al., 1992) and vacuolisation of the cell (Pauchet et al., 2005). Protein binding and insertion detected in Bin-resistant mosquito strains (Schroeder et al., 1989) support that Bin activity depends on post-internalization events. Vacuolisation is associated with the induction of autophagy as a cellular defensive response, but the Bin protein avoids degradative autophagy processes by trafficking in the recycling endosomal pathway (Opota et al., 2011). Further work is needed to fully characterize the significance of these cell responses and their relationship with mosquitocidal activity.

Selected *L. sphaericus* strains active against Bin-resistant mosquito larvae produce crystals containing the Cry48/Cry49 binary protein active against *Culex* spp. (Jones et al., 2007). Interestingly, Cry48 structurally relates to the three-domain Cry proteins, yet it needs Cry49 for activity, constituting a unique example of a three-domain protein acting as part of a binary insecticidal protein. The mode of action of Cry48/Cry49 in *Culex quinquefasciatus* includes features from the Bin and Cry intoxication processes (de Melo et al., 2009), including binding to Bin and Cry receptors (Rezende et al., 2017).

Some strains of *L. sphaericus* also produce mosquitocidal (Mtx) proteins during the vegetative growth stage, although these proteins are not directly associated to high mosquitocidal activity (Thanabalu et al., 1991). In fact, the *mtx* gene in the most commercially used *L. sphaericus* strain (C3-41) appears to be a pseudogene that is no longer active (Hu et al., 2008). Two main types of Mtx proteins are defined based on their mode of action, Mtx1 displaying similarities to ADP-ribosyltransferase bacterial toxins (Thanabalu et al., 1991) and a second group, exemplified by Mtx2, with structural similarities to the *Clostridium* epsilon toxin ETX/MTX2 family of pore forming toxins (Berry, 2012). Individually, each of these proteins can synergize Bin and Cry11Aa (from *Bt* subsp. *israeliensis*) activity against *C. quinquefasciatus* (Wirth et al., 2007), and co-expression of Mtx1 and Mtx2 proteins results in synergistic activity against *A. aegypti* (Rungrod et al., 2009). This synergism can be a valuable tool in enhancing the effectiveness of pesticides based on *L. sphaericus*.

Additional virulence factors occur in *L. sphaericus* strains. The *L. shpaericus* S-layer protein (SlpC) is a structure present on the surface of bacteria displaying toxicity against *Culex spp.* and *A. aegypti* larvae, and synergizing Bin mosquitocidal activity (Allievi et al., 2014). Some *L. sphaericus* strains produce sphaericolysin, a protein active against cockroaches and caterpillars by injection (Nishiwaki et al., 2007).

Similar to the mode of action of Cry, the result of midgut cell killing by Bin protein is the disruption of the gut epithelial barrier, allowing spores and vegetative cells to invade the hemocoel. The host then serves as source of nutrients supporting vegetative growth, which may favor production of Mtx proteins to target unknown cell types. Once nutrient resources are depleted, sporulation ensues and bacteria discharge from collapsing mosquito cadavers to recycle *L. sphaericus* in the environment (Becker et al., 1995). Re-infection by released spores occurs only if release occurs in the larval feeding zone in the water column (Charles and Nicolas, 1986).

3.3 Paenibacillus *spp.*

The first major insect control program based on a bacterial entomopathogen was the use of *Paenibacillus popilliae* (formerly classified as *Bacillus popilliae*) to control larvae of the Japanese beetle, *Popillia japonica* (Klein and Kaya, 1995). In fact, *P. popilliae* was the first entomopathogen registered in the USA as a microbial control agent.

Both *P. popilliae* and the closely related *P. lentimorbus* are characterized by the production of an ellipsoidal spore that deforms the sporangium (Fig. 4a). However, while *P. popilliae* has a spindle-shaped sporangium and forms a parasporal body, *P. lentimorbus* has a lemon-shaped sporangium and lacks a parasporal body in its sporangia (Dutky, 1940). Biopesticides based on *P. popilliae* are used in foliar and soil applications of ornamental and turf grasses to control scarab beetle larvae (grubs).

Being an obligate pathogen, *P. popilliae* is highly fastidious (Dutky, 1940), and the need for *in vivo* production of spores greatly increases manufacturing costs. This inability to grow outside the host distinguishes *P. popilliae* and *P. lentimorbus* from other successfully commercialized spore-forming bacteria, such as *Bacillus* and *Lysinibacillus* spp. In addition, *P. popilliae* strains are also highly specific and although strains have been isolated with activity against different grubs (Chandel et al., 2019), only strains infecting *P. japonica* have been commercialized (Koppenhöfer and Wu, 2017). Direct injection into the hemocoel of vegetative cells helps to overcome this strain specificity (Milner, 1981), suggesting that specificity may be related to the ability of crossing the gut epithelial barrier.

These fastidious features, together with inconsistent performance, have reduced the use of *P. popilliae* pesticides to niche markets. However, there is great potential for commercial expansion once cost-effective *in vitro* production methods or more virulent strains are identified or developed.

Late stages in the infective process by *P. popilliae* result in the characteristic whitish hemolymph coloration due to the high number of refractive bacterial spores present, giving name to the milky disease. At least two crystal insecticidal proteins (Cry18 and Cry43) have been identified in parasporal bodies from *P. popilliae* (Zhang et al., 1997; Yokoyama et al., 2004), yet they are not required for the progression of the milky disease.

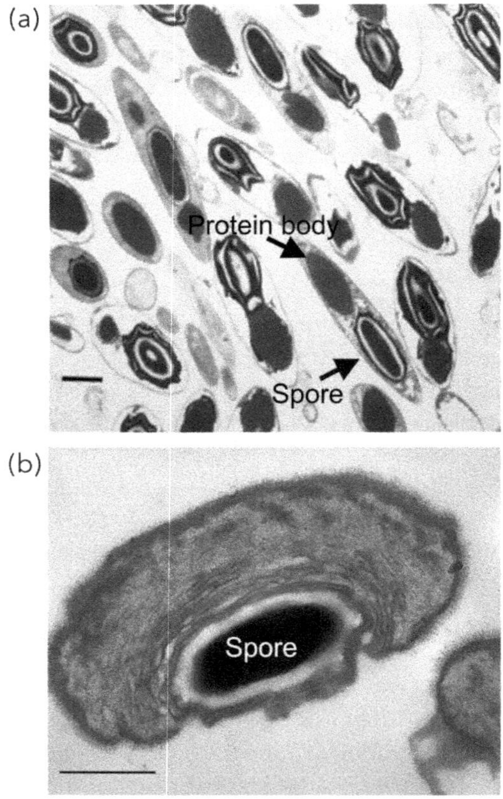

Figure 4 Morphology of spore-forming (a) *Paenibacillus popilliae* and (b) *Brevibacillus laterosporus*. (a) Transmission electron micrograph of *P. popilliae* culture displaying the spindle-shaped sporangia, each containing a spore and associated protein body, as indicated (bar = 1 μm). Source: reprinted from Glare et al. (2017), with permission from Elsevier. (b) Transmission electron micrograph of *B. laterosporus* strain Uniss18 free spore (as indicated) surrounded by the typical canoe-shaped parasporal body, bar = 0.5 μm. Photo courtesy of Dr. Luca Ruiu, University of Sassari, Italy.

Upon ingestion, *P. popilliae* spores germinate in the physicochemical conditions of the scarab midgut and vegetative cells penetrate the midgut cells by phagocytosis (Splittstoesser et al., 1973). Vegetative cells seem to undergo a first round of multiplication in the regenerative nidi on the basal epithelium, and then invade the hemocoel for a second round of multiplication (Kawanishi et al., 1978). Strains of *Paenibacillus* producing high levels of chitinolytic activity display insecticidal activity (Singh et al., 2016), suggesting that chitin degradation may facilitate passage through the gut epithelium.

A proportion of the vegetative cells undergo sporulation in waves until later stages, when spores are largely more abundant and give the typical ivory-white appearance to the hemolymph. The infected larvae become moribund and after depletion of all reserves finally die (Klein, 1981). Upon host death, spores liberate to the soil where they persist for long periods until new hosts become available.

4 Spore-forming bacteria with potential for future commercialization

4.1 Brevibacillus laterosporus

This ubiquitous bacterium is characterized by a canoe-shaped lamellar parasporal body attached to one side of the spore (Fig. 4b) (de Oliveira et al., 2004), and has been identified both as mutualist and pathogen, depending on the insect species considered (Marche et al., 2016). Interest in *B. laterosporus* is fueled by its wide range of activities against diverse coleopteran, dipteran and lepidopteran insects, nematodes and molluscs and the production of broad-spectrum antimicrobials against phytopathogenic bacteria and fungi (Ruiu, 2013).

Much interest has been devoted to its potential for the control of houseflies (*Musca domestica*) (Ferreira et al., 2016; Ruiu et al., 2008, 2014). Experimental *B. laterosporus* formulations are effective in controlling house flies in breeding sites at dairy and poultry farms (Ruiu et al., 2008, 2014), and did not affect beneficial parasitoids (Ruiu et al., 2007). In addition to lethality, significant effects on fitness are reported (Ferreira et al., 2016; Ruiu et al., 2006).

Four highly conserved proteins (ExsC, CHRD, CpbA and CpbB) are reported as main virulence factors against *M. domestica* (Marche et al., 2017). These proteins are part of the spore coat and canoe-shaped parasporal body (SP-CSPB) complex of *B. laterosporus*. Insecticidal activity is at times associated with the production of parasporal crystal inclusions containing insecticidal proteins (Orlova et al., 1998; Zubasheva et al., 2010). Different patents describe some of these insecticidal proteins produced by *B. laterosporus* with activity against coleopteran species, which display high homology to vegetative (Vip) and binary Cry proteins from *B. thuringiensis* (Ruiu, 2013). Genome sequencing projects have also detected many genes encoding predicted insecticidal proteins, including *cry* genes, but their role in *B. laterosporus* activity is yet to be demonstrated (Djukic et al., 2011; Sharma et al., 2012).

4.2 Clostridium bifermentans

Although *C. bifermentans* has been recorded to infect humans, safety of the *C. bifermentans* subsp. *malaysia* has been demonstrated in vertebrates (Thiery et al., 1992a) and non-target dipterans (Yiallouros et al., 1994). This isolate contains an operon encoding a spectrum of Cry insecticidal proteins active against nematocerous Diptera (Thiery et al., 1992b; Yiallouros et al., 1994). Another isolate, *C. bifermentans* subsp. *paraiba* is most active against *Anopheles maculatus* (Seleena et al., 1997).

Proteins included in the operon of *C. bifermentans* subsp. *malaysia* include Cry16A, Cry17A, Cbm17.1 and Cbm17.2 (Qureshi et al., 2014). Interestingly, these proteins are active against *Aedes* rather than *Anopheles* mosquitoes, in contrast to *C. bifermentans* subsp. *malaysia* being more active against *Anopheles*, suggesting that insecticidal proteins targeting *Anopheles* are not located in the cry operon (Qureshi et al., 2014; Juarez-Perez and Delecluse, 2001).

House fly larvae and adults exposed to *C. bifermentans* susbp. *malaysia* display reduced feeding and growth rates, progressive paralysis and death (Ruiu et al., 2012). Histopathology of this process includes midgut microvilli disruption followed by cell lysis. The surrounding muscular and connective tissues also degenerate, which may explain the observed paralysis.

5 Commercialized non-spore-forming entomopathogenic bacteria

5.1 *Serratia entomophila*

This bacterium was isolated as a causative agent of amber disease in larvae of the New Zealand grass grub (*Costelytra zealandica*), a damaging scarab pest of grasses. The bacterium was developed into a biopesticide, but short shelf life and complex application methods for efficacy resulted in initially limited economic success (Jackson, 2007). Continuous improvements in the formulation have greatly extended its shelf life and facilitated application by broadcast methods (Wright et al., 2017). Performance of this improved biopesticide surpasses synthetic alternatives in the long-term control of *C. zealandica* (Zydenbos et al., 2016).

Amber disease has a unique pathogenicity (Jurat-Fuentes and Jackson, 2012). The bacterium attaches using fimbriae to the foregut cuticle, and clearing of the larval midgut within 72 h leads to the characteristic amber coloration of infected hosts. Infection spreads throughout the whole digestive tract while the host remains in a chronic starvation phase consuming storage resources (fat body). Weakening of the gut epithelial barrier leads to rupture, allowing bacterial invasion and deadly septicemia (Jackson et al., 2001).

A 155-kb plasmid (pADAP) isolated from pathogenic *S. entomophila* was found to contain two gene clusters, the *sep* (*Serratia entomophila* pathogenicity) cluster associated with virulence and the *afp* (antifeeding prophage) cluster related to the cessation of feeding (Hurst and Glare, 2002). The pathogenesis of amber disease requires both clusters: Afp factors induce cessation of feeding and Sep proteins prompt clearing of gut contents and amber coloration (Gatehouse et al., 2009). Colonization of the digestive system probably enables continuous production of Afp and Sep factors to prevent purging and enable rupture of the gut epithelium.

The *sep* cluster contains genes producing toxin complex (Tc) proteins (SepA, SepB and SepC) that combine into an insecticidal protein complex similar to Tc proteins from nematode-associated bacteria, *Photorhabdus luminescens* and *Xenorhabdus nematophilus* (Hurst et al., 2007b). On the other hand, the *afp* cluster encodes 18 open reading frames involved in the assembly of a prophage virus-like structure (Hurst et al., 2007a). This anti-feeding prophage includes an inner tube extending into a needle-like puncture device surrounded by a helical sheath, and is hypothesized to serve as a toxin-delivery apparatus (Heymann et al., 2013).

5.2 Chromobacterium subtsugae

Originally isolated from forest soil, *Chromobacterium subtsugae* is a motile gram-negative bacterium with colonies displaying a characteristic violet pigmentation (Martin et al., 2007a). Biopesticides based on *C. subtsugae* display a broad range of activity and diversity of crop applications. The bacterium is active against members of Coleoptera, including the Colorado potato beetle (*Leptinotarsa decemlineata*), small-hive beetle (*Aethina tumida*), pecan weevil (*Curculio caryae*) and *Diabrotica* spp. rootworms (Martin et al., 2007a; Shapiro-Ilan et al., 2013; Stamm et al., 2013); hemipterans such as the Southern green stink bug (*Nezara viridula*) (Martin et al., 2007b); and lepidopteran larvae such as diamondback moth (*Plutella xylostella*) (Martin et al., 2007a).

Commercialized extracts based on strain PRAA4-1^T contain live bacteria and fermentation residues to control thrips, two-spotted spider mites, whiteflies, caterpillars, psyllids and beetles. The bacterium produces a number of stable insecticidal factors. After ingestion, host death can take between 2 and 7 days. This biopesticide is stable when stored in cool conditions, and the formulation with sodium benzoate reduces its environmental degradation by sunlight (Asolkar et al., 2016).

5.3 Burkholderia rinojensis

The genus *Burkholderia* includes resident gut bacteria that often affect fitness of their insect hosts, for oviposition in the bean bug (*Riptortus pedestris*) (Kil et al., 2014). Within this genus, *Burkholderia rinojensis* is a member of the *B. cepacia* complex that displays activity against beet armyworm (*Spodoptera exigua*) and two-spotted spider mite (*Tetranychus urticae*) through both contact and ingestion (Cordova-Kreylos et al., 2013). Cells and spent fermentation media containing active compounds spanning different chemical classes from cultures of *B. rinojensis* strain A396 are commercialized as an environmentally safe biopesticide (Venerate) for foliar and soil applications. This product is activated by exposure and ingestion, inducing exoskeleton degradation and molting interference in lepidopteran larvae and hemipteran pests (https://marronebio.com/products/venerate/). The combination of diverse actives is expected to hinder the evolution of resistance.

6 Non-spore-forming entomopathogenic bacteria with potential for commercialization

6.1 Yersinia entomophaga

Yersinia is a genus consisting of known human pathogens, such as *Y. pestis* (bubonic plague), but also species displaying insecticidal activity. The bacterium *Yersinia entomophaga* was first isolated from diseased larvae of *C. zealandica*, yet it presents activity against a wide spectrum of insect pests in the order of Lepidoptera, Coleoptera and Orthoptera (Hurst et al., 2011a). Pathogenesis involves deterioration of the peritrophic matrix and midgut epithelium, followed by entry into the hemocoel to cause septicemia (Hurst et al., 2014; Marshall et al., 2012).

The main *Y. entomophaga* virulence factor localizes to a pathogenicity island containing seven genes encoding an insecticidal toxin complex (YenTc) and two chitinases (Chi1 and Chi[2]) (Hurst et al., 2011b). The YenTcB and YenTcC components are present on the YenTc surface and are the main toxicity determinants (Landsberg et al., 2011). The YenTcA pentameric component also contains Chi1 and Chi2, and structural analyses suggest its involvement in determining binding and pore formation specificity of the Yen-Tc complex, with a key role for Chi1 in this process (Piper et al., 2019). Other proposed roles for Chi1 and Chi2 in the pathogenic process include degradation of the peritrophic matrix and post-mortem degradation of tissues (Busby et al., 2012).

Experimental preparations of *Y. entomophaga* are efficient in foliar sprays when added to baits to control larvae of porina moth (*Wiseana* spp.) and diamondback moth (*Plutella xylostella*) (Hurst et al., 2019a,b), and adult bronze beetle (*Eucolaspis* spp.) (Hurst et al., 2011c). Although *Y. entomophaga* lacks virulence factors associated with *Yersinia* pathogenicity in humans, its safety needs to be carefully re-evaluated, especially considering isolation in an asymptomatic human patient with urine infection (Le Guern et al., 2018).

6.2 Pseudomonas *spp.*

Members of the *Pseudomonas* genus are strictly aerobic, rod-shaped and highly motile by means of a polar flagellum. Most research on the entomopathogenic members of this genus have focused on *Pseudomonas entomophila*, which was originally isolated as inducing a systemic immune response in both larvae and adults of fruit fly (*Drosophila melanogaster*) (Vodovar et al., 2005). Subsequent tests have detected that *P. entomophila* also presents insecticidal activity against species in three taxonomic orders: Diptera (*Anopheles gambiae*), Lepidoptera (*Bombyx mori* and *Galleria mellonella*) and Coleoptera (*Sitophilus oryzae*) (Dieppois et al., 2015). In addition, it can reduce ethylene levels and allow plants to resist environmental stress and pathogen infection (Kamala-Kannan et al., 2010).

Most available information on *P. entomophila* pathogenesis comes from studies in *Drosophila*. Insecticidal activity is multifactorial (Vodovar et al., 2006) and includes toxin complex (Tc) proteins, secondary metabolites, lipases and proteases that protect against the immune system and help degrade host tissues and probably the peritrophic matrix (Lee et al., 2018; Kuraishi et al., 2011) and cell surface-associated factors allowing adhesion and colonization of the host.

Infection by feeding includes destruction of the gut epithelium and proceeds quickly to host death within 2–3 days. The ingestion of *P. entomophila* and the subsequent action of Tc proteins and secondary metabolites (reactive oxygen species) inhibit protein translation and impede immune and gut epithelial repair programs, hindering the host defensive response to infection (Chakrabarti et al., 2012). The AprA metalloprotease produced by *P. entomophila* is critical for insecticidal activity and contributes to suppress cellular and humoral innate immune responses in the host (Lee et al., 2018).

An insecticidal protein isolated from *P. chlororaphis* (IPD072Aa, 10 kDa) shows activity against coleopteran pests, especially against the Western corn rootworm (*Diabrotica virgifera virgifera*) (Boeckman et al., 2019; Schellenberger et al., 2016). Available data support that IPD072Aa is a unique novel insecticidal protein (Carlson et al., 2019) that acts by disrupting the midgut epithelium (Schellenberger et al., 2016). This protein is currently advancing towards its commercialization as a plant-incorporated protectant in transgenic corn.

7 Microbial control agents (MCAs)

There are a number of examples of commercialized agents with insecticidal activity that were originally isolated from entomopathogenic bacteria. Spinosyns, produced during fermentation by the bacterium *Saccharopolyspora spinosa* (Actinomycetales) are a major insecticidal metabolite group (Kirst, 2010). This group of metabolites disrupt the binding of acetylcholine to postsynaptic nicotinic receptors in the nervous system, leading to paralysis and death. Pesticidal products containing spinosyns have been registered in the USA since 1997 as chemical pesticides to control caterpillars, thrips and other pests of vegetable, fruit and ornamental crops.

Another example of insecticidal MCAs are avermectins and their derivatives, which were originally isolated from the soil bacterium *Streptomyces avermitilis* (Wright, 1987). These metabolites are a family of macrocyclic lactones and they have been commercialized for the control of numerous arthropod pests and parasites (Omura, 2016).

8 Advantages and drawbacks on the use of bacterial biopesticides

Insecticidal products based on strains of *B. thuringiensis* vastly dominate the bacterial biopesticide market, with documented success in diverse systems including forestry, vegetables, ornamentals and major field crops. These biopesticides present advantages and drawbacks, most of which are directly applicable to other bacterial biopesticides and candidate entomopathogenic bacteria for commercialization.

In general, the main strengths of bacterial biopesticides include their higher specificity, resulting in environmental and health safety (Koch et al., 2015), and ease of commercialization compared to synthetic products (Velivelli et al., 2014). Sprays of entomopathogenic bacteria are mostly used in specialty crops, organic agriculture and IPM programs.

Owing to their dependency on the action of insecticidal proteins displaying high specificity for specific receptors, bacterial biopesticides do not have negative effects on non-target and beneficial organisms at concentrations used for field applications (Glare and O'Callaghan, 2000). In addition, this high specificity also supports the lack of environmental and health issues related to the environmental accumulation of bacterial biopesticides, and identifies them as the safest available insecticides (World Health Organization, 1999).

Bacterial biopesticides are exempt from residue limits (Marrone, 2019), presenting a unique opportunity for application shortly before harvest as a clean-up spray. In fact, *Bt* biopesticide residues are commonly found in produce at the point of sale (Frederiksen et al., 2006). This probably results in subsequent dietary exposure to insecticidal proteins (Federici and Siegel, 2008), which is not considered hazardous by regulatory agencies (Koch et al., 2015).

As a drawback resulting from their high specificity, bacterial biopesticides have a limited range of activity, which reduces their competiveness in some markets. In addition, bacterial biopesticides are commonly considered less efficacious and cost-effective when compared to synthetic pesticides. However, increases in the number of field-evolved resistance cases and phasing out of synthetic pesticides motivates interest in bacterial biopesticides as alternative.

Traditional comparisons with synthetic pesticides are commonly limited to comparisons under worst-case or last alternative scenarios. Unfortunately, these comparisons do not consider relevant features of bacterial biopesticides, such as a unique mode of action and amenability to IPM programs, that would promote broader use of biopesticides (Marrone, 2019). For instance, while biopesticides based on *S. entomophila* are limited to a single target host (*C. zealandica*), they are more cost-effective in the longer term compared to synthetic alternatives due to the bacteria's ability to establish in the soil and generate a cycle of infection (Zydenbos et al., 2016). Similarly, the chitinolytic *Paenibacillus* sp. D1 strain reduces rates of chemical pesticides needed to control *Helicoverpa armigera* (Singh et al., 2016).

Short field stability and shelf life are drawbacks that especially affect performance and commercialization of biopesticides based on non-spore-forming bacteria. Even *Bt* spores and insecticidal proteins are susceptible to be broken down by UV rays and can be washed off from plant surfaces by rain or water. In addition, the control of boring and soil pests is difficult as the application of the biopesticide must be conducive to ingestion (Sanahuja et al., 2011).

While not a weakness specific to bacteria, the evolution of resistant insect populations needs consideration when developing new biopesticides. Even though there are several examples of field-evolved resistance (de Bortoli and Jurat-Fuentes, 2019), bacterial entomopathogens producing multiple insecticidal proteins with diverse 'mode of action', defined as recognizing distinct receptors in the host, help delay resistance evolution. Combining or rotating treatments with distinct bacterial species or synthetic pesticides is another strategy to apply distinct modes of action and hinder resistance evolution.

9 Future trends in research

Bacterial biopesticides are currently the most commercially successful microbial pesticides, and market research projects a compounded annual growth rate of 17% in 2020 (Marrone, 2019). The main factors contributing to successful bacterial biopesticide are the availability of a highly virulent strain of the entomopathogen, the development of cost-effective mass production *in vitro* systems, optimized formulation for enhanced persistence and long shelf life, and compatibility with predominant environmental conditions and commonly used pesticides. Research efforts in all these areas are needed to promote commercialization of new, more effective and stable bacterial biopesticides.

Advancements in the application of high throughput molecular methods should contribute to significant increases in virulence of new bacterial biopesticides. For example, reduced costs in DNA sequencing greatly facilitates screening programs to identify candidate bacterial entomopathogens producing multiple virulence factors with new or unique activities (Bi et al., 2015) and genome-level comparisons to characterize the evolution of pathogenicity (Zheng et al., 2017). Screening programs should also greatly benefit from molecular methods providing more sensitive and comprehensive classification of new isolates (Wang et al., 2018) and insecticidal proteins (Berry and Crickmore, 2017).

The development of recombinant strains and chimeric or evolved proteins with expanded and/or unique range of activity should also contribute to resolve the need for strains with higher efficacy (Sansinenea et al., 2010). For instance, continuous evolution of insecticidal Cry proteins can derive proteins with increased activity (Ishikawa et al., 2007) that overcome resistance (Badran et al., 2016) for incorporation in improved recombinant bacteria.

Reports of synergistic interactions between insecticidal proteins from *Bt* are relatively rare (Walters et al., 2018). In contrast, there are numerous examples of proteins and peptides that enhance activity of Cry insecticidal proteins (Pardo-López et al., 2009), with dramatic examples in synergizing activity against lepidopteran (Chen et al., 2007), coleopteran (Park et al., 2009; Gao et al., 2011) and dipteran (Hua et al., 2008) pests. Additional research is needed to determine the potential use of these enhancers in expanding range and level of activity and contribute to reduce the risk of resistance evolution in *Bt* biopesticides.

The increasing rates of resistance and phasing out of synthetic pesticides in some specialty markets should drive research to develop bacterial entomopathogens that historically have not been able to compete. An example is the Buibui strain of *B. thuringiensis* subsp. *japonensis* and its use to control white grubs in turfgrass (Koppenhöfer and Wu, 2017). Commercialization of this strain suffered from complications with formulation and lower performance compared to synthetic pesticides. However, increased virulence and persistence of the Buibui strain induced by soil bacteria (Mashtoly et al., 2010) suggests the potential for increasing competiveness. Similarly, competition from highly cost-effective imidacloprid-based products stalled commercialization of *Bt* subsp. *tenebrionis* to control *Leptinotarsa decemlineata* (Colorado potato beetle) (Gelernter, 2004). However, reduced efficacy of imidacloprid due to resistance (Crossley et al., 2017) represents an opportunity to renew interest in the use of *Bt* subsp. *tenebrionis*.

Innovative formulations may also offer possibilities to increase virulence and stability, and to extend residual effects (Jackson, 2017). As an example, covalent attachment of a polyethylene glycol synergist to BinA increased its thermostability and larvicidal activity (Sharma et al., 2017). Protection of *Bt* from UV degradation by complexing with anti-ultraviolet absorbents extends field stability and efficacy (Zhou et al., 2018). In another example, the stability and activity of pesticides based on *S. entomophila* greatly improved through changes in formulation (Wright et al., 2017).

Even after developing a highly effective and stable bacterial biopesticide, registration disparities among countries and issues related to effective distribution to growers emerge as issues (Marrone, 2019). Innovative harmonized mechanisms for the regulation of bacterial biopesticides focused on assessing environmental safety and ecological risks of the organism under realistic exposure scenarios are needed (Sundh and Goettel, 2013). Mechanisms for better access to distribution channels should be considered, and may include partnering with bigger companies (albeit risking profit margins), or attempting direct sales to the grower (Marrone, 2019).

Current interest in the commercialization of bacterial entomopathogens from taxonomic groups with representative vertebrate pathogens should determine potential risks in humans and other organisms in target environments. Related to the widespread use of successful bacterial biopesticides, harmonized regulatory demands on the use of native/indigenous versus alien/exotic when referred to entomopathogenic bacteria would be beneficial.

Amenability of bacterial biopesticides to IPM should increase interest in the use of these biopesticides to meet farmer effectiveness needs while maintaining environmental safety and sustainability (Ruiu, 2015). More research is also needed to detail benefits from incorporating bacterial biopesticides in combination or rotation with synthetic pesticides for enhanced activity, which would expand commercial interest on biopesticides from current niche markets (Marrone, 2019).

10 Where to look or further information

The following book chapters and articles are not cited in the main text but are good sources where the reader can obtain additional information on aspects of bacterial entomopathogens and their commercial use:

- Special issue edited by Arthurs, S. P. and Dara, S. K. 2019. Global status of microbial control programs and pactices. *J. Invertbr. Pathol.*, 165, 1-88.
- Frederiks, C. and Wesseler, J. H. 2019. A comparison of the EU and US regulatory frameworks for the active substance registration of microbial biological control agents. *Pest Manag. Sci.*, 75, 87-103.
- Malovichko, Y. V., Nizhnikov, A. A. and Antonets, K. S. 2019. Repertoire of the *Bacillus thuringiensis* virulence factors unrelated to major classes of protein toxins and its role in specificity of host-pathogen interactions. *Toxins* 11(6), pii:E347.
- Glare, T. R., Jurat-Fuentes, J. L. and O'Callaghan, M. 2017. Basic and applied research: entomopathogenic bacteria. In: Lacey, L. A. (Ed.), *Microbial Control of Insect and Mite Pests*. Academic Press, San Diego, CA, pp. 47-67.
- Zhang, Q., Hua, G. and Adang, M. J. 2017. Effects and mechanisms of *Bacillus thuringiensis* crystal toxins for mosquito larvae. *Insect Sci.* 24, 714-29.

Key research in the area of bacterial entomopathogens, their commercialization and benefits of adoption can be found at the following organizations:

- The Society for Invertebrate Pathology (SIP) (www.sipweb.org).
- The International Biocontrol Manufacturers Association (IBMA) (http://www.ibma-global.org).
- The Entomological Society of America (ESA) (www.entsoc.org).
- The U.S. Department of Agricuture (USDA), National Institute of Food and Agriculture (NIFA) multistate project S1070: The working group on improving microbial control of arthropod pests (https://www.nimss.org/projects/view/mrp/outline/18417).
- The U.S. Environmental Protection Agency (EPA) (https://www.epa.gov/pesticides/biopesticides).
- The Bioproducts Industry Alliance (BPIA) (http://www.bpia.org).

11 References

Adang, M., Crickmore, N. and Jurat-Fuentes, J. L. 2014. Diversity of *Bacillus thuringiensis* crystal toxins and mechanism of action. In: Dhadialla, T. S. and Gill, S. (Eds), *Advances in Insect Physiology Vol. 47: Insect Midgut and Insecticidal Proteins*. Academic Press, San Diego, CA, pp. 39–87.

Ahmed, H. K., Mitchell, W. J. and Priest, F. G. 1995. Regulation of mosquitocidal toxin synthesis in *Bacillus sphaericus*. *Appl. Microbiol. Biotechnol.* 43(2), 310–4. doi:10.1007/bf00172830.

Ahmed, I., Yokota, A., Yamazoe, A. and Fujiwara, T. 2007. Proposal of *Lysinibacillus boronitolerans* gen. nov. sp. nov., and transfer of *Bacillus fusiformis* to *Lysinibacillus fusiformis* comb. nov. and *Bacillus sphaericus* to *Lysinibacillus sphaericus* comb. nov. *Int. J. Syst. Evol. Microbiol.* 57(5), 1117–25. doi:10.1099/ijs.0.63867-0.

Allievi, M. C., Palomino, M. M., Prado Acosta, M., Lanati, L., Ruzal, S. M. and Sanchez-Rivas, C. 2014. Contribution of S-layer proteins to the mosquitocidal activity of *Lysinibacillus sphaericus*. *PLoS ONE* 9(10), e111114. doi:10.1371/journal.pone.0111114.

Asolkar, R., Marrone, P. and Namnath, J. 2016. *Chromobacterium* formulations, compositions, metabolites and their uses. US Patent 9259007. US Patent and Trademark Office.

Badran, A. H., Guzov, V. M., Huai, Q., Kemp, M. M., Vishwanath, P., Kain, W., Nance, A. M., Evdokimov, A., Moshiri, F., Turner, K. H., Wang, P., Malvar, T. and Liu, D. R. 2016. Continuous evolution of *Bacillus thuringiensis* toxins overcomes insect resistance. *Nature* 533(7601), 58–63. doi:10.1038/nature17938.

Becker, N., Zgomba, M., Petric, D., Beck, M. and Ludwig, M. 1995. Role of larval cadavers in recycling processes of *Bacillus sphaericus*. *J. Am. Mosq. Control Assoc.* 11(3), 329–34.

Berliner, E. 1915. Uber die schlaffsucht der mehlmottenraupe (*Ephestia kuhniella*, Zell.) und ihren Erreger *B. thuringiensis* n. sp. *Z. Angew. Entom.* 2(1), 29–56. doi:10.1111/j.1439-0418.1915.tb00334.x.

Berry, C. 2012. The bacterium, *Lysinibacillus sphaericus*, as an insect pathogen. *J. Invertebr. Pathol.* 109(1), 1–10. doi:10.1016/j.jip.2011.11.008.

Berry, C. and Crickmore, N. 2017. Structural classification of insecticidal proteins - towards an *in silico* characterisation of novel toxins. *J. Invertebr. Pathol.* 142, 16–22. doi:10.1016/j.jip.2016.07.015.

Berry, C., O'Neil, S., Ben-Dov, E., Jones, A. F., Murphy, L., Quail, M. A., Holden, M. T., Harris, D., Zaritsky, A. and Parkhill, J. 2002. Complete sequence and organization of pBtoxis, the toxin-coding plasmid of *Bacillus thuringiensis* subsp. *israelensis*. *Appl. Environ. Microbiol.* 68(10), 5082–95. doi:10.1128/aem.68.10.5082-5095.2002.

Bi, Y., Zhang, Y., Shu, C., Crickmore, N., Wang, Q., Du, L., Song, F. and Zhang, J. 2015. Genomic sequencing identifies novel *Bacillus thuringiensis* Vip1/Vip2 binary and Cry8 toxins that have high toxicity to *Scarabaeoidea* larvae. *Appl. Microbiol. Biotechnol.* 99(2), 753–60. doi:10.1007/s00253-014-5966-2.

Boeckman, C. J., Huang, E., Sturtz, K., Walker, C., Woods, R. and Zhang, J. 2019. Characterization of the spectrum of insecticidal activity for IPD072Aa: a protein derived from *Pseudomonas chlororaphis* with activity against *Diabrotica virgifera virgifera* (Coleoptera: Chrysomelidae). *J. Econ. Entomol.* 112(3), 1190–6. doi:10.1093/jee/toz029.

Boonserm, P., Moonsom, S., Boonchoy, C., Promdonkoy, B., Parthasarathy, K. and Torres, J. 2006. Association of the components of the binary toxin from *Bacillus sphaericus* in solution and with model lipid bilayers. *Biochem. Biophys. Res. Commun.* 342(4), 1273–8. doi:10.1016/j.bbrc.2006.02.086.

Broadwell, A. H., Baumann, L. and Baumann, P. 1990. Larvicidal properties of the 42-Kilodalton and 51-Kilodalton *Bacillus sphaericus* proteins expressed in different bacterial hosts - evidence for a binary toxin. *Curr. Microbiol.* 21(6), 361–6. doi:10.1007/BF02199438.

Busby, J. N., Landsberg, M. J., Simpson, R. M., Jones, S. A., Hankamer, B., Hurst, M. R. and Lott, J. S. 2012. Structural analysis of Chi1 Chitinase from Yen-Tc: the multisubunit insecticidal ABC toxin complex of *Yersinia entomophaga*. *J. Mol. Biol.* 415(2), 359–71. doi:10.1016/j.jmb.2011.11.018.

Carlson, A. B., Mathesius, C. A., Ballou, S., Boeckman, C. J., Gunderson, T. A., Mirsky, H. P., Mukerji, P., Roe, J. C., Schmidt, J. M., Zhang, J. and Delaney, B. 2019. Safety assessment of coleopteran active IPD072Aa protein from *Pseudomonas chlororaphis*. *Food Chem. Toxicol.* 129, 376–81. doi:10.1016/j.fct.2019.04.055.

Chakrabarti, S., Liehl, P., Buchon, N. and Lemaitre, B. 2012. Infection-induced host translational blockage inhibits immune responses and epithelial renewal in the *Drosophila* gut. *Cell Host Microbe* 12(1), 60–70. doi:10.1016/j.chom.2012.06.001.

Chandel, R. S., Soni, S., Vashisth, S., Pathania, M., Mehta, P. K., Rana, A., Bhatnagar, A. and Agrawal, V. K. 2019. The potential of entomopathogens in biological control of white grubs. *Int. J. Pest Manag.* 65(4), 348–62. doi:10.1080/09670874.2018.1524183.

Charles, J. F. and Nicolas, L. 1986. Recycling of *Bacillus sphaericus* 2362 in mosquito larvae: a laboratory study. *Ann. Inst. Pasteur Microbiol.* 137B(1), 101–11. doi:10.1016/s0769-2609(86)80097-7.

Chattopadhyay, P., Banerjee, G. and Mukherjee, S. 2017. Recent trends of modern bacterial insecticides for pest control practice in integrated crop management system. *3 Biotech* 7(1), 60. doi:10.1007/s13205-017-0717-6.

Chen, J., Hua, G., Jurat-Fuentes, J. L., Abdullah, M. A. and Adang, M. J. 2007. Synergism of *Bacillus thuringiensis* toxins by a fragment of a toxin-binding cadherin. *Proc. Natl. Acad. Sci. U. S. A.* 104(35), 13901-6. doi:10.1073/pnas.0706011104.

Cordova-Kreylos, A. L., Fernandez, L. E., Koivunen, M., Yang, A., Flor-Weiler, L. and Marrone, P. G. 2013. Isolation and characterization of *Burkholderia rinojensis.*, a non-*Burkholderia cepacia* complex soil bacterium with insecticidal and miticidal activities. *Appl. Environ. Microbiol.* 79(24), 7669-78. doi:10.1128/AEM.02365-13.

Crickmore, N., Baum, J., Bravo, A., Lereclus, D., Narva, K., Sampson, K., Schnepf, E., Sun, M. and Zeigler, D. R. 2019. *Bacillus thuringiensis* toxin nomenclature. Available at: http://www.btnomenclature.info/.

Crossley, M. S., Chen, Y. H., Groves, R. L. and Schoville, S. D. 2017. Landscape genomics of Colorado potato beetle provides evidence of polygenic adaptation to insecticides. *Mol. Ecol.* 26(22), 6284-300. doi:10.1111/mec.14339.

Czaja, K., Goralczyk, K., Strucinski, P., Hernik, A., Korcz, W., Minorczyk, M., Lyczewska, M. and Ludwicki, J. K. 2015. Biopesticides-towards increased consumer safety in the European Union. *Pest Manag. Sci.* 71(1), 3-6. doi:10.1002/ps.3829.

Davidson, E. W. and Titus, M. 1987. Ultrastructural effects of the *Bacillus sphaericus* mosquito larvicidal toxin on cultured mosquito cells. *J. Invertebr. Pathol.* 50(3), 213-20. doi:10.1016/0022-2011(87)90085-1.

Davidson, E. W., Oei, C., Meyer, M., Bieber, A. L., Hindley, J. and Berry, C. 1990. Interaction of the *Bacillus sphaericus* mosquito larvicidal proteins. *Can. J. Microbiol.* 36(12), 870-8. doi:10.1139/m90-151.

de Bortoli, C. P. and Jurat-Fuentes, J. L. 2019. Mechanisms of resistance to commercially relevant entomopathogenic bacteria. *Curr. Opin. Insect Sci.* 33, 56-62. doi:10.1016/j.cois.2019.03.007.

de Melo, J. V., Jones, G. W., Berry, C., Vasconcelos, R. H., de Oliveira, C. M., Furtado, A. F., Peixoto, C. A. and Silva-Filha, M. H. 2009. Cytopathological effects of *Bacillus sphaericus* Cry48Aa/Cry49Aa toxin on binary toxin-susceptible and -resistant *Culex quinquefasciatus* larvae. *Appl. Environ. Microbiol.* 75(14), 4782-9. doi:10.1128/AEM.00811-09.

de Oliveira, E. J., Rabinovitch, L., Monnerat, R. G., Passos, L. K. J. and Zahner, V. 2004. Molecular characterization of *Brevibacillus laterosporus* and its potential use in biological control. *Appl. Environ. Microbiol.* 70(11), 6657-64. doi:10.1128/AEM.70.11.6657-6664.2004.

Dieppois, G., Opota, O., Lalucat, J. and Lemaitre, B. 2015. *Pseudomonas entomophila*: a versatile bacterium with entomopathogenic properties. In: Ramos, J. L., Goldberg, J. B. and Filloux, A. (Eds), *Pseudomonas*. Springer, Dordrecht, the Netherlands, pp. 25-49.

Djukic, M., Poehlein, A., Thürmer, A. and Daniel, R. 2011. Genome sequence of *Brevibacillus laterosporus* LMG 15441, a pathogen of invertebrates. *J. Bacteriol.* 193(19), 5535-6. doi:10.1128/JB.05696-11.

Donnarumma, F., Paffetti, D., Stotzky, G., Giannini, R. and Vettori, C. 2010. Potential gene exchange between *Bacillus thuringiensis* subsp. *kurstaki* and *Bacillus* spp. in soil *in situ*. *Soil Biol. Biochem.* 42(8), 1329-37. doi:10.1016/j.soilbio.2010.03.014.

Dutky, S. R. 1940. Two new spore-forming bacteria causing milky diseases of Japanese beetle larvae. *J. Agric. Res.* 61, 57-68.

EPA (Environmental Protection Agency). 2019. Part 180 - tolerances and exemptions for pesticide chemical residues in food. In: *Code of Federal Regulations, Title 40*

- *Protection of Environment*. Available at: https://www.govinfo.gov/app/collection/cfr/2019/title40.

Fabrick, J. A. and Wu, Y. 2015. Roles of insect midgut cadherin in Bt intoxication and resistance. In: *Bt Resistance: Characterization and Strategies for GM Crops Producing Bacillus thuringiensis Toxins*. CABI Publishing, Wallingford, UK, pp. 69–86.

Federici, B. A. and Siegel, J. P. 2008. Safety assessment of *Bacillus thuringiensis* and Bt crops used in insect control. In: Hammond, B. G. (Ed.), *Food Safety of Proteins in Agricultural Biotechnology*. CRC Press, Boca Raton, FL, pp. 45–102.

Ferreira, L. M., Romão, T. P., de-Melo-Neto, O. P. and Silva-Filha, M. H. 2010. The orthologue to the Cpm1/Cqm1 receptor in *Aedes aegypti* is expressed as a midgut GPI-anchored a-glucosidase, which does not bind to the insecticidal binary toxin. *Insect Biochem. Mol. Biol.* 40(8), 604–10. doi:10.1016/j.ibmb.2010.05.007.

Ferreira, L. M., Romão, T. P., Nascimento, N. A., Costa Mda C., Rezende, A. M., de-Melo-Neto, O. P. and Silva-Filha, M. H. 2014. Non conserved residues between Cqm1 and Aam1 mosquito alpha-glucosidases are critical for the capacity of Cqm1 to bind the Binary toxin from *Lysinibacillus sphaericus*. *Insect Biochem. Mol. Biol.* 50, 34–42. doi:10.1016/j.ibmb.2014.04.004.

Ferreira, V. D. S. B., Barcellos, I. D. S., Carramaschi, I. N., Santos-Mallet, J. R., Queiroz, M. M. C. and Zahner, V. 2016. Larvicidal activity and effects on post embrionary development of laboratory reared *Musca domestica* (Linnaeus, 1758) (Diptera: Muscidae), treated with *Brevibacillus laterosporus* (Laubach) spore suspensions. *J. Invertebr. Pathol.* 137, 54–7. doi:10.1016/j.jip.2016.04.009.

Frederiks, C. and Wesseler, J. H. 2019. A comparison of the EU and US regulatory frameworks for the active substance registration of microbial biological control agents. *Pest Manag. Sci.* 75(1), 87–103. doi:10.1002/ps.5133.

Frederiksen, K., Rosenquist, H., Jorgensen, K. and Wilcks, A. 2006. Occurrence of natural *Bacillus thuringiensis* contaminants and residues of *Bacillus thuringiensis*-based insecticides on fresh fruits and vegetables. *Appl. Environ. Microbiol.* 72(5), 3435–40. doi:10.1128/AEM.72.5.3435-3440.2006.

Gao, Y., Jurat-Fuentes, J. L., Oppert, B., Fabrick, J. A., Liu, C., Gao, J. and Lei, Z. 2011. Increased toxicity of *Bacillus thuringiensis* Cry3Aa against *Crioceris quatuordecimpunctata*, *Phaedon brassicae* and *Colaphellus bowringi* by a *Tenebrio molitor* cadherin fragment. *Pest Manag. Sci.* 67(9), 1076–81. doi:10.1002/ps.2149.

Gatehouse, H. S., Tan, B., Christeller, J. T., Hurst, M. R., Marshall, S. D. and Jackson, T. A. 2009. Phenotypic changes and the fate of digestive enzymes during induction of amber disease in larvae of the New Zealand grass grub (*Costelytra zealandica*). *J. Invertebr. Pathol.* 101(3), 215–21. doi:10.1016/j.jip.2009.05.010.

Gelernter, W. 2004. The rise and fall of *Bacillus thuringiensis tenebrionis*. *Phytoparasitica* 32(4), 321–4. doi:10.1007/BF02979840.

Glare, T. R. and O'Callaghan, M. 2000. *Bacillus thuringiensis: Biology, Ecology and Safety*. John Wiley & Sons Ltd, Chichester, UK.

Glare, T. R., Jurat-Fuentes, J. L. and O'Callaghan, M. 2017. Basic and applied research: entomopathogenic bacteria. In: Lacey, L. A. (Ed.), *Microbial Control of Insect and Mite Pests*. Academic Press, San Diego, CA, pp. 47–67.

Guo, W., Kain, W. and Wang, P. 2019. Effects of disruption of the peritrophic membrane on larval susceptibility to Bt toxin Cry1Ac in cabbage loopers. *J. Insect Physiol.* 117, 103897. doi:10.1016/j.jinsphys.2019.103897.

Hadapad, A. B., Hire, R. S., Vijayalakshmi, N. and Dongre, T. K. 2009. UV protectants for the biopesticide based on *Bacillus sphaericus* Neide and their role in protecting the binary toxins from UV radiation. *J. Invertebr. Pathol.* 100(3), 147–52. doi:10.1016/j.jip.2008.12.003.

Heymann, J. B., Bartho, J. D., Rybakova, D., Venugopal, H. P., Winkler, D. C., Sen, A., Hurst, M. R. and Mitra, A. K. 2013. Three-dimensional structure of the toxin-delivery particle antifeeding prophage of *Serratia entomophila*. *J. Biol. Chem.* 288(35), 25276–84. doi:10.1074/jbc.M113.456145.

Hu, X., Fan, W., Han, B., Liu, H., Zheng, D., Li, Q., Dong, W., Yan, J., Gao, M., Berry, C. and Yuan, Z. 2008. Complete genome sequence of the mosquitocidal bacterium *Bacillus sphaericus* C3-41 and comparison with those of closely related *Bacillus* species. *J. Bacteriol.* 190(8), 2892-902. doi:10.1128/JB.01652-07.

Hua, G., Zhang, R., Abdullah, M. A. and Adang, M. J. 2008. *Anopheles gambiae* cadherin AgCad1 binds the Cry4Ba toxin of *Bacillus thuringiensis israelensis* and a fragment of AgCad1 synergizes toxicity. *Biochemistry* 47(18), 5101–10. doi:10.1021/bi7023578.

Hurst, M. R. and Glare, T. R. 2002. Restriction map of the *Serratia entomophila* plasmid pADAP carrying virulence factors for *Costelytra zealandica*. *Plasmid* 47(1), 51–60. doi:10.1006/plas.2001.1551.

Hurst, M. R., Beard, S. S., Jackson, T. A. and Jones, S. M. 2007a. Isolation and characterization of the *Serratia entomophila* antifeeding prophage. *FEMS Microbiol. Lett.* 270(1), 42–8. doi:10.1111/j.1574-6968.2007.00645.x.

Hurst, M. R., Jones, S. M., Tan, B. and Jackson, T. A. 2007b. Induced expression of the *Serratia entomophila* Sep proteins shows activity towards the larvae of the New Zealand grass grub *Costelytra zealandica*. *FEMS Microbiol. Lett.* 275(1), 160–7. doi:10.1111/j.1574-6968.2007.00886.x.

Hurst, M. R., Becher, S. A., Young, S. D., Nelson, T. L. and Glare, T. R. 2011a. *Yersinia entomophaga* sp. nov., isolated from the New Zealand grass grub *Costelytra zealandica*. *Int. J. Syst. Evol. Microbiol.* 61(4), 844–9. doi:10.1099/ijs.0.024406-0.

Hurst, M. R., Jones, S. A., Binglin, T., Harper, L. A., Jackson, T. A. and Glare, T. R. 2011b. The main virulence determinant of *Yersinia entomophaga* MH96 is a broad host-range toxin complex active against insects. *J. Bacteriol.* 193(8), 1966–80. doi:10.1128/JB.01044-10.

Hurst, M. R. H., Rogers, D. J., Wright, D. A., Townsend, R. J., Bruening, R. and Cole, L. M. 2011c. Effect of the bacterium *Yersinia entomophaga* on adult bronze beetle. *N. Z. Plant Protect.* 64, 209–14. doi:10.30843/nzpp.2011.64.5956.

Hurst, M. R., van Koten, C. and Jackson, T. A. 2014. Pathology of *Yersinia entomophaga* MH96 towards *Costelytra zealandica* (Coleoptera; Scarabaeidae) larvae. *J. Invertebr. Pathol.* 115, 102–7. doi:10.1016/j.jip.2013.11.004.

Hurst, M. R., Swaminathan, J., Wright, D. A., Hardwick, S., Ferguson, C. M., Beattie, A., Richards, N. K., Harper, L., Moss, R. A., Cave, V. M., van Koten, C. and McNeill, M. R. 2019a. Development of a *Yersinia entomophaga* bait for control of larvae of the porina moth (*Wiseana* spp.), a pest of New Zealand improved grassland systems. *Pest Manag. Sci.* 76(1), 350–9. doi:10.1002/ps.5521.

Hurst, M. R. H., Jones, S. A., Beattie, A., van Koten, C., Shelton, A. M., Collins, H. L. and Brownbridge, M. 2019b. Assessment of *Yersinia entomophaga* as a control agent of the diamondback moth *Plutella xylostella*. *J. Invertebr. Pathol.* 162, 19–25. doi:10.1016/j.jip.2019.02.002.

Icoz, I., Andow, D., Zwahlen, C. and Stotzky, G. 2009. Is the Cry1Ab protein from *Bacillus thuringiensis* (Bt) taken up by plants from soils previously planted with bt corn and by carrot from hydroponic culture? *Bull. Environ. Contam. Toxicol.* 83(1), 48–58. doi:10.1007/s00128-009-9760-2.

Ishikawa, H., Hoshino, Y., Motoki, Y., Kawahara, T., Kitajima, M., Kitami, M., Watanabe, A., Bravo, A., Soberon, M., Honda, A., Yaoi, K. and Sato, R. 2007. A system for the directed evolution of the insecticidal protein from *Bacillus thuringiensis*. *Mol. Biotechnol.* 36(2), 90–101. doi:10.1007/s12033-007-0001-9.

Jackson, T. A. 2007. A novel bacterium for control of grass grub. In: Vincent, C., Goettel, M. S. and Lazarovits, G. (Eds), *Biological Control: a Global Perspective*. CABI Publishing, Wallingford, UK, pp. 160–8.

Jackson, T. A. 2017. Entomopathogenic bacteria: mass production, formulation, and quality control. In: Lacey, L. A. (Ed.), *Microbial Control of Insect and Mite Pests*. Academic Press, pp. 125–39. Chapter 8.

Jackson, T. A., Boucias, D. G. and Thaler, J. O. 2001. Pathobiology of amber disease, caused by *Serratia* spp., in the New Zealand grass grub, *Costelytra zealandica*. *J. Invertebr. Pathol.* 78(4), 232–43. doi:10.1006/jipa.2002.5078.

Jones, G. W., Nielsen-Leroux, C., Yang, Y., Yuan, Z., Dumas, V. F., Monnerat, R. G. and Berry, C. 2007. A new Cry toxin with a unique two-component dependency from *Bacillus sphaericus*. *FASEB J.* 21(14), 4112–20. doi:10.1096/fj.07-8913com.

Juarez-Perez, V. and Delecluse, A. 2001. The Cry toxins and the putative hemolysins of *Clostridium bifermentans* ser. *malaysia* are not involved in mosquitocidal activity. *J. Invertebr. Pathol.* 78(1), 57–8. doi:10.1006/jipa.2001.5042.

Jurat-Fuentes, J. L. and Jackson, T. A. 2012. Bacterial entomopathogens. In: Vega, F. E. and Kaya, H. K. (Eds), *Insect Pathology* (2nd edn.). Academic Press, San Diego, CA, pp. 265–349.

Kabaluk, J. T., Svircev, A. M., Goettel, M. S. and Woo, S. G. 2010. The use and regulation of microbial pesticides in representative jurisdictions worldwide. Available at: www.IOBC-Global.org.

Kamala-Kannan, S., Lee, K. J., Park, S. M., Chae, J. C., Yun, B. S., Lee, Y. H., Park, Y. J. and Oh, B. T. 2010. Characterization of ACC deaminase gene in *Pseudomonas entomophila* strain PS-PJH isolated from the rhizosphere soil. *J. Basic Microbiol.* 50(2), 200–5. doi:10.1002/jobm.200900171.

Kawanishi, C. Y., Splittstoesser, C. M. and Tashiro, H. 1978. Infection of the European chafer, *Amphimallon majalis*, by *Bacillus popilliae*: ultrastructure. *J. Invertebr. Pathol.* 31(1), 91–102. doi:10.1016/0022-2011(78)90113-1.

Kil, Y. J., Seo, M. J., Kang, D. K., Oh, S. N., Cho, H. S., Youn, Y. N., Yasunaga-Aoki, C. and Yu, Y. M. 2014. Effects of Enterobacteria (*Burkholderia* sp.) on development of *Riptortus pedestris*. *J. Fac. Agric. Kyushu Univ.* 59, 77–84.

Kirst, H. A. 2010. The spinosyn family of insecticides: realizing the potential of natural products research. *J. Antibiot. (Tokyo)* 63(3), 101–11. doi:10.1038/ja.2010.5.

Klein, M. G. 1981. Advances in the use of *Bacillus popillae* for pest control. In: Burges, H. D. (Ed.), *Microbial Control of Pests and Plant Diseases 1970–1980*. Academic Press, London, UK, pp. 183–92.

Klein, M. G. and Kaya, H. 1995. *Bacillus* and *Serratia* species for scarab control. *Mem. Inst. Oswaldo Cruz* 90(1), 87–95. doi:10.1590/S0074-02761995000100019.

Koch, M. S., Ward, J. M., Levine, S. L., Baum, J. A., Vicini, J. L. and Hammond, B. G. 2015. The food and environmental safety of Bt crops. *Front. Plant Sci.* 6, 283. doi:10.3389/fpls.2015.00283.

Koppenhöfer, A. M. and Wu, S. 2017. Microbial control of insect pests of turfgrass. In: Lacey, L. A. (Ed.), *Microbial Control of Insect and Mite Pests*. Academic Press, pp. 331–41.

Kuraishi, T., Binggeli, O., Opota, O., Buchon, N. and Lemaitre, B. 2011. Genetic evidence for a protective role of the peritrophic matrix against intestinal bacterial infection in *Drosophila melanogaster*. *Proc. Natl. Acad. Sci. U. S. A.* 108(38), 15966–71. doi:10.1073/pnas.1105994108.

Landsberg, M. J., Jones, S. A., Rothnagel, R., Busby, J. N., Marshall, S. D., Simpson, R. M., Lott, J. S., Hankamer, B. and Hurst, M. R. 2011. 3D structure of the *Yersinia entomophaga* toxin complex and implications for insecticidal activity. *Proc. Natl. Acad. Sci. U. S. A.* 108(51), 20544–9. doi:10.1073/pnas.1111155108.

Lecadet, M. M., Frachon, E., Dumanoir, V. C., Ripouteau, H., Hamon, S., Laurent, P. and Thiery, I. 1999. Updating the H-antigen classification of *Bacillus thuringiensis*. *J. Appl. Microbiol.* 86(4), 660–72. doi:10.1046/j.1365-2672.1999.00710.x.

Lee, S. A., Jang, S. H., Kim, B. H., Shibata, T., Yoo, J., Jung, Y., Kawabata, S. I. and Lee, B. L. 2018. Insecticidal activity of the metalloprotease AprA occurs through suppression of host cellular and humoral immunity. *Dev. Comp. Immunol.* 81, 116–26. doi:10.1016/j.dci.2017.11.014.

Le Guern, A. S., Savin, C., Bremont, S., Payro, G., Bon, D., Carniel, E. and Pizarro-Cerda, J. 2018. First isolation of *Yersinia entomophaga* in human urinary tract. *New Microbes New Infect.* 26, 3–7. doi:10.1016/j.nmni.2018.08.002.

Marche, M. G., Mura, M. E. and Ruiu, L. 2016. *Brevibacillus laterosporus* inside the insect body: beneficial resident or pathogenic outsider? *J. Invertebr. Pathol.* 137, 58–61. doi:10.1016/j.jip.2016.05.002.

Marche, M. G., Mura, M. E., Falchi, G. and Ruiu, L. 2017. Spore surface proteins of *Brevibacillus laterosporus* are involved in insect pathogenesis. *Sci. Rep.* 7, 43805. doi:10.1038/srep43805.

Marrone, P. G. 2019. Pesticidal natural products - status and future potential. *Pest Manag. Sci.* 75(9), 2325–40. doi:10.1002/ps.5433.

Marshall, S. D., Hares, M. C., Jones, S. A., Harper, L. A., Vernon, J. R., Harland, D. P., Jackson, T. A. and Hurst, M. R. 2012. Histopathological effects of the Yen-Tc toxin complex from *Yersinia entomophaga* MH96 (*Enterobacteriaceae*) on the *Costelytra zealandica* (Coleoptera: Scarabaeidae) larval midgut. *Appl. Environ. Microbiol.* 78(14), 4835–47. doi:10.1128/AEM.00431-12.

Martin, P. A. 2007. USDA finds broader uses for bacterial insecticide. *Ind. Bioproces* 29, 6–7.

Martin, P. A., Gundersen-Rindal, D., Blackburn, M. and Buyer, J. 2007a. *Chromobacterium subtsugae* sp. nov., a betaproteobacterium toxic to Colorado potato beetle and other insect pests. *Int. J. Syst. Evol. Microbiol.* 57(5), 993–9. doi:10.1099/ijs.0.64611-0.

Martin, P. A., Hirose, E. and Aldrich, J. R. 2007b. Toxicity of *Chromobacterium subtsugae* to southern green stink bug (Heteroptera: Pentatomidae) and corn rootworm (Coleoptera: Chrysomelidae). *J. Econ. Entomol.* 100(3), 680–4. doi:10.1603/0022-0493(2007)100[680:tocsts]2.0.co;2.

Mashtoly, T. A., Abolmaaty, A., Thompson, N., El-Zemaity, Mel-S., Hussien, M. I. and Alm, S. R. 2010. Enhanced toxicity of *Bacillus thuringiensis japonensis* strain Buibui toxin to oriental beetle and northern masked chafer (Coleoptera: Scarabaeidae) larvae with *Bacillus* sp. NFD2. *J. Econ. Entomol.* 103(5), 1547–54. doi:10.1603/ec10028.

Milner, R. J. 1981. A novel milky disease organism from Australian scarabaeids: field occurrence, isolation, and infectivity. *J. Invertebr. Pathol.* 37(3), 304–9. doi:10.1016/0022-2011(81)90091-4.

Mnif, I. and Ghribi, D. 2015. Potential of bacterial derived biopesticides in pest management. *Crop Protect.* 77, 52–64. doi:10.1016/j.cropro.2015.07.017.

Mulligan, F. S., Schaefer, C. H. and Miura, T. 1978. Laboratory and field evaluation of *Bacillus sphaericus* as a mosquito (Diptera-Culicidae) control agent. *J. Econ. Entomol.* 71(5), 774–7. doi:10.1093/jee/71.5.774.

Nicolas, L., Lecroisey, A. and Charles, J. F. 1990. Role of the gut proteinases from mosquito larvae in the mechanism of action and the specificity of the *Bacillus sphaericus* toxin. *Can. J. Microbiol.* 36(11), 804–7. doi:10.1139/m90-138.

Nishiwaki, H., Nakashima, K., Ishida, C., Kawamura, T. and Matsuda, K. 2007. Cloning, functional characterization, and mode of action of a novel insecticidal pore-forming toxin, sphaericolysin, produced by *Bacillus sphaericus*. *Appl. Environ. Microbiol.* 73(10), 3404–11. doi:10.1128/AEM.00021-07.

Oei, C., Hindley, J. and Berry, C. 1992. Binding of purified *Bacillus sphaericus* binary toxin and its deletion derivatives to *Culex quinquefasciatus* gut: elucidation of functional binding domains. *J. Gen. Microbiol.* 138(7), 1515–26. doi:10.1099/00221287-138-7-1515.

Olson, S. 2015. An analysis of the biopesticide market now and where it is going. *Outlk Pest Manag.* 26(5), 203–6. doi:10.1564/v26_oct_04.

Omura, S. 2016. A splendid gift from the Earth: the origins and impact of the avermectins (Nobel lecture). *Angew. Chem. Int. Ed. Engl.* 55(35), 10190–209. doi:10.1002/anie.201602164.

Opota, O., Charles, J. F., Warot, S., Pauron, D. and Darboux, I. 2008. Identification and characterization of the receptor for the *Bacillus sphaericus* binary toxin in the malaria vector mosquito, *Anopheles gambiae*. *Comp. Biochem. Physiol. B Biochem. Mol. Biol.* 149(3), 419–27. doi:10.1016/j.cbpb.2007.11.002.

Opota, O., Gauthier, N. C., Doye, A., Berry, C., Gounon, P., Lemichez, E. and Pauron, D. 2011. *Bacillus sphaericus* binary toxin elicits host cell autophagy as a response to intoxication. *PLoS ONE* 6(2), e14682. doi:10.1371/journal.pone.0014682.

Orlova, M. V., Smirnova, T. A., Ganushkina, L. A., Yacubovich, V. Y. and Azizbekyan, R. R. 1998. Insecticidal activity of *Bacillus laterosporus*. *Appl. Environ. Microbiol.* 64(7), 2723–5. doi:10.1128/AEM.64.7.2723-2725.1998.

Pardo-López, L., Muñoz-Garay, C., Porta, H., Rodríguez-Almazán, C., Soberón, M. and Bravo, A. 2009. Strategies to improve the insecticidal activity of Cry toxins from *Bacillus thuringiensis*. *Peptides* 30(3), 589–95. doi:10.1016/j.peptides.2008.07.027.

Park, Y., Abdullah, M. A., Taylor, M. D., Rahman, K. and Adang, M. J. 2009. Enhancement of *Bacillus thuringiensis* Cry3Aa and Cry3Bb toxicities to coleopteran larvae by a toxin-binding fragment of an insect cadherin. *Appl. Environ. Microbiol.* 75(10), 3086–92. doi:10.1128/AEM.00268-09.

Parte, A. C. 2018. LPSN – List of Prokaryotic names with Standing in Nomenclature (bacterio.net), 20 years on. *Int. J. Syst. Evol. Microbiol.* 68, 1825–9. doi:10.1099/ijsem.0.002786.

Pauchet, Y., Luton, F., Castella, C., Charles, J. F., Romey, G. and Pauron, D. 2005. Effects of a mosquitocidal toxin on a mammalian epithelial cell line expressing its target receptor. *Cell. Microbiol.* 7(9), 1335–44. doi:10.1111/j.1462-5822.2005.00560.x.

Piper, S. J., Brillault, L., Rothnagel, R., Croll, T. I., Box, J. K., Chassagnon, I., Scherer, S., Goldie, K. N., Jones, S. A., Schepers, F., Hartley-Tassell, L., Ve, T., Busby, J. N., Dalziel, J. E., Lott, J. S., Hankamer, B., Stahlberg, H., Hurst, M. R. H. and Landsberg, M. J. 2019. Cryo-EM structures of the pore-forming A subunit from the *Yersinia entomophaga* ABC toxin. *Nat. Commun.* 10(1), 1952. doi:10.1038/s41467-019-09890-8.

Qureshi, N., Chawla, S., Likitvivatanavong, S., Lee, H. L. and Gill, S. S. 2014. The cry toxin operon of *Clostridium bifermentans* subsp. *malaysia* is highly toxic to *Aedes* larval mosquitoes. *Appl. Environ. Microbiol.* 80(18), 5689–97. doi:10.1128/AEM.01139-14.

Rasko, D. A., Altherr, M. R., Han, C. S. and Ravel, J. 2005. Genomics of the *Bacillus cereus* group of organisms. *FEMS Microbiol. Rev.* 29(2), 303–29. doi:10.1016/j.femsre.2004.12.005.

Raymond, B., Johnston, P. R., Nielsen-LeRoux, C., Lereclus, D. and Crickmore, N. 2010. *Bacillus thuringiensis*: an impotent pathogen? *Trends Microbiol.* 18(5), 189–94. doi:10.1016/j.tim.2010.02.006.

Rezende, T. M. T., Romão, T. P., Batista, M., Berry, C., Adang, M. J. and Silva-Filha, M. H. N. L. 2017. Identification of Cry48Aa/Cry49Aa toxin ligands in the midgut of *Culex quinquefasciatus* larvae. *Insect Biochem. Mol. Biol.* 88, 63–70. doi:10.1016/j.ibmb.2017.08.001.

Romao, T. P., de Melo Chalegre, K. D., Key, S., Ayres, C. F., Fontes de Oliveira, C. M., de-Melo-Neto, O. P. and Silva-Filha, M. H. 2006. A second independent resistance mechanism to *Bacillus sphaericus* binary toxin targets its alpha-glucosidase receptor in *Culex quinquefasciatus*. *FEBS J.* 273(7), 1556–68. doi:10.1111/j.1742-4658.2006.05177.x.

Ruan, L., Crickmore, N., Peng, D. and Sun, M. 2015. Are nematodes a missing link in the confounded ecology of the entomopathogen *Bacillus thuringiensis*? *Trends Microbiol.* 23(6), 341–6. doi:10.1016/j.tim.2015.02.011.

Ruiu, L. 2013. *Brevibacillus laterosporus*, a pathogen of invertebrates and a broad-spectrum antimicrobial species. *Insects* 4(3), 476–92. doi:10.3390/insects4030476.

Ruiu, L. 2015. Insect pathogenic bacteria in integrated pest management. *Insects* 6(2), 352–67. doi:10.3390/insects6020352.

Ruiu, L., Delrio, G., Ellar, D. J., Floris, I., Paglietti, B., Rubino, S. and Satta, A. 2006. Lethal and sublethal effects of *Brevibacillus laterosporus* on the housefly (*Musca domestica*). *Entomol. Exp. Appl.* 118(2), 137–44. doi:10.1111/j.1570-7458.2006.00370.x.

Ruiu, L., Satta, A. and Floris, I. 2007. Susceptibility of the house fly pupal parasitoid *Muscidifurax raptor* (Hymenoptera: Pteromalidae) to the entomopathogenic bacteria *Bacillus thuringiensis* and *Brevibacillus laterosporus*. *Biol. Control* 43(2), 188–94. doi:10.1016/j.biocontrol.2007.08.005.

Ruiu, L., Satta, A. and Floris, I. 2008. Immature house fly (*Musca domestica*) control in breeding sites with a new *Brevibacillus laterosporus* formulation. *Environ. Entomol.* 37(2), 505–9. doi:10.1603/0046-225x(2008)37[505:ihfmdc]2.0.co;2.

Ruiu, L., Satta, A. and Floris, I. 2012. Observations on house fly larvae midgut ultrastructure after *Brevibacillus laterosporus* ingestion. *J. Invertebr. Pathol.* 111(3), 211–6. doi:10.1016/j.jip.2012.08.005.

Ruiu, L., Satta, A. and Floris, I. 2014. Administration of *Brevibacillus laterosporus* spores as a poultry feed additive to inhibit house fly development in feces: a new eco-sustainable concept. *Poult. Sci.* 93(3), 519–26. doi:10.3382/ps.2013-03418.

Rungrod, A., Tjahaja, N. K., Soonsanga, S., Audtho, M. and Promdonkoy, B. 2009. *Bacillus sphaericus* Mtx1 and Mtx2 toxins co-expressed in *Escherichia coli* are synergistic against *Aedes aegypti* larvae. *Biotechnol. Lett.* 31(4), 551–5. doi:10.1007/s10529-008-9896-x.

Salazar-Marroquín, E. L., Galán-Wong, L. J., Moreno-Medina, V. R., Reyes-López, M. Á. and Pereyra-Alférez, B. 2016. Bacteriocins synthesized by *Bacillus thuringiensis*: generalities and potential applications. *Rev. Med. Microbiol.* 27(3), 95–101. doi:10.1097/MRM.0000000000000076.

Sanahuja, G., Banakar, R., Twyman, R. M., Capell, T. and Christou, P. 2011. *Bacillus thuringiensis*: a century of research, development and commercial applications. *Plant Biotechnol. J.* 9(3), 283–300. doi:10.1111/j.1467-7652.2011.00595.x.

Sansinenea, E., Vázquez, C. and Ortiz, A. 2010. Genetic manipulation in *Bacillus thuringiensis* for strain improvement. *Biotechnol. Lett.* 32(11), 1549–57. doi:10.1007/s10529-010-0338-1.

Sato, R., Adegawa, S., Li, X., Tanaka, S. and Endo, H. 2019. Function and role of ATP-binding cassette transporters as receptors for 3D-cry toxins. *Toxins* 11(2). doi:10.3390/toxins11020124.

Schellenberger, U., Oral, J., Rosen, B. A., Wei, J. Z., Zhu, G., Xie, W., McDonald, M. J., Cerf, D. C., Diehn, S. H., Crane, V. C., Sandahl, G. A., Zhao, J. Z., Nowatzki, T. M., Sethi, A., Liu, L., Pan, Z., Wang, Y., Lu, A. L., Wu, G. and Liu, L. 2016. A selective insecticidal protein from *Pseudomonas* for controlling corn rootworms. *Science* 354(6312), 634–7. doi:10.1126/science.aaf6056.

Schroeder, J. M., Chamberlain, C. and Davidson, E. W. 1989. Resistance to the *Bacillus sphaericus* toxin in cultured mosquito cells. *In Vitro Cell. Dev. Biol.* 25(10), 887–91. doi:10.1007/bf02624000.

Seiber, J. N., Coats, J., Duke, S. O. and Gross, A. D. 2014. Biopesticides: state of the art and future opportunities. *J. Agric. Food Chem.* 62(48), 11613–9. doi:10.1021/jf504252n.

Seleena, P., Lee, H. L. and Lecadet, M. M. 1997. A novel insecticidal serotype of *Clostridium bifermentans*. *J. Am. Mosq. Control Assoc.* 13(4), 395–7.

Shapiro-Ilan, D. I., Cottrell, T. E., Jackson, M. A. and Wood, B. W. 2013. Control of key pecan insect pests using biorational pesticides. *J. Econ. Entomol.* 106(1), 257–66. doi:10.1603/ec12302.

Sharma, V., Singh, P. K., Midha, S., Ranjan, M., Korpole, S. and Patil, P. B. 2012. Genome sequence of *Brevibacillus laterosporus* strain GI-9. *J. Bacteriol.* 194(5), 1279. doi:10.1128/JB.06659-11.

Sharma, M., Hire, R. S., Hadapad, A. B., Gupta, G. D. and Kumar, V. 2017. PEGylation enhances mosquito-larvicidal activity of *Lysinibacillus sphaericus* binary toxin. *Bioconjug. Chem.* 28(2), 410–8. doi:10.1021/acs.bioconjchem.6b00565.

Silva-Filha, M. H. 2017. Resistance of mosquitoes to entomopathogenic bacterial-based larvicides: current status and strategies for management. In: Fiuza, L. M., Polanczyk, R. A. and Crickmore, N. (Eds), *Bacillus thuringiensis and Lysinibacillus sphaericus: Characterization and Use in the Field of Biocontrol*. Springer International Publishing, Cham, pp. 239–57.

Silva-Filha, M. H., Nielsen-Leroux, C. and Charles, J. F. 1999. Identification of the receptor for *Bacillus sphaericus* crystal toxin in the brush border membrane of the mosquito *Culex pipiens* (Diptera: Culicidae). *Insect Biochem. Mol. Biol.* 29(8), 711–21. doi:10.1016/s0965-1748(99)00047-8.

Silva Filha, M. H. N. L., Berry, C. and Regis, L. 2014. *Lysinibacillus sphaericus*: toxins and mode of action, applications for mosquito control and resistance management. In: Dhadialla, T. S. and Gill, S. S. (Eds), *Advances in Insect Physiology Vol. 47*. Academic Press, pp. 89–176.

Singh, A. K., Singh, A. and Joshi, P. 2016. Combined application of chitinolytic bacterium *Paenibacillus* sp. D1 with low doses of chemical pesticides for better control of *Helicoverpa armigera*. *Int. J. Pest Manag.* 62(3), 222–7. doi:10.1080/09670874.2016.1167267.

Smith, A. W., Cámara-Artigas, A., Brune, D. C. and Allen, J. P. 2005. Implications of high-molecular-weight oligomers of the binary toxin from *Bacillus sphaericus*. *J. Invertebr. Pathol.* 88(1), 27–33. doi:10.1016/j.jip.2004.10.005.

Splittstoesser, C. M., Tashiro, H., Lin, S. L., Steinkraus, K. H. and Fiori, B. J. 1973. Histopathology of the European chafer, *Amphimallon majalis*, infected with *Bacillus popilliae*. *J. Invertebr. Pathol.* 22(2), 161–7. doi:10.1016/0022-2011(73)90128-6.

Stamm, M. D., Prochaska, T. J., Matz, N. A. and Baxendale, R. W. 2013. Efficacy of *Chromobacterium subtsugae* for control of southern masked chafers, 2012. *Arthropod Manag. Tests* 38(1). doi:10.4182/amt.2013.G16.

Steinhaus, E. A. 1975. *Disease in a Minor Chord*. Ohio State University Press, Columbus, OH.

Su, T., Thieme, J. L. and Cheng, M. L. 2018. Impact of storage and handling temperatures on the activities of mosquito larvicides. *J. Am. Mosq. Control Assoc.* 34(3), 244–8. doi:10.2987/18-6770.1.

Sundh, I. and Goettel, M. S. 2013. Regulating biocontrol agents: a historical perspective and a critical examination comparing microbial and macrobial agents. *BioControl* 58(5), 575–93. doi:10.1007/s10526-012-9498-3.

Thanabalu, T., Hindley, J., Jackson-Yap, J. and Berry, C. 1991. Cloning, sequencing, and expression of a gene encoding a 100-kilodalton mosquitocidal toxin from *Bacillus sphaericus* SSII-1. *J. Bacteriol.* 173(9), 2776–85. doi:10.1128/jb.173.9.2776-2785.1991.

Thiery, I., Hamon, S., Dumanoir, V. C. and de Barjac, H. 1992a. Vertebrate safety of *Clostridium bifermentans* serovar *malaysia*, a new larvicidal agent for vector control. *J. Econ. Entomol.* 85(5), 1618–23. doi:10.1093/jee/85.5.1618.

Thiery, I., Hamon, S., Gaven, B. and De Barjac, H. 1992b. Host range of *Clostridium bifermentans* serovar. *Malaysia*, a mosquitocidal anaerobic bacterium. *J. Am. Mosq. Control Assoc.* 8(3), 272–7.

Vachon, V., Laprade, R. and Schwartz, J. L. 2012. Current models of the mode of action of *Bacillus thuringiensis* insecticidal crystal proteins: a critical review. *J. Invertebr. Pathol.* 111(1), 1–12. doi:10.1016/j.jip.2012.05.001.

van Frankenhuyzen, K. 2009. Insecticidal activity of *Bacillus thuringiensis* crystal proteins. *J. Invertebr. Pathol.* 101(1), 1–16. doi:10.1016/j.jip.2009.02.009.

Velivelli, S. L., De Vos, P., Kromann, P., Declerck, S. and Prestwich, B. D. 2014. Biological control agents: from field to market, problems, and challenges. *Trends Biotechnol.* 32(10), 493–6. doi:10.1016/j.tibtech.2014.07.002.

Verplaetse, E., Slamti, L., Gohar, M. and Lereclus, D. 2015. Cell differentiation in a *Bacillus thuringiensis* population during planktonic growth, biofilm formation, and host infection. *mBio* 6(3), e00138–15. doi:10.1128/mBio.00138-15.

Vettori, C., Paffetti, D., Saxena, D., Stotzky, G. and Giannini, R. 2003. Persistence of toxins and cells of *Bacillus thuringiensis* subsp. *kurstaki* introduced in sprays to Sardinia soils. *Soil Biol. Biochem.* 35(12), 1635–42. doi:10.1016/j.soilbio.2003.08.009.
Vimala Devi, P. S., Duraimurugan, P. and Chandrika, K. S. V. P. 2019. *Bacillus thuringiensis*-based nanopesticides for crop protection. In: Koul, O. (Ed.), *Nano-Biopesticides Today and Future Perspectives*. Academic Press, pp. 249–60. Chapter 10.
Vodovar, N., Vinals, M., Liehl, P., Basset, A., Degrouard, J., Spellman, P., Boccard, F. and Lemaitre, B. 2005. *Drosophila* host defense after oral infection by an entomopathogenic *Pseudomonas* species. *Proc. Natl. Acad. Sci. U. S. A.* 102(32), 11414–9. doi:10.1073/pnas.0502240102.
Vodovar, N., Vallenet, D., Cruveiller, S., Rouy, Z., Barbe, V., Acosta, C., Cattolico, L., Jubin, C., Lajus, A., Segurens, B., Vacherie, B., Wincker, P., Weissenbach, J., Lemaitre, B., Médigue, C. and Boccard, F. 2006. Complete genome sequence of the entomopathogenic and metabolically versatile soil bacterium *Pseudomonas entomophila*. *Nat. Biotechnol.* 24(6), 673–9. doi:10.1038/nbt1212.
Walters, F. S., Graser, G., Burns, A. and Raybould, A. 2018. When the whole is not greater than the sum of the parts: a critical review of laboratory bioassay effects testing for insecticidal protein interactions. *Environ. Entomol.* 47(2), 484–97. doi:10.1093/ee/nvx207.
Wang, K., Shu, C., Soberon, M., Bravo, A. and Zhang, J. 2018. Systematic characterization of Bacillus Genetic Stock Center *Bacillus thuringiensis* strains using Multi-Locus Sequence Typing. *J. Invertebr. Pathol.* 155, 5–13. doi:10.1016/j.jip.2018.04.009.
Wirth, M. C., Yang, Y., Walton, W. E., Federici, B. A. and Berry, C. 2007. Mtx toxins synergize *Bacillus sphaericus* and Cry11Aa against susceptible and insecticide-resistant *Culex quinquefasciatus* larvae. *Appl. Environ. Microbiol.* 73(19), 6066–71. doi:10.1128/AEM.00654-07.
World Health Organization (WHO) & International Programme on Chemical Safety. 1999. Microbial pest control agent: *Bacillus thuringiensis*. World Health Organization, Geneva. Available at: https://apps.who.int/iris/handle/10665/42242.
Wraight, S. P., Molloy, D. P. and Singer, S. 1987. Studies on the culicine mosquito host range of *Bacillus sphaericus* and *Bacillus thuringiensis* var. *israelensis* with notes on the effects of temperature and instar on bacterial efficacy. *J. Invertebr. Pathol.* 49(3), 291–302. doi:10.1016/0022-2011(87)90061-9.
Wright, D. J. 1987. Avermectins: action on target pest species. *Biochem. Soc. Trans.* 15(1), 65–7. doi:10.1042/bst0150065.
Wright, D. A., Zydenbos, S. M., Wessman, P., O'Callaghan, M., Townsend, R. J., Jackson, T. A., van Koten, C. and Mansfield, S. 2017. Surface coating aids survival of *Serratia entomophila* (Enterobacteriaceae) in granules for surface application. *Biocontrol Sci. Technol.* 27(12), 1383–99. doi:10.1080/09583157.2017.1402861.
Yiallouros, M., Storch, V., Thiery, I. and Becker, N. 1994. Efficacy of *Clostridium bifermentans* serovar *Malaysia* on target and nontarget organisms. *J. Am. Mosq. Control Assoc.* 10(1), 51–5.
Yokoyama, T., Tanaka, M. and Hasegawa, M. 2004. Novel cry gene from *Paenibacillus lentimorbus* strain Semadara inhibits ingestion and promotes insecticidal activity in *Anomala cuprea* larvae. *J. Invertebr. Pathol.* 85(1), 25–32. doi:10.1016/j.jip.2003.12.009.
Zhang, J., Hodgman, T. C., Krieger, L., Schnetter, W. and Schairer, H. U. 1997. Cloning and analysis of the first cry gene from *Bacillus popilliae*. *J. Bacteriol.* 179(13), 4336–41. doi:10.1128/jb.179.13.4336-4341.1997.

Zhang, X., Candas, M., Griko, N. B., Rose-Young, L. and Bulla, L. A. 2005. Cytotoxicity of *Bacillus thuringiensis* Cry1Ab toxin depends on specific binding of the toxin to the cadherin receptor BT-R(1) expressed in insect cells. *Cell Death Differ.* 12(11), 1407–16. doi:10.1038/sj.cdd.4401675.

Zheng, J., Gao, Q., Liu, L., Liu, H., Wang, Y., Peng, D., Ruan, L., Raymond, B. and Sun, M. 2017. Comparative genomics of *Bacillus thuringiensis* reveals a path to specialized exploitation of multiple invertebrate hosts. *mBio* 8(4), e00822-00817. doi:10.1128/mBio.00822-17.

Zhou, X., Li, H., Liu, Y., Hao, J., Liu, H. and Lu, X. 2018. Improvement of stability of insecticidal proteins from *Bacillus thuringiensis* against UV-irradiation by adsorption on sepiolite. *Adsorpt. Sci. Technol.* 36(5–6), 1233–45. doi:10.1177/0263617418759777.

Zubasheva, M. V., Ganushkina, L. A., Smirnova, T. A. and Azizbekyan, R. R. 2010. Larvicidal activity of crystal-forming strains of *Brevibacillus laterosporus*. *Appl. Biochem. Microbiol.* 46(8), 755–62. doi:10.1134/S0003683810080041.

Zydenbos, S. M., Townsend, R. J., Lane, P. M. S., Mansfield, S., Callaghan, M. O., van Koten, C. and Jackson, T. A. 2016. Effect of *Serratia entomophila* and diazinon applied with seed against grass grub populations on the North Island volcanic plateau. *N. Z. Plant Protect.* 69, 86–93.

Chapter 6

Advances in the use of Bt genes in insect-resistant crops

Salvatore Arpaia, ENEA Research Centre Trisaia-Rotondella (MT), Italy

1 Introduction

Genetically modified (GM) crops have often been referred to as one of the agricultural innovations with the fastest adoption rate by farmers worldwide. Indeed in 2018, about 17 million farmers in 26 countries planted 191.7 million ha with GM crops, an increase of 1% or 1.9 million hectares from 2017 (ISAAA, 2018). This fast pace of adoption has led in the last two decades to the unchallenged primacy of GM varieties on worldwide production of cotton (80% of the invested surface) and soybean (77%), while about one-third of cultivated maize and canola is currently GM. This large adoption of GM varieties, especially in the Americas, has created several problems in international trading (Parisi et al., 2016), since there are major importers of staple food and feed, for example the European Union, that still have a precautionary attitude towards GM products and, therefore, require imported GM food and feed to be specifically authorized and their processed products on the market, labelled. The negative attitude of the EU towards GM crops is even more obvious in case of cultivation and as a consequence the current practice of growing GM crops in Europe is limited to the Iberian peninsula where the 'historical' GM event maize MON810 resistant to the corn borers *Ostrinia nubilalis* and *Sesamia nonagrioides* was grown on ca 121 000 ha in 2018. The gene conferring resistance to Lepidoptera

http://dx.doi.org/10.19103/AS.2020.0073.07

in the MON810 maize is a *cry1Ab* gene derived by the soil bacterium *Bacillus thuringiensis* (Bt) and which is selectively toxic to Lepidoptera. This gene can be considered the founder of a series of synthetically reproduced traits mimicking the natural array of genes codifying for insecticidal toxins present in the genome of different strains of *B. thuringiensis*, each characterized by a rather narrow range of target organisms (Palma et al., 2014). These genes still constitute the large majority of genes expressed by GM crops resistant to insects, alone or in combination with other traits expressed in the so-called stacked events, which represent to date the majority of GM crops used for cultivation. Even though insect-resistant traits are considered as belonging to the first generation of GM crops, the role of Bt-derived genes is still very relevant in the currently grown biotech crops worldwide. In fact, the only available commercial event expressing a gene to induce resistance to insects not derived from Bt is the MON87411 maize developed by Monsanto (authorized for cultivation by the US Environmental Protection Agency in 2015) which, in addition to a Cry3Bb delta-endotoxin, encodes for a double-stranded RNA transcript containing a 240 bp fragment of the *Diabrotica virgifera virgifera Snf7* gene to confer resistance to this beetle. This maize event has been then crossed to several other GM events and it is now included in several stacked GM maize hybrids.

2 *Bacillus thuringiensis* (Bt) insecticidal toxins

The list of Bt-derived genes currently contains 858 entries (http://www.btnomenclature.info/; accessed 5 July 2019), although only less than a dozen are currently expressed in commercially available GM crops (http://www.isaaa.org/gmapprovaldatabase/geneslist/default.asp; accessed on 5 July 2019). The majority of the proteins produced by these genes are δ-endotoxin formed during the sporulating phase of the bacterium life cycle and can be grouped in two families: cry and cyt. The different cry toxins are specifically noxious to insects of the orders Lepidoptera, Coleoptera, Hymenoptera and Diptera. In contrast, cyt toxins are mostly found in *B. thuringiensis* strains active against Diptera, although a few exceptions of coleopteran active strains containing Cyt proteins have been documented (Bravo and Soberon, 2010). In addition, strains of *B. thuringiensis* active against insect orders of Hymenoptera, Homoptera, Orthoptera and Mallophaga and to other invertebrates such as nematodes and mites have also been identified (http://www.btnomenclature.info/; accessed on 5 July 2019). Another class of toxins is synthesized by this bacterium during the vegetative phase of the life cycle (Vip toxins) and their use has been exploited recently for obtaining GM plants resistant to insects (e.g. MIR162 maize, Bollgard III cotton). Bt toxins are known to be quite specific, though some cross-order activity has been reported (Van Frankenhuyzen, 2013). Bt-based microbial pesticides have been used for the biological control of insect pests for more than 80 years, and are currently included in over 100 commercial products (Glare and O'Callaghan, 2000). Bt microbial pesticides represent one of the main pillars of pest management for organic farmers, since their remarkable selectivity, their short persistence in the environment and their harmlessness for mammals make this class of pesticides quite innocuous for non-target organisms.

Due to all the positive characteristics of these molecules, it is not surprising that Bt genes have been one of the first practical applications of genetic engineering in agriculture, when the successful transformation of crop plants was achieved in the 1980s. The efficiency of these GM plants in tolerating the attack of insect pests spurred seed companies to produce new varieties which reached the market in the late 1990s upon the first approval by the US EPA of the StarLink maize expressing the toxin Cry9C as a plant incorporating pesticide (https://archive.epa.gov/pesticides/biopesticides/web/html/starlink_corn.html; accessed 23 July 2019).

There are some differences in the functioning of Bt toxins between microbial pesticides and GM plant-expressing Bt toxins (Arpaia et al., 2017) that trigger specifically targeted risk assessment in the two different cases. Some practical implications for the different action of Bt toxins in the two different delivery methods are summarized in Table 1.

The mechanism of action of cry toxins originating from *B. thuringiensis* occurring in the environment requires the cleavage of the protoxin (the natural form of cry proteins) upon ingestion, under favourable pH conditions in the midgut of target insects. This activation is not necessary in GM plants, since the

Table 1 Practical differences between the characteristics of Bt toxins in microbial pesticides and in GM plants due to the different delivery method

Microbial pesticides	GM plants
Release controlled by applicator: timing, point location, and so on	Under the regulation of constitutive promoters (e.g. 35s) toxin is released continuously and in all plant parts, independently from the infestation level
Product is evenly distributed over whole plants	There is a different tissue-specific toxin production (usually highest in leaf tissues), which also varies along the cropping season
Degradation begins immediately after application	Degradation only occurs when plant-expressed compounds are released in the environment (i.e. via root exudates and cultivation residues)
Possible dispersal by spray drift at the time of application	Possible dispersal by plant tissue transport during (e.g. with pollen) and after (e.g. via plant residues) the cropping season
Limited interactions with plant or other naturally occurring metabolites in the environment	Possible interactions with other plant produced compounds

newly expressed genes codify for the shorter form, already active, toxin. The way the toxic effect can be enhanced in microbial pesticides or in plants is also sensibly different. Laboratory studies have demonstrated that when grown in specific media, *B. thuringiensis* colonies produce higher amount of chitinases, which act synergistically with cry toxins (Vu et al., 2009); the toxic effect under these conditions can be 1.5–2 times higher as demonstrated in laboratory experiments targeting the larvae of *Aedes aegypti* (Ramirez-Suero et al., 2011).

The range of organisms and the modality of exposure to Bt toxins induced by the two delivery methods can also be quite different. For example, aerial transportation of Bt-pollen can expose non-target caterpillar larvae living on wild plants at some distance from the GM-cultivated areas (Perry et al., 2010; Holst et al., 2013), while this pathway seems more remote for microbial pesticides. The use of microbial pesticides against corn borers is usually recommended before maize flowering, thus pollen feeding organisms are normally not exposed to the toxins in this form. On the contrary, cry toxins are commonly present in pollen of GM plants, although they are less abundant compared to other tissues, therefore exposure of flower visiting organisms is still possible. Another difference, with potentially relevant biosafety implications, is due to the exposure pathway, which is expected to be largely bitrophic when Bt toxins are distributed on plant canopy with sprays since they will have a limited persistence. Possibilities of exposure for non-target organisms can instead follow several direct and indirect paths when these toxins are expressed in plants (Andow et al., 2006).

3 Incorporating Bt-expressing GM plants in integrated pest management (IPM)

The report 'Agriculture at a Crossroads' (IAASTD, 2009) indicated that agricultural knowledge, science and technology are fundamental to meeting the Millennium development goals. The biggest challenge of modern agriculture is ensuring enough food and feed production for a fast-growing world population while safeguarding economic, social and environmental sustainability of cropping practices. This scenario is further complicated by the evidence that climate change is altering the distribution, incidence and intensity of animal and plant pests and diseases (Taylor et al., 2018; Deutsch et al., 2018). GM crops represent a valuable new tool to support these actions and their new traits have the potential to improve crop production especially under heavy pest, disease and weed pressure (Gomez-Barbero et al., 2008; Areal et al., 2013). In particular, insecticidal and virus-resistant crops could help to keep pests and diseases in check, to lower the chemical pesticide load in the environment, to support complementary IPM tactics such as the active use of biocontrol agents and, therefore, to increase reliance on natural pest control. However, the reduction of pesticide use alone does not guarantee that natural pest control in the presence of GM crops is maintained or enhanced, and the need to assess the role that GM plants might play in different IPM systems was highlighted by Menozzi et al. (2017). In order to verify how likely Bt crops might fit well in IPM programmes, we need to consider if these varieties can be successfully managed in the context of the main principles of IPM (i.e. monitoring population density, establishing action thresholds, selection of the most appropriate method(s) of pest control). Particularly, it is paramount to investigate if the ecological functions provided by the organisms living in an agro-ecosystem (e.g. pollinators, natural enemies, detritivores) are maintained to the same extent as in 'traditional' non-GM-based agro-ecosystems.

GM crops resistant to insects, as well as traditionally bred insect-resistant plants, work independently of any action threshold established for target pests, since they produce a kind of preventive defense even before insects attack cropped plants. Monitoring population levels of the target insects will, therefore, not be necessary when GM varieties are being cultivated, though monitoring for possible onset of resistance is regularly conducted by the applicants. On the other hand, the monitoring of other insect pests and natural enemies (which remains a necessary activity in any IPM programme) is not hampered in any way by the presence of GM plants. Therefore, the main issue to be considered is the role of this germplasm in combination with other means of pest control in the agro-ecosystem, including the use of chemical pesticides.

A meta-analysis conducted by Klümper and Qaim (2014) has revealed that on average GM technology adoption has reduced chemical pesticide use by 37%, and increased crop yields by 22%. The extent of these effects vary especially according to the modified crop traits and geographic regions, both yield gains and pesticide reductions are larger for insect-resistant crops than for other foreign traits expressed in GM plants. Not surprisingly, cotton - a crop on which the use of insecticides is quite heavy - scores as the agro-ecosystem where reduction of pesticide load is highest when GM varieties are adopted. A reduced use of insecticides will obviously favour the activity of natural enemies, especially generalist predators that can find secondary preys among non-target herbivores in the agro-ecosystem and maintain their valuable contribution to ecosystem services (Arpaia et al., 2007). Even in some cases where particular species showed a reduced abundance in Bt stands compared to the untransformed control plants, the redundancy in the agro-ecosystem allowed other species of the same guild to concur in maintaining the level of natural pest control, therefore guaranteeing the ecological service (Naranjo, 2005). However, ecological relationships regulating food webs in the field are most commonly not linear ones and the possible issue of secondary pests needs to be carefully monitored (Wang et al., 2008; Hagenbucher et al., 2013; Catarino et al., 2016).

Several meta-analyses have investigated comparatively the impact of different pest control approaches, including Bt-expressing GM crops, on natural enemies (e.g. Marvier et al., 2007; Naranjo, 2009). The general outcomes of these studies indicate that non-target invertebrates are more abundant in Bt cotton and Bt maize fields compared with non-transgenic fields managed with insecticides. However, in comparison with insecticide-free control fields, certain non-target taxa are less abundant in Bt fields. The organisms which experience a negative effect when reared on Bt-based diets are mostly specialized parasitoids linked to target pests, which suffer most likely for the poor quality of available hosts (e.g. Ramirez-Romero et al., 2007; Walker et al., 2007; Naranjo, 2009). This is not surprising, considering the strict link with the host species that this group of natural enemies requires for successfully completing their life cycle. Even though direct toxic effects are not expected to occur, the reduced abundance or quality of hosts in Bt fields may hamper the efficacy of parasitoid-based natural biological control in some agro-ecosystems.

The concepts of IPM are usually successfully realized over an extended spatial and temporal scale. Approaching the problem from an agro-ecosystems angle is best to design less vulnerable cropping systems, moving towards longer-term prevention and suppression of key pests in the system over multiple fields, farms and seasons (Barzman et al., 2015). To analyse the effects of Bt maize adoption on European corn borer (ECB), *O. nubilalis* populations, Hutchison et al. (2010) estimated the annual per capita growth rates in maize-growing areas in the Midwest of the United States where maize constitutes the most relevant cropping system. Possible confounding factors, such as previous population levels in different regions, and density-dependent mortality were accounted for in the model when time series were analysed. Reduction of populations of the ECB ranged from 23% to 73% when historical data from the pre-Bt era (i.e. 1963–1996) where compared with data collected from 1996 to 2009. This clearly indicates that benefits from the adoption of Bt maize were extended to all maize-growing farmers, including non-adopters of GM maize. A more recent paper (Dively et al., 2018) addressed the issue of area-wide pest management at a regional scale, considering also mixed cropping systems where several crops share relevant surfaces in the area. The case study concerns the Mid-Atlantic United States where crops such as peppers, green beans and sweet corn are cultivated on large acreages and are all susceptible to the attack of ECB and *Helicoverpa zea*. Using data spanning from 1976 to 2016 through a regional network of blacklight traps in three states, Dively et al. (2018) demonstrated that vegetable growers benefit via decreased crop damage and insecticide applications due to the large decline of populations of the two pests. Their model also includes new features and it is able to account for different concurring biological phenomena linked to climate change. The number of recommended insecticide sprays to control ECB in peppers, green beans and sweet corn, and *H. zea* in sweet corn, significantly decreased as a function of average Bt corn adoption rates. Time series regarding field damage reports for ECB in the area are available, and the comparisons from the pre-Bt maize introduction and the current surveys indicate their decline in pepper reached 78%, and 70% in sweet corn.

Positive examples of area-wide pest management achieved with the contribution of Bt crops are not limited to maize. For instance, adoption of Bt cotton in China suppressed pink bollworm *Pectinophora gossypiella* and cotton bollworm *Helicoverpa armigera* populations in non-Bt cotton as well as in other crops (Wu et al., 2008). Regional reductions in the pink bollworm, which is fairly specialized to cotton, have also been reported from the use of Bt cotton in the United States (Carrière et al., 2003).

Another positive aspect regarding the action of natural enemies and GM crops was elucidated by using population models which indicated that natural enemies can support the effectiveness of GM cultivations, by delaying the possible adaptation of target pests to Bt toxins (Gould 1994; Arpaia et al., 1997). The possible effect is linked to the behaviour of the natural enemies occurring in the agro-ecosystem, namely if natural enemies prey on the pest in a density-dependent fashion, this could lead to faster adaptation by the pest to the toxin, since the difference in fitness between resistant and susceptible individuals will be further enhanced in favour of the former. In contrast, inverse density-dependent predation is expected to slow the rate of adaptation by imposing more fitness costs to the few expected resistant survivors.

Theoretical considerations and some experimental evidence suggest that Bt-expressing GM crops might profitably be included in an IPM framework (Lundgren et al., 2009; Romeis et al., 2019); however, Hokkanen (2015) warned that in reality these benefits seldom seem to be achieved, as GM crops are still perceived by the growers as a stand-alone technology without any real attempts to integrate them as a component in IPM. For example, not conducting field sampling for secondary pests and natural enemies may cause a failure in estimating predator-prey relationships in the agro-ecosystem with the consequence that intense insecticide use might become necessary to control secondary pests, which are not sensitive to Bt toxins.

4 Insecticide resistance

Insecticide resistance is a well-known phenomenon regarding almost 600 insect species and more than 300 active ingredients (Sparks and Nauen, 2015). Population models indicate that the higher the selection pressure against a trait (i.e. susceptibility to a chemical compound), the quicker the spread of a resistant allele will be in the population. This principle applies equally to exposed insect populations when a Bt toxin is expressed in a GM plant in the field. The possible build-up of resistance to Bt toxins can be mainly considered as an agronomic and commercial problem; however, the main regulatory agencies (e.g. EPA in the USA, EFSA in the EU) demand that developers of GM varieties also monitor in a proactive manner the risk of insurgence of resistance in commercial cultivations, due to its possible environmental consequences (e.g. spread of resistant strains that could damage one or several crop species and then require additional use of pesticides). That's why insecticide resistance management measures have been deemed necessary from the very beginning of commercialization of Bt crops.

Current insect resistance management strategies rely on the combined use of GM events with high expression levels of cry toxins, able to kill all susceptible and heterozygote individuals and implementing refuge areas (i.e. non-GM stands) close to GM cultivations to favour mating between susceptible and resistant individuals to further dilute resistance allele frequency in the population. Indeed, to date Bt crops have remained effective against most pest populations after many years of commercial cultivation (Castañera et al., 2016; Tabashnik and Carrière, 2017); however, cases of field-evolved resistance with consequent increased crop damage to GM events have been reported in some populations of major target insect pests (Luttrell et al., 2004; Van Rensburg, 2007; Tabashnik et al., 2008, 2009; Bagla, 2010; Storer et al., 2010; Dhurua and Gujar, 2011; Gassmann et al., 2011; Kruger et al., 2012). The main cause of these failures has been the non-compliance to the high expression–refuge strategy, but lately a few newly occurring cases of previously susceptible pests damaging Bt crops also when compliance to the strategy was ensured (US EPA, FIFRA Scientific Advisory Panel, 2018), reopened the debate on how to improve monitoring methods and continue to develop strategies on how to preserve resistant germplasm over time.

Mathematical models have been largely used by regulatory authorities to establish a system of early warning to prevent the onset of resistance by the target pests. Though with some differences all models concur in indicating that increase of allele frequency over time is a natural, unavoidable, phenomenon that in the long run will lead to the onset of resistance. To proactively prevent the insurgence of resistance, common approaches suggest the possible introduction of novel toxins against the target pest either replacing those originally used or by pyramiding more toxins in a single stacked GM event. The latter tactic is now quite popular among biotech companies who have on the market GM events expressing several insect resistance and herbicide-tolerant genes in the same varieties (http://www.isaaa.org/gmapprovaldatabase). However, the strong asynchrony in the approval of GM cultivations between countries renders this approach not currently feasible in some areas, especially in Europe.

5 Conclusion and future trends

The possible environmental impacts of GM crops and the way to assess the relative risks have been quite extensively debated (Arpaia, 2010). A recent survey indicated that biodiversity preservation has been the predominant sub-topic (58%) in the projects studying implications of GM crop cultivations in Europe (Menozzi et al., 2017). The effect on non-target organisms was specifically analysed in 25% of those projects. In spite of the large body of data accumulating on this subject, the dichotomy regarding the approval of commercial cultivations of GM crops between geographical regions remains quite obvious. In order to support the production of scientifically sound data during environmental risk assessment (ERA), the European Food Safety Authority (EFSA) has renewed its guidelines for environmental risk assessment of GM plants (EFSA, 2010), making clear the need for sound ecological data on which ERA of GM crops should be rooted. The practical application of such guidelines has been implemented in a EU-funded research project (Arpaia et al., 2014) which has collected relevant baseline ecological data regarding active food webs in maize and potato agro-ecosystems and the surrounding landscapes in different European areas. Such datasets represent an important benchmark against which ERA for introduction of new GM germplasm could be realistically built. Surveys conducted in commercial fields in the framework of this project also triggered the recommendation that region-specific non-target indicator species should be included in ERA protocols to better account for the relevance of the receiving environment where Bt crops are to be released (van Capelle et al., 2016). The rationale on which the ERA is conducted by several regulatory agencies involve the monitoring of cultivated land (on-crop area), but sometimes of the surrounding landscape (off-crop area) and their interactions in terms of ecological functionality. This type of monitoring should include possibilities to study combined effects of different stressors on selected focal species and the extent of ecosystem services they provide (e.g. pollination, natural pest control). In the context of IPM, natural enemies represent obviously the most relevant category of non-target organisms to be studied; however, developing pest control systems, which are compatible with the normal activities of other functional guilds and which are very relevant for agriculture (e.g. pollinators, detritivores), should be a primary goal for the sustainability of modern agriculture.

The maintenance of ecosystem services is a declared principle of ERA in Europe, not only in the field of GM organisms (EFSA, 2016). However, analysing ecosystem services is not a simple exercise since it involves the measurement of the contribution of service-providing species, the scale at which they operate and determine the possible effects of stressors on these species (Kremen and Ostfeld, 2005).

GM crops expressing *B. thuringiensis* toxins represent a considerable part of the first generation of GM crops, characterized by a single newly expressed trait (either insect resistance or herbicide tolerance). After more than two decades of cultivation of GM crops worldwide, the current trend is represented by stacked events, containing more than one transgene. Additionally, some new characteristics are being introduced by seed companies into elite varieties (e.g. drought resistance, improved quality, etc.). Bt genes have been so far almost the only source of resistance to insects engineered into plant varieties which were brought to the market by industry; approved GM events contain only a handful of them: *Cry1Ab*, *Cry1Ac*, *Cry1F*, *Cry3Bb*. In spite of the progress made in discovering new Bt toxins in the last few years (Adang et al., 2014), only more recently new Bt-derived genes have been exploited for the production of Cry34/Cry35 binary toxins which act in combination against Coleoptera, or vip3A active against a broad range of lepidopteran pests and they have been expressed in a few GM commercialized events.

Even within this limited group of insecticidal toxins, some aspects of their mode of action are not completely understood (Hernàndez-Martinez et al., 2017). When some of these new traits were expressed in GM events aimed at commercial cultivation, the European Food Safety Authority argued that to resolve the remaining scientific uncertainty on the potential toxicity of the binary Cry34Ab1/Cry35Ab1 proteins on some predators, for example ladybirds, additional laboratory studies had to be performed prior to authorization (EFSA, 2013). A potential drastic change of this scenario may occur when more insect-resistant GM events expressing dsRNA will become available (Christiaens et al., 2018), if the request for authorization of these events will increase, therefore the possibilities of using genetically engineered plants for insect pest control will become less dependent on Bt toxins.

The possible onset of resistance to insecticidal toxins expressed in plants might also constitute a driving force towards searching new traits or combining more than one trait in a single variety to conduct a successful 'arms race' against new insect populations locally. These new scenarios will also trigger priority to new research questions in order to assess the possible effects due to the interactions of more than one toxin used in combination, on non-target organisms. The possible effects of multiple stressors, especially in valued species active as service providers in agro-ecosystems, is certainly a topic which has received limited attention so far. Combinatorial effects of Cry toxins with other proteins or chemicals are actually widely recognized and reported in the literature (Hilbeck and Otto, 2015). In Europe, ERA requirement for stacked GM varieties could be less extensive than the assessment of the parental GM events if the latter have already been considered to be safe, additional environmental studies are deemed relevant on a case-by-case basis (De Schrijver et al., 2007). Especially stacked events combining more than one insect-resistant toxin should be risk assessed on non-target organisms, since potential for additive interactions between different Bt toxin classes exists (Pérez et al., 2005; Sharma et al., 2010).

A final consideration should be given to the potentially different environmental impacts in different receiving environments. Non-target organisms are known to show quite a different sensitivity to Bt toxins even within the same insect order (Van Frenkenhuizen, 2009). Review of laboratory studies have shown that experiments aimed at evaluating activity of different cry toxins on natural enemies are still performed on a relatively small number of surrogate species mostly in America and Europe and many relevant taxonomic groups which are involved in providing important ecological services are poorly represented (Lövei et al., 2009). In the last few years, however, efforts are ongoing in collecting new information in countries where only recently GM crops have been authorized. Environmental risk assessment needs, at a certain stage of the process, to collect information on regional ecologically relevant focal species that have been judged to have an important role in providing ecological services in a particular agro-ecosystem. Conducting monitoring to develop enough information for a thorough estimate of ecological services may represent a massive effort, which goes well beyond the scope and possibilities of an ERA specifically aimed at the approval of a new variety or a new insecticide to be used in agriculture. Local resources and expertise should then be involved, also after the approval of varieties resistant to insects, in order to confirm that biotech products are exploiting their potential and are managed in a way to support the sustainability of agriculture.

6 Acknowledgements

I am indebted to Nicholas A. E. Birch for his useful comments on an earlier version of this manuscript.

7 Where to look for further information

Regulatory agencies provide information on the legal framework, the current status of GMO authorized for commercialization and details about the opinions given on each dossier for risk assessment of GM crops in their territories.

- United States Environmental Protection Agency - https://www.epa.gov/regulation-biotechnology-under-tsca-and-fifra/epas-regulation-biotechnology-use-pest-management.
- European Food Safety Authority - http://www.efsa.europa.eu/en/topics/topic/gmo.

The ICGEB has a web-based collection of reviews on different aspects of biosafety of GM organisms:

- International Centre for genetic Engineering and Biotechnology, Collection of Biosafety Reviews – https://www.icgeb.org/biosafety-publications-collections/.

The ISAAA web site publishes annual reports concerning worldwide use of GM crops:

- International Service for the Acquisition of Agri-biotech Applications – www.isaaa.org.

The following website is a good source of teaching materials for students wishing to improve their knowledge on the scientific aspects and the socio-economic implications of growing GM crops:

- University of Reading, teaching module on GMOs – http://www.reading.ac.uk/AMIGA/ScienceofGM/ami-WhatareGMcrops.aspx.

8 References

Adang, M. J., Crickmore, N. and Jurat-Fuentes, J. L. 2014. Diversity of *Bacillus thuringiensis* crystal toxins and mechanism of action. *Adv. Insect Physiol.* 47, 39–87. doi:10.1016/B978-0-12-800197-4.00002-6.

Andow, D. A., Lovei, G. L. and Arpaia, S. 2006. Ecological risk assessment for Bt crops. *Nat. Biotechnol.* 24(7), 749–51; author reply 751. doi:10.1038/nbt0706-749.

Areal, F. J., Riesgo, L. and Rodriguez-Cerezo, E. 2013. Economic and agronomic impact of commercialized GM crops: a meta-analysis. *J. Agric. Sci.* 151(1), 7–33. doi:10.1017/S0021859612000111.

Arpaia, S. 2010. Genetically modified plants and "non-target" organisms: analysing the functioning of the agro-ecosystem. *Coll. Biosafety Rev.* 5, 12–80.

Arpaia, S., Gould, F. and Kennedy, G. G. 1997. Potential impact of *Coleomegilla maculata* DeGeer (Coleoptera: Coccinellidae) predation on adaptation of *Leptinotarsa decemlineata* Say (Coleoptera: Chrysomelidae) to Bt-transgenic potatoes. *Entomol. Exper. Appl.* 82(1), 91–100. doi:10.1046/j.1570-7458.1997.00117.x.

Arpaia, S., Di Leo, G. M., Fiore, M. C., Schmidt, J. E. U. and Scardi, M. 2007. Composition of arthropod species assemblages in Bt-expressing and near isogenic eggplants in experimental fields. *Environ. Entomol.* 36(1), 213–27. doi:10.1603/0046-225x(2007)36[213:coasai]2.0.co;2.

Arpaia, S., Messéan, A., Birch, N. A., Hokannen, H., Härtel, S., Van Loon, J. J., Lovei, G. L., Park, J., Spreafico, H., Squire, G. R., Steffan-Dewenter, I., Tebbe, C. and Van Der Voet, H. 2014. Assessing and monitoring impacts of genetically modified plants on agro-ecosystems: the approach of AMIGA project. *Entomologia* 2, 154. doi:10.4081/entomologia.2014.154.

Arpaia, S., Birch, A. N. E., Kiss, J., Van Loon, J. J. A., Messéan, A., Nuti, M., Perry, J. N., Sweet, J. B. and Tebbe, C. C. 2017. Assessing environmental impacts of genetically modified plants on non-target organisms: the relevance of in planta studies. *Sci. Total Environ.* 583, 123–32. doi:10.1016/j.scitotenv.2017.01.039.

Bagla, P. 2010. Hardy cotton-munching pests are latest blow to GM crops. *Science* 327, 1439.

Barzman, M., Bàrberi, P., Birch, A. N. E., Boonekamp, P., Dachbrodt-Saaydeh, S., Graf, B., Hommel, B., Jensen, J. E., Kiss, J., Kudsk, P., Lamichhane, J. R., Messéan, A., Moonen, A. C., Ratnadass, A., Ricci, P., Sarah, J. L. and Sattin, M. 2015. Eight principles of integrated pest management. *Agron. Sustain. Dev.* 35(4), 1199–215. doi:10.1007/s13593-015-0327-9.

Bravo, A. and Soberon, M. 2010. *Bacillus thuringiensis*: mechanisms and use. In: Gilbert, L. I. and Gill, S. S. (Eds), *Insect Control Biological and Synthetic Agents*. Academic Press, Amsterdam.

Carrière, Y., Ellers-Kirk, C., Sisterson, M., Antilla, L., Whitlow, M., Dennehy, T. J. and Tabashnik, B. E. 2003. Long-term regional suppression of pink bollworm by *Bacillus thuringiensis* cotton. *Proc. Natl. Acad. Sci. U.S.A.* 100(4), 1519–23. doi:10.1073/pnas.0436708100.

Castañera, P., Farinós, G. P., Ortego, F. and Andow, D. A. 2016. Sixteen years of Bt maize in the EU hotspot: why has resistance not evolved? *PLoS ONE* 11(5), e0154200. doi:10.1371/journal.pone.0154200.

Catarino, R., Ceddia, G., Areal, F., Parisey, N. and Park, J. 2016. Managing maize under pest species competition: is Bt (*Bacillus thuringiensis*) maize the solution? *Ecosphere* 7(6), e01340. doi:10.1002/ecs2.1340.

Christiaens, O., Dzhambazova, T., Kostov, K., Arpaia, S., Joga, M. R., Urru, I., Sweet, J. J. and Smagghe, G. 2018. Literature review of baseline information on RNAi to support the environmental risk assessment of RNAi-based GM plants. EFSA supporting publication 2018: EN-1424, 173pp. doi:10.2903/sp.efsa.2018.EN-1424.

De Schrijver, A., Devos, Y., Van den Bulcke, M., Cadot, P., De Loose, M., Reheul, D. and Sneyers, M. 2007. Risk assessment of GM stacked events obtained from crosses between GM events. *Trends Food Sci. Technol.* 18(2), 101–9. doi:10.1016/j.tifs.2006.09.002.

Deutsch, C. A., Tewksbury, J. J., Tigchelaar, M., Battisti, D. S., Merrill, S. C., Huey, R. B. and Naylor, R. L. 2018. Increase in crop losses to insect pests in a warming climate. *Science* 361(6405), 916–9. doi:10.1126/science.aat3466.

Dhurua, S. and Gujar, G. T. 2011. Field-evolved resistance to Bt toxin Cry1Ac in the pink bollworm, *Pectinophora gossypiella* (Saunders) (Lepidoptera: Gelechiidae) from India. *Pest Manag. Sci.* 67, 898–903.

Dively, G. P., Venugopal, P. D., Bean, D., Whalen, J., Holmstrom, K., Kuhar, T. P., Doughty, H. B., Patton, T., Cissel, W. and Hutchison, W. D. 2018. Regional pest suppression associated with widespread Bt maize adoption benefits vegetable growers. *Proc. Natl. Acad. Sci. U.S.A.* 115(13), 3320–5. doi:10.1073/pnas.1720692115.

EFSA Panel on Genetically Modified Organisms (GMOs). 2010. Guidance on the environmental risk assessment of genetically modified plants. *EFSA J.* 8(11), 1879. doi:10.2903/j.efsa.2010.1879.

EFSA Panel on Genetically Modified Organisms (GMOs). 2013. Statement supplementing the environmental risk assessment conclusions and risk management recommendations on genetically modified insect-resistant maize 59122 for cultivation in the light of new scientific information on non-target organisms and regionally sensitive areas. *EFSA J.* 11(11), 3443.

EFSA Scientific Committee. 2016. Guidance to develop specific protection goals options for environmental risk assessment at EFSA, in relation to biodiversity and ecosystem services. *EFSA J.* 14(6), e04499.

Gassmann, A. J., Petzold-Maxwell, J. L., Keweshan, R. S. and Dunbar, M. W. 2011. Field-evolved resistance to Bt maize by western corn rootworm. *PLoS One* 6, e22629.

Glare, T. R. and O'Callaghan, M. 2000. *Bacillus thuringiensis: Biology, Ecology and Safety*. John Wiley and Sons Ltd, Chichester UK.

Gomez-Barbero, M., Berbel, J. and Rodriguez-Cerezo, E. 2008. Bt corn in Spain–the performance of the EU's first GM crop. *Nat. Biotechnol.* 26(4), 384–6. doi:10.1038/nbt0408-384.

Gould, F. 1994. Potential and problems with high dose strategies for pesticidal, engineered crops. *Biocontr. Sci. Technol.* 4, 451–61.

Hagenbucher, S., Wäckers, F. L., Wettstein, F. E., Olson, D. M., Ruberson, J. R. and Romeis, J. 2013. Pest trade-offs in technology: reduced damage by caterpillars in Bt cotton benefits aphids. *Proc. Royal Soc. B: Biol. Sci.* 280(1758), 20130042. doi:10.1098/rspb.2013.0042.

Hernández-Martínez, P., Gomis-Cebolla, J., Ferré, J. and Escriche, B. 2017. Changes in gene expression and apoptotic response in *Spodoptera exigua* larvae exposed to sublethal concentrations of Vip3 insecticidal proteins. *Sci. Rep.* 7(1), 16245.

Hilbeck, A. and Otto, M. 2015. Specificity and combinatorial effects of *Bacillus thuringiensis* Cry toxins in the context of GMO environmental risk assessment. *Front. Environ. Sci.* 3, 71. doi:10.3389/fenvs.2015.00071.

Hokkanen, H. M. 2015. Integrated pest management at the crossroads: science, politics, or business (as usual)? *Arthrop. Plant Interact.* 9(6), 543–5.

Holst, N., Lang, A., Lövei, G. and Otto, M. 2013. Increased mortality is predicted of *Inachis io* larvae caused by Bt-maize pollen in European farmland. *Ecol. Modell.* 250, 126–33. doi:10.1016/j.ecolmodel.2012.11.006.

Hutchison, W. D., Burkness, E. C., Mitchell, P. D., Moon, R. D., Leslie, T. W., Fleischer, S. J., Abrahamson, M., Hamilton, K. L., Steffey, K. L., Gray, M. E., Hellmich, R. L., Kaster, L. V., Hunt, T. E., Wright, R. J., Pecinovsky, K., Rabaey, T. L., Flood, B. R. and Raun, E. S. 2010. Areawide suppression of European corn borer with Bt maize reaps savings to non-Bt maize growers. *Science* 330(6001), 222–5. doi:10.1126/science.1190242.

IAASTD. 2009. Agriculture at a crossroads. Synthesis Report. Island Press, Washington DC.

ISAAA. 2018. Global Status of Commercialized Biotech/GM Crops in 2018: Biotech Crops Continue to Help Meet the Challenges of Increased Population and Climate Change. ISAAA Brief No. 54. ISAAA, Ithaca, NY.

Klümper, W. and Qaim, M. 2014. A meta-analysis of the impacts of genetically modified crops. *PLoS ONE* 9(11), e111629. doi:10.1371/journal.pone.0111629.

Kremen, C. and Ostfeld, R. S. 2005. A call to ecologists: measuring, analyzing, and managing ecosystem services. *Front. Ecol. Environ.* 3(10), 540–8. doi:10.1890/1540-9295(2005)003[0540:ACTEMA]2.0.CO;2.

Kruger, M., Van Rensburg, J. B. J. and Van den Berg, J. 2012. Transgenic Bt maize: farmers perceptions, refuge compliance and reports of stem borer resistance in South Africa. *J. Appl. Entomol.* 136, 38–50.

Lövei, G. L., Andow, D. A. and. Arpaia, S. 2009. Transgenic insecticidal crops and natural enemies: a detailed review of laboratory studies. *Environ. Entomol.* 38(2), 293–306. doi:10.1603/022.038.0201.

Lundgren, J. G., Gassmann, A. J., Bernal, J., Duan, J. J. and Ruberson, J. 2009. Ecological compatibility of GM crops and biological control. *Crop Protect.* 28(12), 1017–30. doi:10.1016/j.cropro.2009.06.001.

Luttrell, R. G., Ali, I., Allen, K. C., Young III, S. Y., Szalanski, A., Williams, K., Lorenz, G., Parker Jr, C. D. and Blanco, C. 2004. Resistance to Bt in Arkansas populations of cotton bollworm. In: Richter, D. A. (Ed.), *Proceedings of the 2004 Beltwide Cotton Conferences*, San Antonio, Texas, 5–9 January 2004, pp. 1373–83.

Marvier, M., McCreedy, C., Regetz, J. and Kareiva, P. 2007. A meta-analysis of effects of Bt cotton and maize on nontarget invertebrates. *Science* 316(5830), 1475–7. doi:10.1126/science.1139208.

Menozzi, D., Kostov, K., Sogari, G., Arpaia, S., Moyankova, D. and Mora, C. 2017. A stakeholder engagement approach for identifying future research directions in the evaluation of current and emerging applications of GMOs. *Bio-Based Appl. Econom. J.* 6, 57–79.

Naranjo, S. E. 2005. Long-term assessment of the effects of transgenic *Bt* cotton on the function of the natural enemy community. *Environ. Entomol.* 34(5), 1211–23. doi:10.1093/ee/34.5.1211.

Naranjo, S. E. 2009. Impacts of Bt crops on non-target invertebrates and insecticide use patterns. *CAB Rev.* 4(11), 011. doi:10.1079/PAVSNNR20094011.

Palma, L., Muñoz, D., Berry, C., Murillo, J. and Caballero, P. 2014. *Bacillus thuringiensis* toxins: an overview of their biocidal activity. *Toxins* 6(12), 3296–325. doi:10.3390/toxins6123296.

Parisi, C., Tillie, P. and Rodriguez-Cerezo, E. 2016. The global pipeline of GM crops out to 2020. *Nat. Biotechnol.* 34(1), 31–6. doi:10.1038/nbt.3449.

Pérez, C., Fernandez, L. E., Sun, J., Folch, J. L., Gill, S. S., Soberón, M. and Bravo, A. 2005. *Bacillus thuringiensis* subsp. israelensis Cyt1Aa synergizes Cry11Aa toxin by functioning as a membrane-bound receptor. *Proc. Natl. Acad. Sci. U.S.A.* 102(51), 18303–8. doi:10.1073/pnas.0505494102.

Perry, J. N., Devos, Y., Arpaia, S., Bartsch, D., Gathmann, A., Hails, R. S., Kiss, J., Lheureux, K., Manachini, B., Mestdagh, S., Neemann, G., Ortego, F., Schiemann, J. and Sweet, J. B. 2010. A mathematical model of exposure of non-target Lepidoptera to Bt-maize pollen expressing Cry1Ab within Europe. *Proc. R. Soc. B.* 277(1686), 1417–25.

Ramirez-Romero, R., Bernal, J. S., Chaufaux, J. and Kaiser, L. 2007. Impact assessment of Bt-maize on a moth parasitoid, *Cotesia marginiventris* (Hymenoptera: Braconidae), via host exposure to purified Cry1Ab protein or Bt-plants. *Crop Protect.* 26(7), 953–62. doi:10.1016/j.cropro.2006.09.001.

Ramírez-Suero, M., Valerio-Alfaro, G., Bernal, J. S. and Ramírez-Lepe, M. 2011. Synergistic effect of chitinases and *Bacillus thuringiensis* israelensis spore-toxin complex against *Aedes aegypti* larvae. *Can. Entomol.* 143(2), 157–64. doi:10.4039/n10-051.

Romeis, J., Naranjo, S. E., Meissle, M. and Shelton, A. M. 2019. Genetically engineered crops help support conservation biological control. *Biol. Contr.* 130, 136–54. doi:10.1016/j.biocontrol.2018.10.001.

Sharma, P., Nain, V., Lakhanpaul, S. and Kumar, P. A. 2010. Synergistic activity between *Bacillus thuringiensis* Cry1Ab and Cry1Ac toxins against maize stem borer (*Chilo partellus* Swinhoe). *Lett. Appl. Microbiol.* 51(1), 42–7. doi:10.1111/j.1472-765X.2010.02856.x.

Sparks, T. C. and Nauen, R. 2015. IRAC: mode of action classification and insecticide resistance management. *Pestic. Biochem. Physiol.* 121, 122–8. doi:10.1016/j.pestbp.2014.11.014.

Storer, N. P., Babcock, J. M., Schlenz, M., Meade, T., Thompson, G. D., Bing, J. W. and Huckaba, R. M. 2010. Discovery and characterization of field resistance to Bt maize: *Spodoptera frugiperda* (Lepidoptera: Noctuidae) in Puerto Rico. *J. Econom. Entomol.* 103, 1031–8.

Tabashnik, B. E. and Carrière, Y. 2017. Surge in insect resistance to transgenic crops and prospects for sustainability. *Nat. Biotechnol.* 35(10), 926.

Tabashnik, B. E., Gassmann, A. J., Crowder, D. W. and Carrière, Y. 2008. Insect resistance to Bt crops: evidence versus theory? *Nat. Biotechnol.* 26, 199–202.

Tabashnik, B. E., Van Rensburg, J. B. J. and Carriere, Y. 2009. Field-evolved insect resistance to Bt crops: definition, theory, and data. *J. Econom. Entomol.* 102, 2011–25.

Taylor, R., Herms, D., Cardina, J. and Moore, R. 2018. Climate change and pest management: unanticipated consequences of trophic dislocation. *Agronomy* 8(1), 7. doi:10.3390/agronomy8010007.

US EPA, FIFRA Scientific Advisory Panel. 2018. Meeting Minutes and Final Report No. 2018-06, 17–19 July 2018. Available at: https://www.regulations.gov/document?D=EPA-HQ-OPP-2017-0617-0032.

Van Capelle, C., Schrader, S. and Arpaia, S. 2016. Selection of focal earthworm species as non-target soil organisms for environmental risk assessment of genetically modified plants. *Sci. Total Environ.* 548–549, 360–9.

Van Frankenhuyzen, K. 2009. Insecticidal activity of Bacillus thuringiensis crystal proteins. *J. Invertebr. Pathol.* 101(1), 1–16. doi:10.1016/j.jip.2009.02.009.

Van Frankenhuyzen, K. 2013. Cross-order and cross-phylum activity of *Bacillus thuringiensis* pesticidal proteins. *J. Invertebr. Pathol.* 114(1), 76–85. doi:10.1016/j.jip.2013.05.010.

Van Rensburg, J. B. J. 2007. First report of field resistance by the stem borer, *Busseola fusca* (Fuller) to Bt-transgenic maize. *S. A. J. Plant Soil* 24(3), 147–51.

Vu, K. D., Yan, S., Tyagi, R. D., Valéro, J. R. and Surampalli, R. Y. 2009. Induced production of chitinase to enhance entomotoxicity of *Bacillus thuringiensis* employing starch industry wastewater as a substrate. *Bioresour. Technol.* 100(21), 5260–9. doi:10.1016/j.biortech.2009.03.084.

Walker, G. P., Cameron, P. J., MacDonald, F. M., Madhusudhan, V. V. and Wallace, A. R. 2007. Impacts of *Bacillus thuringiensis* toxins on parasitoids (Hymenoptera:Braconidae) of *Spodoptera litura* and *Helicoverpa armigera* (Lepidoptera: Noctuidae). *Biol. Contr.* 40(1), 142–51. doi:10.1016/j.biocontrol.2006.09.008.

Wang, S., Just, D. R. and Pinstrup-Andersen, P. P. 2008. Bt-cotton and secondary pests. *Int. J. Biotechnol.* 10(2/3), 113. doi:10.1504/IJBT.2008.018348.

Wu, K. M., Lu, Y. H., Feng, H. Q., Jiang, Y. Y. and Zhao, J. Z. 2008. Suppression of cotton bollworm in multiple crops in China in areas with Bt toxin-containing cotton. *Science* 321(5896), 1676–8. doi:10.1126/science.1160550.

Chapter 7

Plant growth-promoting bacteria (PGPBs) as biocontrol agents against invertebrate pests

Luca Ruiu, Università degli Studi di Sassari, Italy

1 Introduction

Rhizobacterial communities play a leading role in favoring soil health and productivity. While many microbial species are directly involved in basic biogeochemical cycles ensuring substance turnover in the ecosystem and the availability of nutrients for plants and other living organisms, a variety of bacteria have evolved specific ways to interact with members of their biocenotic environment. From a plant perspective, some species have developed as pathogens while others have established special relationships determining beneficial effects on the plant growth, vigor, and productivity.

In evolutionary terms, bacterial genomes have evolved and reshaped over time from common ancestors with decomposing functions in the ecosystem, to develop specific adaptations leveraging an arsenal of genes and their products regulating the complex of relations with the external environment.

Historically, most of the scientific community's efforts and industrial investments have been focused on the study of plant pathogenic bacteria, with the aim of understanding the physical and molecular interactions with the host and, consequently, developing strategies and products for their containment.

http://dx.doi.org/10.19103/AS.2020.0073.08

Today, however, there is a growing interest in bacterial species spontaneously associated with crop plants and able to interact with them in a beneficial way, making them more productive. Such benefits for the plant derive from biocenotic interactions that may favor the acquisition of nutrients from the external environment or may increase their resistance to pathogens and parasites. Nature therefore offers the opportunity to exploit these sophisticated biological mechanisms to support agricultural crops. Many of these plant growth-promoting bacteria (PGPB) are cultivable, and the same benefits observed in nature are reproducible through their application in agricultural soil. The result is a strong industrial interest pursuing the principles of eco-sustainable crop management involving the reduction of chemical inputs in the agroecosystem in favor of beneficial microbial agents making plants less susceptible to biotic adversities (Glick, 2012).

In an initial view, PGPBs were mainly considered for their role in making plants' use of resources more efficient, but by defining the mechanisms that govern their interactions with the plant and the external environment it was understood how they can contribute to other functions. According to an extreme simplification, the following types of overlapping benefits could be included:

- optimizing the use of environmental resources;
- improving plant health and resistance to biotic and abiotic factors;
- acting directly against plant pathogens; and
- acting directly against invertebrate pests.

The effects of PGPB species against invertebrates such as insects and nematodes may derive from the action of bacterial toxins and virulence factors or by other mechanisms causing an increased plant resistance. In summary, the effects on invertebrates can be classified as:

- direct (toxicity or pathogenicity); and
- indirect (plant resistance induction).

This chapter includes the analysis of a case study on a PGPB species, *Pseudomonas protegens*, behaving as a biocontrol agent against both plant phytopathogens and invertebrate pests.

2 Plant growth promotion

Association of PGPBs with plants may lead to an increased plant health and growth rate as a result of a facilitated and optimized use of soil resources (i.e. nutrients, water), and an improved physiological equilibrium. It is well known that in order to make efficient use of the nutrient supply in the soil, it is necessary to favor chemical forms that improve bioavailability, that is, that allow and facilitate absorption by the roots. So, for example, abundant elements in nature such as nitrogen, phosphorus, and iron need to be chemically converted in the soil to be made available to plants (Gamalero and Glick, 2011).

Accordingly, atmospheric nitrogen fixation is essentially left to the enzymatic activity of plant-associated *Rhizobia* species and only to a lesser extent to free-living bacteria such as *Azospirillum* spp. *Rhizobia* and leguminous plants have developed a symbiotic relationship leading to the formation of nodules on the root. These special structures host bacteria that convert atmospheric nitrogen into ammonia (Somasegaran and Hoben, 2012).

Another capability associated with PGPB is the enzymatic activity of certain bacterial species contributing to solubilization and mineralization of phosphorous, an essential nutrient for plant that is normally immobilized in the soil (Zaidi et al., 2009). Such an important function is, however, more often covered by specialized fungi such as mychorrizae (Bolan, 1991).

Plants can also benefit from soil bacteria secreting siderophores, that is, chelating compounds forming a complex with iron, and making this element soluble and available for root absorption. Siderophore-producing bacteria (SPB) include several *Pseudomonas fluorescens* strains (Kloepper et al., 1980).

PGPBs may also play a role in modulating plant physiology, thus promoting its growth and supporting its response to stress. Among bacterially synthesized phytohormones, some species were found to produce cytokinins or gibberellins, and indoleacetic acid (IAA). Some of these bacteria were also found to produce ACC-deaminase that is known to mitigate the inhibitory action of stress-associated ethylene (Glick, 2014).

3 Action against plant pathogens

In addition to promoting plant growth through direct mechanisms facilitating the access to environmental resources, bacteria may protect the plant against pathogens, thus adding further support to growth and health preservation (Compant et al., 2005a). While many mechanisms of action against plant pathogens still need to be clarified, there is a growing interest in using PGPBs as biocontrol agents, in line with the principles of reducing chemical inputs in agroecosystems (Ruiu, 2018).

A first mechanism of plant protection is related to the root surface colonization by these bacteria, often represented by biofilm formers, which determines a competition for both space and nutrients with soil borne phytopathogens (Kloepper et al., 1999). In addition, some endophytic species enter the root establishing a more intimate relationship with the host plant (Gray and Smith, 2005).

On the other hand, the biocontrol activity may depend on the synthesis and release of antibiotics and other compounds like amphisin, 2,4-diacetylphloroglucinol (DAPG), hydrogen cyanide, oomycin A, phenazine, pyoluteorin, pyrrolnitrin, tensin, tropolone, cyclic lipopeptides, oligomycin A, kanosamine, and zwittermicin A, produced by different bacterial species belonging to diverse genera including *Pseudomonas*, *Bacillus*, and *Streptomyces* (Compant et al., 2005a).

Several enzymes variably produced by different species and strains of these bacteria can also act directly against pathogens. These include proteases, chitinases, and beta-glucanases, acting against a variety of phytopathogenic fungi in different genera including *Fusarium*, *Phytophthora*, *Pythium*, and *Botrytis* (Frankowski et al., 2001; Ordentlich et al., 1988).

Some PGPB species are capable of producing ACC-deaminase that interferes with the ethylene released by a plant attacked by phytopathogens, thus reducing its expected deleterious effects (Glick, 2005). Other bacteria may employ siderophores to bind iron and indirectly lower its availability in the rhizosphere for potential plant pathogen development (Leong, 1986).

On the other hand, PGPBs may produce signals that induce a systemic resistance triggering plant immune system. Such induced systemic resistance (ISR), can be an effective tool to control the action of plant pathogens and it represent a feature associated with several PGPBs (Pieterse et al., 2014).

4 Potential against invertebrate pests

Several bacterial species that live in the soil and interact with plants promoting their access to nutrients, colonizing the phylloplane, or containing phytopathogens, might also be able to exercise an antagonistic action against invertebrate pests. This effect is often the result of the production and release of specific toxins and virulence factors. The versatility of these multivalent bacterial strains makes them particularly useful, including from a commercial point of view, ensuring diverse benefits for the plant. Among a wide variety of species and strains with such features, good examples are represented by species in the genera *Bacillus, Paenibacillus, Brevibacillus, Pseudomonas, Serratia, Burkholderia,* and *Streptomyces*.

Among Bacillaceae, some strains of *Bacillus firmus* have been commercially developed into products used to control plant-parasitic nematodes (PPNs). While the mechanism of action of these strains is not completely understood, recent studies have highlighted the involvement of a bacterial serine protease (Sep1) degrading the nematode physical barriers including intestinal tissues and cuticle proteins (Geng et al., 2016). Similarly, *Bacillus subtilis* strains were shown to have insecticidal potential related to the production of specific metabolites like the lipopeptides (i.e. surfactins and iturins) produced by strain S499 (Assie et al., 2002).

Several *Bacillus* species involved at different levels in promoting plant growth and in antagonism against pathogens and pests could be mentioned. Among these are strains of *B. cereus sensu lato* (Herman et al., 2008) and *B. pumilus* (Garcia-Ramon et al., 2016).

Within the same bacterial order, different *Paenibacillus* species have shown to act as PGPB. Some species in this genus behave as insect pathogens, and this insecticidal activity is directly related to the production of specific toxins and virulence factors (Zhao et al., 2011).

Plant growth promotion properties are also attributed to the spore-former *Brevibacillus laterosporus*, a species involved in soil fertilization, antagonism against plant pathogens, insecticidal and nematocidal activity (Ruiu, 2013). This species has also been shown to have endophytic ability in host plants (Ormskirk et al., 2019).

Gammaproteobacteria enhancing plant growth includes *Pseudomonas* and *Serratia* species. The insecticidal potential of *P. entomophila* was initially studied on *Drosophila melanogaster*, and the pathogenicity was found to be associated with a specific toxin secretion system (Vodovar et al., 2006). The *Pseudomonas fluorescens* group involves different species involved in both plant growth promotion and antagonism against phytopathogens and invertebrate pests (Ruffner et al., 2013).

The entomopathogenic properties of *Serratia* species, like *Serratia nematodiphila*, against insects and nematodes derive from the production of diverse virulence factors, among which a significant part is played by protein toxin complexes, in addition to various enzymes such as chitinases, proteases, and lipases (Kwak et al., 2015).

The betaproteobacteria in genus *Burkholderia* were found to have epiphytic and endophytic plant colonization patterns determining several benefits, inducing a local host defence effect (Compant et al., 2005b). This genus is also represented by species acting as biocontrol agents against phytopathogens (Janisiewicz and Roitman, 1988), while some soil species were reported to have a broad spectrum insecticidal and miticidal activity (Cordova-Kreylos et al., 2013).

The soil-dwelling actinobacteria *Streptomyces* spp. include several species establishing beneficial interactions with plants, favoring their growth and indirectly protecting against pathogens and parasites. Such properties are related to the production of a variety of bioactive compounds (de Jesus Sousa and Olivares, 2016). These secondary metabolites include the macrocyclic lactone derivatives, avermectins, produced by *S. avermitilis* and acting as potent insecticides inhibiting neurotransmission, which have led to the development of commercially available insecticidal products (Bloomquist, 1996).

5 Applications in agriculture and forestry

The need to reduce the use of chemical inputs in agricultural and forest ecosystems supports the application of low environmental impact solutions to promote plant growth and health. The employment of suitable plant growth-promoting microorganisms, including bacteria, is therefore in line with modern legislative frameworks fostering the reduction in the use of conventional chemical substances for plant nutrition and protection.

Agriculture represents the field in which PGPBs have been more studied and commercialized with products exploiting the growth promotion and biocontrol properties of several species in the genera *Azospirillum*, *Azobacter*, *Bacillus*, *Burkholderia*, *Enterobacter*, *Pseudomonas*, *Serratia*, and *Xanthomonas*. Several commercial products include a single or a consortium of different species that find use as soil amendments benefiting a variety of agriculture crops (Lucy et al., 2004). Additionally, many of these PGPB species can have secondary beneficial effects protecting plants from pathogens and pests through the previously described direct and indirect mechanisms. The result is an important regulatory distinction between soil amendments and biological control agents, where the placing on the market of the latter requires compliance with specific procedures and regulations implying the approval and risk assessment of any new active substance and formulated product (Ruiu, 2015).

While research and use of PGPBs in forestry is much less developed, there are several successful application examples, including inoculation with root growth-promoting strains belonging to species such as *Bacillus polymyxa* and *Pseudomonas fluorescens*, which, among other things, have the characteristic of overcoming adverse winter environmental conditions in the plant rhizosphere (Chanway et al., 2000).

The global market for biofertilizers, worth more than US$ 1.8 billion in 2018, is characterized by a significant growth at an estimated rate of 14.3%, with the segment of nitrogen-fixing species (i.e. *Rhizobium*, *Azobacter*, *Azospirillum*) being the most represented (ResearchAndMarkets, 2019). This trend is in line with the growing attention to employ safer and environmentally responsible biological products.

6 Case study: *Pseudomonas protegens*

Fluorescent *Pseudomonas* species, involved in plant growth promotion and protection against phytopathogens and invertebrate pests, are characterized by the production of a yellowish-green siderophore called pyoverdine. These soil-dwelling bacteria are able to induce plant resistance and produce metabolites directly acting against pathogens (Raaijmakers et al., 2009). A taxonomic study published in 2011 based on genome and phenotypic analyses, led to the proposal of *P. protegens* as a new species (Ramette et al., 2011). As a result, some strains previously included in the *P. fluorescens* species complex were moved to the novel pseudomonad species and new isolates are being collected worldwide. Beneficial properties in favor of plants include stimulation of root growth, biofilm formation and competition with potential microbial pathogens, resistance induction, and the production of bioactive metabolites (Table 1).

The colonization of rhizosphere by *P. protegens* optimizes the plant use of environmental resources in favor of its growth and development leveraging specific mechanisms like nitrogen fixation (Fox et al., 2016). In addition, some strains, like the well-studied Pf-5, secrete siderophores and are also able to use heterologous siderophores released by other rhizobials (Sexton et al., 2017). Such behavior reduces the availability of iron for phytopathogens, thus indirectly protecting the plant from their infective action. *Pseudomonas* species are also involved in phosphate mobilization in soil, which contributes to their biofertilization properties (Fernández et al., 2012).

The resistance to pathogens induced in the plant colonized by *P. protegens* can be classified as:

- Systemic acquired resistance (SAR); and
- Induced systemic resistance (ISR).

Systemic resistance is naturally induced by plant pathogens and parasites, but the production of signaling molecules by beneficial soil microbes, such as

Table 1 Representative secondary metabolites produced by *Pseudomonas protegens* strains and their main activity against phytopathogens and invertebrate pests

Metabolite	Main activity
2,4-diacetylphloroglucinol (DAPG)	Antimicrobial
Hydrogen cyanide (HCN)	Antimicrobial
Pyoluteorin	Antimicrobial
Pyrrolnitrin	Antimicrobial
Toxoflavin	Antimicrobial
Phenazine	Antimicrobial
Orfamide A	Antimicrobial, insecticidal
Rhizoxin analogs	Antimicrobial, insecticidal
Protein toxin FitD	Insecticidal
Chitinases ChiC	Antimicrobial, nematocidal, insecticidal
Phospholipase PlcN	Insecticidal
Extracellular proteases	Insecticidal

ethylene, jasmonic acid, and salicylic acid may confer on the plant resistance to future pathogenic actions (Ramamoorthy et al., 2001).

Though SAR and ISR are often used as synonyms, a more accurate definition indicates that SAR refers to resistance induced by pathogens or salicylic acid, while ISR derives from the action of beneficial microbes or does not depend on salicylic acid (Pieterse et al., 2014). Resistance induction is related to the overexpression of plant defense-related genes (PR) and with physiological changes leading to structural barriers (i.e. cell wall) increasing (Benhamou et al., 1996).

In addition to free bacterial cells moving in the soil, pseudomonad cells can aggregate on the root surface to form biofilms, represented by bacterial cells embedded in an extracellular polysaccharide (EPS) matrix including extracellular DNAs (eDNA) and proteins (Ueda and Saneoka, 2015). The colonizing bacterial community in the biofilm is protected from the external environment increasing its tolerance to antimicrobials and adheres persistently to the substrate finding favorable conditions to grow. Biofilm formation by beneficial microbes can result in an advantage for the plant through two main mechanisms that keep pathogens away:

- Competition for space (due to a limited space on root surface); and
- Competition for nutrients (due to limited nutrient availability in the rhizosphere).

On the other hand, the action of beneficial *Pseudomonas* species against plant pathogens is not limited in creating an environment less favorable to them for their development and pathogenic activity. Accordingly, beneficial species are capable of producing secondary metabolites with direct action against pathogens. For instance, different *P. protegens* strains produce several antifungal compounds including 2,4-diacetylphloroglucinol (DAPG), hydrogen cyanide (HCN), pyoluteorin, pyrrolnitrin, toxoflavin, and orfamide A, derivatives rhizoxin, chitinases and other extracellular proteases (Philmus et al., 2015). The production of secondary metabolites is regulated by specific signal transduction mechanisms like the two global regulators GacA and RpoS, highly conserved in *Pseudomonas* species (Hassan et al., 2010; Kidarsa et al., 2013).

Alongside this action against plant pathogens, *P. protegens* can also produce an inhibitory or toxic action against invertebrate pests.

Several members of the *Pseudomonas* genus are implicated in pathogenic interactions with invertebrates, particularly insects. One of the most studied insecticidal species is *P. entomophila*, the ingestion of whose cells causes epithelial cells disruption in the midgut of susceptible targets.

Such effects derive from a toxin-mediated process that appears to be related to a specific secretion system (Vodovar et al., 2006). Sequencing and analysis of the genome of different pseudomonad species revealed conserved traits associated with the insecticidal potential. However, while several species in the *P. fluorescens* group may cause varying degrees of insecticidal effects, strains of the sub-clade 1, specifically including *P. protegens*, were found to harbor a proper arsenal of genes associated with the insecticidal action (Flury et al., 2016).

Main types of toxin genes found in *P. fluorescens* group species shows significant similarity with insecticidal genes of entomopathogenic bacteria such as *Xenorhabdus* and *Photorhabdus* species that produce an insecticidal toxin complex (Tc), including three protein components (A, B, and C), acting against insects by ingestion (Waterfield et al., 2001). In the case of *P. protegens*, a significant contribution to insect pathogenicity is given by a large protein toxin called Fit (which stand for fluorescent insecticidal toxin) showing homology with the Makes caterpillars floppy (Mcf) toxin produced by the entomopathogenic nematode symbiont *Photorhabdus luminescens* (Péchy-Tarr et al., 2008). More specifically, the toxin product called FitD is a protein with a size of approximately 320 kDa. Gene *fitD* is flanked by *fitABC* (upstream) and *fitE* (downstream) encoding for a protein secretion system. Other associated genes (i.e. *fitG* and *fitH*) play a regulatory function. The toxicity of the Fit protein is observed after injection or ingestion by susceptible insects, but it seems not to be the sole insecticidal metabolite this bacterium can produce. Accordingly, full oral toxicity of *P. protegens* mutants lacking *fitD* gene was observed in treated *Drosophila melanogaster*, suggesting the action of other insecticidal toxins. These were identified in rhizoxin analogs, orfamide A and chitinase that are also known to be involved in the action against phytopathogenic microorganisms (Loper et al., 2016). In other experiments, a contribution to insecticidal activity has been directly attributed to chitinase ChiC and phospholipase PlcN (Flury et al., 2016).

While several studies are underway to determine the mechanism of action of different strains with insecticidal properties, current knowledge leads to the conclusion that *P. protegens* acts against insects through a complex mechanism.

The potential of *P. protegens* to act as PGPB with a dual action of control against plant pathogens and invertebrate pests have recently been confirmed by genome sequencing and annotation studies on different strains including reference strain CHA0 (Smits et al., 2019).

Leveraging this potential, some root-associated *Pseudomonas protegens* strains and other *Pseudomonas* species have been developed as commercial products available for use in agriculture (Kupferschmied et al., 2013).

7 Conclusion and future trends

Free-living soil microorganisms represent a natural resource supporting plant growth and counteracting the potential negative action of pathogens and invertebrate pests. While PGPBs have found application especially for promoting soil bioavailability of nutrients for plants, their interest is growing in relation to employment prospects for the protection of agricultural crops. Several PGPB-based products are commercialized worldwide as soil amendment, but are also expected to have additional containment effects against plant pathogens, and sometimes against invertebrate pests, such as insects and nematodes. Accordingly, the scientific and industrial interest in bacterial strains with multiple benefits on the plants is particularly sought after. This trend is in line with the commercial objective of identifying bacterial strains that can cover multiple market segments. This approach contrasts one of the limits that has long been associated with microbiological control agents, whose use was relegated to niche contexts because of a high specificity and therefore a potential market limited to one or a few target pests or pathogens (Glare et al., 2012).

There is a clear economic and practical advantage in using a single microorganism that can facilitate the nutrition of the plant, induce system resistance mechanisms, and protect it from the deleterious action of pathogens and parasites.

Screening programs are in progress to discover strains with multiple properties and to study the mechanisms of action (Ruiu, 2015). Next generation sequencing, genomic analyses, and molecular studies on the interactions between soil microorganisms and between microorganisms and the plant are also providing new insights on the actual potential of various PGPBs in biocontrol (Compant et al., 2005a). On the other hand, this trend is corroborated by a growing interest of consumers and legislators for effective and safer bio-based solutions. While risk assessment remains an essential prerequisite, the market launch of new products based on beneficial microorganisms is expected in the near future.

8 Where to look for further information

8.1 Further reading

- A good introduction to PGPB mechanisms and potential is Glick (2012).
- A comprehensive introduction to PGPB as biofertilizers is Vessey (2003).
- A good review on the mechanisms of induced plant systemic resistance is Pieterse et al. (2014).
- Good reviews of PGPB as biocontrol agents are Beneduzi et al. (2012) and Compant et al. (2005a).
- An overview on the applications of beneficial soil microorganisms is Lucy et al. (2004).

8.2 Key international organizations and conferences on biocontrol

- SIP (Society for Invertebrate Pathology) promoting scientific knowledge in microbial entomopathogens.
- IOBC (International Organisation for Biological Control) supporting environmentally safe methods of pest and disease control in plant protection.
- IBMA (International Biocontrol Manufacturers Association) supporting the introduction of new biocontrol technologies.
- ABIM (Annual Biocontrol Industry Meeting) internationally recognized meeting place for the biocontrol industry.

9 References

Assie, L. K., Deleu, M., Arnaud, L., Paquot, M., Thonart, P., Gaspar, C. H. and Haubruge, E. 2002. Insecticide activity of surfactins and iturins from a biopesticide *Bacillus subtilis* Cohn (S499 strain). *Mededelingen (Rijksuniversiteit te Gent. Fakulteit van de Landbouwkundige en Toegepaste Biologische Wetenschappen)* 67(3), 647–55.

Beneduzi, A., Ambrosini, A. and Passaglia, L. M. 2012. Plant growth-promoting rhizobacteria (PGPR): their potential as antagonists and biocontrol agents. *Genetics and Molecular Biology* 35(4 Suppl.), 1044–51. doi:10.1590/s1415-47572012000600020.

Benhamou, N., Bélanger, R. R. and Paulitz, T. C. 1996. Pre-inoculation of Ri T-DNA-transformed pea roots with *Pseudomonas fluorescens* inhibits colonization by *Pythium ultimum* Trow: an ultrastructural and cytochemical study. *Planta* 199(1), 105–17. doi:10.1007/BF00196887.

Bolan, N. S. 1991. A critical review on the role of mycorrhizal fungi in the uptake of phosphorus by plants. *Plant and Soil* 134(2), 189–207. doi:10.1007/BF00012037.

Bloomquist, J. R. 1996. Ion channels as targets for insecticides. *Annual Review of Entomology* 41(1), 163–90. doi:10.1146/annurev.en.41.010196.001115.

Chanway, C. P., Shishido, M., Nairn, J., Jungwirth, S., Markham, J., Xiao, G. and Holl, F. B. 2000. Endophytic colonization and field responses of hybrid spruce seedlings after inoculation with plant growth-promoting rhizobacteria. *Forest Ecology and Management* 133(1–2), 81–8. doi:10.1016/S0378-1127(99)00300-X.

Compant, S., Duffy, B., Nowak, J., Clément, C. and Barka, E. A. 2005a. Use of plant growth-promoting bacteria for biocontrol of plant diseases: principles, mechanisms of action, and future prospects. *Applied and Environmental Microbiology* 71(9), 4951–9. doi:10.1128/AEM.71.9.4951-4959.2005.

Compant, S., Reiter, B., Sessitsch, A., Nowak, J., Clément, C. and Barka, E. A. 2005b. Endophytic colonization of *Vitis vinifera* L. by plant growth-promoting bacterium *Burkholderia* sp. strain PsJN. *Applied and Environmental Microbiology* 71(4), 1685–93. doi:10.1128/AEM.71.4.1685-1693.2005.

Cordova-Kreylos, A. L., Fernandez, L. E., Koivunen, M., Yang, A., Flor-Weiler, L. and Marrone, P. G. 2013. Isolation and characterization of *Burkholderia rinojensis* sp. nov., a non-*Burkholderia cepacia* complex soil bacterium with insecticidal and miticidal activities. *Applied and Environmental Microbiology* 79(24), 7669–78. doi:10.1128/AEM.02365-13.

de Jesus Sousa, J. A. and Olivares, F. L. 2016. Plant growth promotion by streptomycetes: ecophysiology, mechanisms and applications. *Chemical and Biological Technologies in Agriculture* 3(1), 24. doi:10.1186/s40538-016-0073-5.

Fernández, L., Agaras, B., Zalba, P., Wall, L. G. and Valverde, C. 2012. *Pseudomonas* spp. isolates with high phosphate-mobilizing potential and root colonization properties from agricultural bulk soils under no-till management. *Biology and Fertility of Soils* 48(7), 763–73. doi:10.1007/s00374-012-0665-6.

Flury, P., Aellen, N., Ruffner, B., Péchy-Tarr, M., Fataar, S., Metla, Z., Dominguez-Ferreras, A., Bloemberg, G., Frey, J., Goesmann, A., Raaijmakers, J. M., Duffy, B., Höfte, M., Blom, J., Smits, T. H., Keel, C. and Maurhofer, M. 2016. Insect pathogenicity in plant-beneficial pseudomonads: phylogenetic distribution and comparative genomics. *The ISME Journal* 10(10), 2527–42. doi:10.1038/ismej.2016.5.

Fox, A. R., Soto, G., Valverde, C., Russo, D., Lagares Jr., A., Zorreguieta, Á., Alleva, K., Pascuan, C., Frare, R., Mercado-Blanco, J., Dixon, R. and Ayub, N. D. 2016. Major cereal crops benefit from biological nitrogen fixation when inoculated with the nitrogen-fixing bacterium *Pseudomonas protegens* Pf-5 X940. *Environmental Microbiology* 18(10), 3522–34. doi:10.1111/1462-2920.13376.

Frankowski, J., Lorito, M., Scala, F., Schmid, R., Berg, G. and Bahl, H. 2001. Purification and properties of two chitinolytic enzymes of *Serratia plymuthica* HRO-C48. *Archives of Microbiology* 176(6), 421–6. doi:10.1007/s002030100347.

Gamalero, E. and Glick, B. R. 2011. Mechanisms used by plant growth-promoting bacteria. In: Maheshwari, D. K. (Ed.), *Bacteria in Agrobiology: Plant Nutrient Management*. Springer, Berlin, Heidelberg, pp. 17–46.

Garcia-Ramon, D. C., Molina, C. A., Osuna, A. and Vílchez, S. 2016. An in-depth characterization of the entomopathogenic strain *Bacillus pumilus* 15.1 reveals that it produces inclusion bodies similar to the parasporal crystals of *Bacillus thuringiensis*. *Applied Microbiology and Biotechnology* 100(8), 3637–54. doi:10.1007/s00253-015-7259-9.

Geng, C., Nie, X., Tang, Z., Zhang, Y., Lin, J., Sun, M. and Peng, D. 2016. A novel serine protease, Sep1, from *Bacillus firmus* DS-1 has nematicidal activity and degrades multiple intestinal-associated nematode proteins. *Scientific Reports* 6, 25012. doi:10.1038/srep25012.

Glare, T., Caradus, J., Gelernter, W., Jackson, T., Keyhani, N., Köhl, J., Marrone, P., Morin, L. and Stewart, A. 2012. Have biopesticides come of age? *Trends in Biotechnology* 30(5), 250–8. doi:10.1016/j.tibtech.2012.01.003.

Glick, B. R. 2005. Modulation of plant ethylene levels by the bacterial enzyme ACC deaminase. *FEMS Microbiology Letters* 251(1), 1–7. doi:10.1016/j.femsle.2005.07.030.

Glick, B. R. 2012. Plant growth-promoting bacteria: mechanisms and applications. *Scientifica* 2012, 963401. doi:10.6064/2012/963401.

Glick, B. R. 2014. Bacteria with ACC deaminase can promote plant growth and help to feed the world. *Microbiological Research* 169(1), 30–9. doi:10.1016/j.micres.2013.09.009.

Gray, E. J. and Smith, D. L. 2005. Intracellular and extracellular PGPR: commonalities and distinctions in the plant-bacterium signaling processes. *Soil Biology and Biochemistry* 37(3), 395–412. doi:10.1016/j.soilbio.2004.08.030.

Hassan, K. A., Johnson, A., Shaffer, B. T., Ren, Q., Kidarsa, T. A., Elbourne, L. D., Hartney, S., Duboy, R., Goebel, N. C., Zabriskie, T. M., Paulsen, I. T. and Loper, J. E. 2010. Inactivation of the GacA response regulator in *Pseudomonas fluorescens* Pf-5 has far-reaching transcriptomic consequences. *Environmental Microbiology* 12(4), 899–915. doi:10.1111/j.1462-2920.2009.02134.x.

Herman, M. A. B., Nault, B. A. and Smart, C. D. 2008. Effects of plant growth-promoting rhizobacteria on bell pepper production and green peach aphid infestations in New York. *Crop Protection* 27(6), 996–1002. doi:10.1016/j.cropro.2007.12.004.

Janisiewicz, W. J. and Roitman, J. 1988. Biological control of blue mold and gray mold on apple and pear with *Pseudomonas cepacia*. *Phytopathology* 78(12), 1697–700. doi:10.1094/Phyto-78-1697.

Kidarsa, T. A., Shaffer, B. T., Goebel, N. C., Roberts, D. P., Buyer, J. S., Johnson, A., Kobayashi, D. Y., Zabriskie, T. M., Paulsen, I. and Loper, J. E. 2013. Genes expressed by the biological control bacterium *Pseudomonas protegens* Pf-5 on seed surfaces under the control of the global regulators GacA and RpoS. *Environmental Microbiology* 15(3), 716–35. doi:10.1111/1462-2920.12066.

Kloepper, J. W., Leong, J., Teintze, M. and Schroth, M. N. 1980. Enhanced plant growth by siderophores produced by plant growth-promoting rhizobacteria. *Nature* 286(5776), 885–6. doi:10.1038/286885a0.

Kloepper, J. W., Rodriguez-Kabana, R., Zehnder, G. W., Murphy, J. F., Sikora, E. and Fernandez, C. 1999. Plant root-bacterial interactions in biological control of soilborne diseases and potential extension to systemic and foliar diseases. *Australasian Plant Pathology* 28(1), 21–6. doi:10.1071/AP99003.

Kupferschmied, P., Maurhofer, M. and Keel, C. 2013. Promise for plant pest control: root-associated pseudomonads with insecticidal activities. *Frontiers in Plant Science* 4, 287. doi:10.3389/fpls.2013.00287.

Kwak, Y., Khan, A. R. and Shin, J. H. 2015. Genome sequence of *Serratia nematodiphila* DSM 21420T, a symbiotic bacterium from entomopathogenic nematode. *Journal of Biotechnology* 193, 1–2. doi:10.1016/j.jbiotec.2014.11.002.

Leong, J. 1986. Siderophores: their biochemistry and possible role in the biocontrol of plant pathogens. *Annual Review of Phytopathology* 24(1), 187–209. doi:10.1146/annurev.py.24.090186.001155.

Loper, J. E., Henkels, M. D., Rangel, L. I., Olcott, M. H., Walker, F. L., Bond, K. L., Kidarsa, T. A., Hesse, C. N., Sneh, B., Stockwell, V. O. and Taylor, B. J. 2016. Rhizoxin analogs, orfamide A and chitinase production contribute to the toxicity of *Pseudomonas protegens* strain Pf-5 to *Drosophila melanogaster*. *Environmental Microbiology* 18(10), 3509–21. doi:10.1111/1462-2920.13369.

Lucy, M., Reed, E. and Glick, B. R. 2004. Applications of free living plant growth-promoting rhizobacteria. *Antonie van Leeuwenhoek* 86(1), 1–25. doi:10.1023/B:ANTO.0000024903.10757.6e.

Ordentlich, A., Elad, Y. and Chet, I. 1988. The role of chitinase of *Serratia marcescens* in biocontrol of *Sclerotium rolfsii*. *Phytopathology* 78(1), 84–8.

Ormskirk, M. M., Narciso, J., Hampton, J. G. and Glare, T. R. 2019. Endophytic ability of the insecticidal bacterium *Brevibacillus laterosporus* in *Brassica*. *PLoS ONE* 14(5), e0216341. doi:10.1371/journal.pone.0216341.

Péchy-Tarr, M., Bruck, D. J., Maurhofer, M., Fischer, E., Vogne, C., Henkels, M. D., Donahue, K. M., Grunder, J., Loper, J. E. and Keel, C. 2008. Molecular analysis of a novel gene cluster encoding an insect toxin in plant-associated strains of *Pseudomonas fluorescens*. *Environmental Microbiology* 10(9), 2368–86. doi:10.1111/j.1462-2920.2008.01662.x.

Philmus, B., Shaffer, B. T., Kidarsa, T. A., Yan, Q., Raaijmakers, J. M., Begley, T. P. and Loper, J. E. 2015. Investigations into the biosynthesis, regulation, and self-resistance of toxoflavin in *Pseudomonas protegens* Pf-5. *Chembiochem: a European Journal of Chemical Biology* 16(12), 1782–90. doi:10.1002/cbic.201500247.

Pieterse, C. M., Zamioudis, C., Berendsen, R. L., Weller, D. M., Van Wees, S. C. and Bakker, P. A. 2014. Induced systemic resistance by beneficial microbes. *Annual Review of Phytopathology* 52, 347–75. doi:10.1146/annurev-phyto-082712-102340.

Raaijmakers, J. M., Paulitz, T. C., Steinberg, C., Alabouvette, C. and Moënne-Loccoz, Y. 2009. The rhizosphere: a playground and battlefield for soilborne pathogens and beneficial microorganisms. *Plant and Soil* 321(1-2), 341–61. doi:10.1007/s11104-008-9568-6.

Ramamoorthy, V., Viswanathan, R., Raguchander, T., Prakasam, V. and Samiyappan, R. 2001. Induction of systemic resistance by plant growth promoting rhizobacteria in crop plants against pests and diseases. *Crop Protection* 20(1), 1–11. doi:10.1016/S0261-2194(00)00056-9.

Ramette, A., Frapolli, M., Fischer-Le Saux, M., Gruffaz, C., Meyer, J. M., Défago, G., Sutra, L. and Moënne-Loccoz, Y. 2011. *Pseudomonas protegens* sp. nov., widespread plant-protecting bacteria producing the biocontrol compounds 2,4-diacetylphloroglucinol and pyoluteorin. *Systematic and Applied Microbiology* 34(3), 180–8. doi:10.1016/j.syapm.2010.10.005.

ResearchAndMarkets. 2019. Biofertilizer market: global industry trends, share, size, growth, opportunity and forecast 2019–2024. Available at: https://www.researchandmarkets.com/r/ucz1gg.

Ruffner, B., Péchy-Tarr, M., Ryffel, F., Hoegger, P., Obrist, C., Rindlisbacher, A., Keel, C. and Maurhofer, M. 2013. Oral insecticidal activity of plant-associated pseudomonads. *Environmental Microbiology* 15(3), 751–63. doi:10.1111/j.1462-2920.2012.02884.x.

Ruiu, L. 2013. *Brevibacillus laterosporus*, a pathogen of invertebrates and a broad-spectrum antimicrobial species. *Insects* 4(3), 476–92. doi:10.3390/insects4030476.

Ruiu, L. 2015. Insect pathogenic bacteria in integrated pest management. *Insects* 6(2), 352–67. doi:10.3390/insects6020352.

Ruiu, L. 2018. Microbial biopesticides in agroecosystems. *Agronomy* 8(11), 235. doi:10.3390/agronomy8110235.

Sexton, D. J., Glover, R. C., Loper, J. E. and Schuster, M. 2017. *Pseudomonas protegens* Pf-5 favours self-produced siderophore over free-loading in interspecies competition for iron. *Environmental Microbiology* 19(9), 3514–25. doi:10.1111/1462-2920.13836.

Smits, T. H. M., Rezzonico, F., Frasson, D., Vesga, P., Vacheron, J., Blom, J., Pothier, J. F., Keel, C., Maurhofer, M. and Sievers, M. 2019. Updated genome sequence and annotation for the full genome of *Pseudomonas protegens* CHA0. *Microbiology Resource Announcements* 8(39), e01002-19. doi:10.1128/MRA.01002-19.

Somasegaran, P. and Hoben, H. J. 2012. *Handbook for Rhizobia: Methods in Legume-Rhizobium Technology*. Springer Science & Business Media, Berlin (DE).

Ueda, A. and Saneoka, H. 2015. Characterization of the ability to form biofilms by plant-associated *Pseudomonas* species. *Current Microbiology* 70(4), 506–13. doi:10.1007/s00284-014-0749-7.

Vessey, J. K. 2003. Plant growth promoting rhizobacteria as biofertilizers. *Plant and Soil* 255(2), 571–86. doi:10.1023/A:1026037216893.

Vodovar, N., Vallenet, D., Cruveiller, S., Rouy, Z., Barbe, V., Acosta, C., Cattolico, L., Jubin, C., Lajus, A., Segurens, B., Vacherie, B., Wincker, P., Weissenbach, J., Lemaitre, B., Médigue, C. and Boccard, F. 2006. Complete genome sequence of the entomopathogenic and metabolically versatile soil bacterium *Pseudomonas entomophila*. *Nature Biotechnology* 24(6), 673–9. doi:10.1038/nbt1212.

Waterfield, N. R., Bowen, D. J., Fetherston, J. D., Perry, R. D. and ffrench-Constant, R. H. 2001. The *tc* genes of *Photorhabdus*: a growing family. *Trends in Microbiology* 9(4), 185–91. doi:10.1016/s0966-842x(01)01978-3.

Zaidi, A., Khan, M. S., Ahemad, M. and Oves, M. 2009. Plant growth promotion by phosphate solubilizing bacteria. *Acta microbiologica et immunologica Hungarica* 56(3), 263–84. doi:10.1556/AMicr.56.2009.3.6.

Zhao, L., Yang, X., Li, X., Mu, W. and Liu, F. 2011. Antifungal, insecticidal and herbicidal properties of volatile components from *Paenibacillus polymyxa* strain BMP-11. *Agricultural Sciences in China* 10(5), 728–36. doi:10.1016/S1671-2927(11)60056-4.

Chapter 8

Advances in the use of entomopathogenic viruses as biopesticides in suppressing crop insect pests

Martin Erlandson, Agriculture and Agri-Food Canada, Canada

1 Introduction

Insect pests have an immense impact on agricultural and forestry production and are also vectors of plant and animal diseases including, disease organisms of plants and animals of agricultural importance and humans. Over the last seven decades, insect pest control has relied heavily on the use of broad-spectrum chemical pesticides. Over-reliance on chemical pesticides has led to the development of resistance in insect pest populations to chemical insecticides and the emergence of secondary pest problems due to the loss of arthropod natural enemies. In addition, negative human health and environmental effects, including negative impacts on insect pollinators, have been noted as a result of the broad target spectrum of these chemicals and their persistence in the environment. This has become an increasingly important issue, and regulatory agencies in many jurisdictions worldwide are currently re-evaluating the registration of many key chemical insecticides. As well, increasingly stringent regulations concerning chemical insecticide residues levels in food and feed have had significant impact with regard to restricting the current and potential expansion of the chemical insecticide market. As a result more environmentally sustainable pest control options such as integrated pest management (IPM) systems have been proposed. IPM incorporates various control strategies

http://dx.doi.org/10.19103/AS.2020.0073.09
Published by Burleigh Dodds Science Publishing Limited, 2020.

including, cultural methods, chemical insecticides and biological control agents to suppress insect pest populations below levels that cause economic damage. Insect biological control agents include microbial pathogens that have been developed as biopesticides and their implementation using chemical insecticide application methodology (Lacey et al., 2015). Although microbial biopesticides still constitute a relatively small proportion of the insecticide market globally (Lacey et al., 2015; references therein), 1–3%, they have the potential to become increasingly important in the future for the reasons outlined above. Indeed the overall biopesticides market has been projected to grow from USD 3.0 billion in 2018 to USD 6.4 billion by 2023 representing a CAGR of 16% during the forecast period due to increasing support for biological solutions as alternatives to chemical pesticides in both organic farming and IPM systems (MarketsAndMarkets, 2018). Currently, the microbial biopesticide market targeting insects is dominated by *Bacillus thuringiensis* (Bt)-based products; however, the use of transgenic crops expressing one or more Bt-toxin genes for insect pest suppression has significantly curtailed the use of Bt as a biopesticide (James, 2014).

Much of the developmental research on microbial pathogens of insects has focussed on the incorporation of pest-control strategies and application in the manner that has historically been used for chemical insecticides. However, the potential for microbial pathogens to amplify in hosts and persist in pest populations following application allows for the implementation of alternate use strategies for insect control. Four potential use strategies have been suggested for insect-specific viruses including implementation as: (i) classical biological control agents, (ii) conservation, (iii) inoculative and (iv) inundative strategies. The suitability of a virus for any of these strategies depends on the basic biology of the host–pathogen interaction and the specific attributes of the agricultural or forest production system.

Classical biological control involves the introduction of a microbial pathogen from a foreign source into a new ecosystem with an aim of permanent establishment and long-term suppression of an insect pest, most commonly an introduced pest. A good example is the European spruce sawfly, *Gilpinia hercyniae*, a major forest pest of European origin that became established in eastern North America in the 1930s through early 1940s. Following the introduction of a highly specific and efficacious baculovirus originating from northern Europe the pest population was reduced to below an economically significant level and remains so until now.

Conservation aims to conserve the natural enemies already present. There are few examples of the implementation of this strategy with microbial pathogens of insects in general and no good examples can be cited for viruses.

The inoculative release strategy involves the application of a microbial pathogen to an insect pest population at the early stage of an outbreak with the expectation of a season-long or a multi-year control. There are a number of examples of this approach in forestry, including the baculovirus product TM Biocontrol-1 (OpMNPV) for control of douglas-fir tussock moth. For this and the classical approach to work effectively, the virus needs a mechanism for efficient spread through the host population and is often most successful for pest species that exhibit a colonial or gregarious behaviour that increases the interaction between individuals.

The inundative strategy is similar to the use of conventional chemical insecticides in that the virus or bioinsecticide is applied to a pest insect population that has exceeded its economic threshold with the expectation of rapid control. This approach has been used mostly in greenhouse and horticultural crops, as for example, the use of codling moth granulovirus in orchards or the Spodoptera exigua nucleopolyhedrovirus (SPOD-X) for control of beet armyworm in greenhouse crops in Europe. A virus, as a biopesticide, must be sufficiently infectious and virulent to produce a high rate of infection when applied in the field and kill insects in a timely fashion so as to limit economic plant damage or the vectoring potential of the insect pest. As well, the virus needs to be produced relatively cheaply, in quantities required for the market, and be in a form that can be formulated for storage and application. The host specificity must be sufficiently broad to target pest(s) that constitutes an economically viable market and yet have enough target specificity to be advantageous in an IPM system. Finally a virus's target specificity, biology and ecology need to be well understood in order for governmental regulatory agencies to be convinced of their environmental and human-health safety as a prerequisite for registration as a bioinsecticide.

Viruses are associated with insects in a wide range of ecological relationships: as pathogens, as plant and animal viruses relying on insect vectors for transmission, and as specialized symbionts that allow hymenopteran parasitoids to fully develop in their hosts (e.g. polydnaviruses). The current taxonomy of viruses includes 17 families whose hosts include an insect or other arthropod for at least some members of the group (Table 1) (Lefkowitz et al., 2017). At least 14 virus families have representatives that are pathogenic in invertebrates (Table 1) and at least some of these families have representative viruses that have been investigated for potential as biological control agents of insects.

The virus families that are most-extensively investigated for the potential as insect control agents will be described in the following sections. Historically research has focussed on virus groups with readily observable symptomology (e.g. baculoviruses and iridescent viruses), those producing impressive epizootics in insect populations (baculoviruses), and those easily detected by light microscopy as a consequence of being occluded in relatively large protein crystals (baculoviruses, cypoviruses and entomopoxviruses) within the infected

Table 1 Virus families associated with insects

Family/genus	Nucleic acid	Virion shape	Occlusion body	Related Viruses in Vertebrates	Related Viruses in Plants
Reoviridae***	dsRNA, linear	Icosahedral		Orbivirus	Rice
Dwarf	10 segments			(e.g. blue tongue)	Wound tumour
Spinareovirinae					
Cypovirus	dsRNA, linear	Icosahedral	+	None	None
Birnaviridae	dsRNA, linear	Icosahedral	–	Infectious pancreatic necrosis virus	None
Tetraviridae					
Betatetravirus	ssRNA(+), 1-linear	Icosahedral	–	None	None
Omegatetravirus	ssRNA(+), 2-linear	Icosahedral	–	None	None
Dicistroviridae	ssRNA(+), linear	Icosahedral	–	None	None
	Cripavirus				
	Aparavirus				
Iflaviridae	ssRNA(+), linear	Icosahedral	–	None	None
Iflavirus	"	"	"	"	"
Nodaviridae	ssRNA(+), 2-linear	Icosahedral	–		
Alphanodavirus					
Flaviviridae***	ssRNA(+), linear	Icosahedral	–	Yellow fever	None
Flavivirus	"	"	"	"	"
Rhabdoviridae***	ssRNA(–)	Bullet-shaped	–	Rabies	
Bunyaviridae***	ssRNA(+/–)	Spherical/ pleomorphic	–	Hantavirus	None
Togaviridae	ssRNA(+)	Spherical	–	Rubella	None
Baculoviridae	dsDNA, circular	Rod	+	None	None
Alphabaculovirus	"	"	"	"	"
Betabaculovirus	"	"	"	"	"
Gammabaculovirus	"	"	"	"	"
Deltabaculovirus	"	"	"	"	"
Poxviridae	dsDNA, linear		+	Smallpox	None
Entomopoxvirinae					
Alphaentomopoxvirus	dsDNA, linear	Ovoid	+	None	None
Betaentomopoxvirus	dsDNA, linear	Ovoid	+	None	None
Gammaentomopoxvirus	dsDNA, linear	Brick-shaped	+	None	None
Iridoviridae	dsDNA, linear	Icosahedral	–	African swine fever virus	None
Iridovirus and *Chloridovirus*	dsDNA, linear	Icosahedral	–	None	None

Family/genus	Nucleic acid	Virion shape	Occlusion body	Related Viruses in Vertebrates	Related Viruses in Plants
Ascoviridae					
Ascovirus	dsDNA, circular	Rod/ovoid	–	None	None
Nudiviridae	dsDNA, circular	Rod	–	None	None
Alphanudivirus	"	"	"	"	"
Betanudivirus	"	"	"	"	"
Polydnaviridae					
Bracovirus	dsDNA, m-circular	Rod	–	None	None
Ichnovirus	dsDNA, m-circular	Fusiform	-	None	None
Parvoviridae	ssDNA linear	Icosahedral	–	Canine distemper	None
Densovirinae	ssDNA, linear	Icosahedral	–	None	None
Iteravirus	ssDNA, monosense	"	"	"	"
Brevidensovirus	ssDNA, monosense	"	"	"	"
Densovirus	ssDNA, ambisense	"	"	"	"
Pefudensovirus	ssDNA, ambisense	"	"	"	"

* Virus family association with arthropods largely as vectors of animal and plant viruses.

host cells. As well, insect groups such as Lepidoptera, Diptera, Coleoptera, Hymenoptera and Orthoptera that are of economic importance as pests, vectors of disease or exploited by man (e.g. silkworm, honey bees etc.) have been more thoroughly screened for viruses than other arthropod and insect groups.

Although a number of virus groups have been investigated, those in the family *Baculoviridae* are viewed as the most promising for development as biopesticides.

2 RNA viruses

2.1 Tetraviridae

Tetraviruses are small icosahedral, non-enveloped viruses with linear ssRNA genomes, that fall into two genera, *Betatetravirus* and *Omegatetravirus* (Dorrington et al., 2012). Tetraviruses have been isolated only from Lepidoptera and infection is restricted to the midgut and horizontal transmission by ingestion of infectious virus is likely the major route of infection. The pathogenicity of tetravirus isolates ranges from non-symptomatic to acutely lethal infections. Helicoverpa armigera stunt virus (HaSV) is the most extensively studied virus in this group and has been explored as a potential biocontrol agent for pests in the *Helicoverpa* group (Dorrington and Short, 2010). The *Helicoverpa* group contains some of the most damaging pests of agricultural crops, including the cotton bollworm, tobacco budworm and corn earworm. HaSV is highly pathogenic for early instar larvae but has almost no impact on the last two larval stages. HaSV infection produces rapid cessation of feeding, significant delays in larval growth and a very characteristic stunting or shrinkage of the larval body.

In a field test, HaSV applied against *H. armigera* in sorghum achieved control comparable to that of commercial preparations of Gemstar (a registered strain of the baculovirus HzSNPV) and standard chemical insecticides (Christian et al., 2005). Tetraviruses are difficult to produce in the quantities required for insect control, however, HaSV virion assembly has been demonstrated in plant protoplasts co-transformed with a plasmid expressing the HaSV capsid gene and plasmids carrying the complete cDNAs of the two HaSV genomic RNAs (Gordon et al., 2001). Thus this novel method of HaSV production suggests that tetraviruses may have potential as viral insecticides or in more novel approaches comparable to transgenic crops expressing *B. thurinigensis* σ-endotoxins. However to date no commercial insect control product based on a tetravirus has been registered.

2.2 Reoviridae

Reoviruses are characterized by icosahedral virions with genomes consisting of 10–12 linear dsRNA molecules. The family *Reoviridae* contains a number of genera that replicate in both vertebrate hosts and in arthropod vectors such as ticks and mosquitoes, as well as plant viruses that replicate both in plant hosts and in arthropod vectors such as leafhoppers. The family also contains a single genus, *Cypovirus,* that exclusively infects insects (Attoui et al., 2012).

Cypoviruses (CPV), like baculoviruses and entomopoxviruses, are occluded within crystalline proteinaceous structures referred to as occlusion bodies (OBs) or polyhedra. The OB provides some stability for the virus particles in the environment. Cypoviruses are transmitted orally and upon ingestion the OB dissolves in the alkaline conditions of the host insect gut releasing virions to infect gut cells. Typical CPV infections are limited to midgut epithelial cells and disease symptoms are often chronic in nature, mimicking those of starvation. The major symptoms include reduced feeding and larval growth, increased development times and often malformed adults that have reduced longevity and fecundity.

Cypoviruses have been isolated most frequently from Lepidoptera but a few isolates from Diptera and Hymenoptera have been described (Mori and Metcalf, 2010). Although cypoviruses are highly infectious and persist in insect populations, they have received limited attention for development as viral insecticides largely due to their generally chronic rather than acute symptomology (Belloncik and Mori, 1998). A CPV product, Matsukemin, based on Dendrolimus spectabilis CPV 1 (DsCPV-1) was registered in Japan in 1944 for use against the pine caterpillar but was deregistered in 1995 (Kunimi, 2007). Dendrolimus punctatus 1 (DpCPV-1) is registered for the control of a number of *Dendrolimus* species in pine forests in China (Sun, 2015). It is perhaps significant that these viruses target pests of forests where substantial defoliation can be tolerated without economic damage. The other proposed use for CPV is in combination with chemical insecticides for control of insect populations resistant to these specific chemicals. There is evidence that CPV infection of the insect gut is able to suppress chemical insecticide detoxification based on cytochrome P450 pathways (Mori and Metcalf, 2010).

3 DNA viruses

The most studied insect-specific DNA viruses are those readily detected due to the presence of OB and those producing observable epizootics in insect pests and thus have been considered to have good potential as biological control agents of insects. In the following sections those virus families with species showing the best potential will be discussed including *Parvoviridae*, *Poxviridae*, *Baculoviridae* and *Nudivirdae*.

3.1 Subfamily Densovirinae*:* Parvoviridae

Members of the subfamily *Densovirinae* (DNV) include five genera that exclusively infect arthropods, and three of these genera, *Ambidensovirus*, *Brevidensovirus* and *Iteradensovirus,* have members that infect insects in the orders Blattodea, Diptera, Hemiptera, Homoptera, Hymenoptera, Lepidoptera and Orthoptera (Tijssen et al., 2018). These viruses are small, non-enveloped, icosahedral virions with a single linear 4–6 kb ssDNA genome. Densoviruses were first discovered in laboratory colonies of the wax moth, *Galleria mellonella*, and in insects reared commercially such as wax moth and silk worm (Bergoin and Tijssen, 1998). Densoviruses typically infect insect larvae and produce a range of symptoms that include alterations in cuticular pigmentation, progressive paralysis and death. Some densoviruses appear to infect a single host species while others have broad experimental host ranges, and their virulence varies considerably depending on the virus species.

Some of the densoviruses infecting economically important pests are quite virulent and host-specific, and have been investigated as biocontrol agents (Bergoin and Tijssen, 1998 and references therein). Early studies showed that introductions of Galleria mellonella DNV into beehives controlled heavy infestations of wax moth. The potential of several densoviruses for the control of the oil palm insect pests, *Sibine fusca* and *Casphalia extranea*, was tested in spray trials in South America, Africa and Egypt using homogenates of infected insects (Bergoin and Tijssen, 1998). The Aedes aegypti DNV was investigated for mosquito control in the Soviet Union, and a commercial product called Viroden was developed. More recently, a densovirus isolate (PfDNV) from the smoky brown cockroach, *Periplaneta fulginosa*, has been investigated as a viral insecticide for cockroach control (Hu et al., 1994; Sun, 2015). A product has been registered in China based on PfDNV particles formulated as a paste along with the cockroach pheromone bait (Sun, 2015).

3.2 Subfamily Entomopoxvirinae*:* Poxviridae

Entomopoxviruses have a typical poxvirus morphology, being large brick-shaped or oval-enveloped virions containing a large (229–307 kilobase pair), linear double-stranded DNA genome. These large complex viruses replicate in the cytoplasm of infected cells and mature virions are typically occluded within proteinaceous occlusion bodies called spheroids. There are three recognized genera in the *Entomopoxvirinae*, distinguished by virion morphology and size, as well as by host range. Entomopoxviruses in the genus *Alphaentomopoxvirus* have been isolated exclusively from Coleoptera, *Betaentomopoxvirus* from Lepidoptera and Orthoptera, and *Gammaentomopoxvirus* from Diptera (Perera et al., 2010; Skinner et al., 2012).

Entomopoxviruses are infectious orally upon ingestion of spheroids that dissolve in the alkaline conditions of the host midgut aided by the presence of spheroid-associated alkaline proteases (Perera et al., 2010). The released occluded virions enter midgut cells, possibly via membrane fusion, and primary replication occurs with production of non-occluded virions, and limited production of the occluded virions spheroids occurs in the gut (King et al., 1997). Secondary infection of hemocytes appears to be an important conduit for virus spread to the major site of replication which is the fat body in most hosts. Late in infection the fat body often becomes hypertrophied by virtue of the massive accumulation of spheroids in infected cells. Often, infected insects lack their normal dark pigmentation and the abdomen of heavily infected hosts becomes quite distended. The time frame of infectivity, symptomology and mortality is typically protracted compared to other groups of insect pathogens but this can be quite variable depending on the entomopoxvirus genus and specific host interaction involved. The betaentomopoxvirues infecting lepidopteran hosts are the most efficacious in terms of time to death, taking 3–5 weeks to kill while betaentomopoxviruses infecting orthopteran hosts kill more slowly (reviewed in Perera et al., 2010). It has been noted that for alphaentomopoxvirues infecting coleopteran hosts the infection can be quite protracted taking up to 37 weeks for mortality to occur (reviewed in Perera et al., 2010). One of the interesting features that has been noted in betaentomopoxvirus infections is the substantial delay in larval/nymphal development as, for example, in Melanoplus sanguinipes entomopoxvirus (MSEV) infection of grasshoppers (Olfert and Erlandson, 1991; Woods et al., 1992) and Choristoneura fumiferana entomopoxvirus infection of spruce budworm larvae (Palli et al., 2000). In the later study it was determined that the entomopoxvirus infection impact on delaying or preventing metamorphosis was due to the modulation of the insect host's juvenile hormone and ecdysteroid titres.

A number of betaentomopoxvirues infecting lepidopteran larvae as well as the alphaentomopoxviruses contain a gene encoding the protein fusolin which often forms 'spindle bodies' in infected cells or is occluded along with virions into the occlusion bodies, spheroids (King et al., 1997). Entomopoxvirus-spindle bodies have been shown to increase the infectivity of baculoviruses when co-fed to lepidopteran hosts. The infection-enhancing impact of the spindle bodies has been linked to perturbations of the insect's peritrophic membrane (PM) which allows increased access of occluded virions to the midgut epithelial cells (Mitsuhashi et al., 2007; Takemoto et al., 2008). The N-terminal amino acid sequence of fusolin contains a globular domain that resembles the lytic polysaccharide monoxygenase domain of the AA10 proteins that target chitin (Chiu et al., 2015). These proteins have chitin-binding activity that is thought to, in part, explain the fusolin's ability to disrupt PM structure and function, and it has been suggested that these proteins could be useful in enhancing the activity of baculoviruses as biopesticides (Hukuhara et al., 1999; Takemoto et al., 2008).

Entomopoxviruses have not been investigated extensively for their potential use as biological control agents due to several general features of their infectivity and pathogenesis in their insect hosts. First, as stated above, entomopoxviruses tend to produce somewhat chronic infections that can take significant periods of time to kill their hosts. In addition, the morphological and biochemical similarities of entomopoxviruses to poxviruses infecting mammals have raised potential safety concerns thus limiting their appeal as possible virus biopesticides due to perceived regulatory hurdles. However, although there are still a limited number of complete genome sequence for entomopoxviruses, the gene-content results suggest that entomopoxviruses have a set of co-linear core genes that distinguish them from vertebrate poxviruses and suggest that they are fundamentally distinct virus groups (Perera et al., 2010).

A few entomopoxviruses have been investigated for their potential use as biological control agents against orthopteran insects from which other insect virus groups, including baculoviruses, have not been described and therefore this virus group may yet have some utility. Melanoplus sanguinipes entomopoxvirus (MSEV) and Oedaleius senegalensis entomopoxvirus are of interest because they infect many of the major grasshopper and locust pest species (Erlandson and Streett, 1997; Streett et al., 1997). MSEV, in particular, has a broad host range among grasshoppers and locusts. This virus is infectious upon ingestion; it initially infects midgut epithelium cells and eventually produces a systemic infection in the host with the major site of virus replication being the fat body. MSEV infections develop slowly in grasshoppers with mortality occurring only after three or more weeks, though other symptoms such as inhibition of pigment production, developmental delay and reduced food consumption occur more quickly (Olfert and Erlandson, 1991; Woods et al., 1992). Preliminary field trials in which 1.0×10^{10} MSEV spheroids per ha, formulated in starch granules or on wheat bran, were applied to rangeland plots produced 25–30% infection levels among grasshoppers within 2 weeks (Streett and Woods, 1990; McGuire et al., 1991). Streett et al. (1997) have reviewed studies on the economic feasibility of producing MSEV in grasshoppers and concluded that production of this virus is too costly to be practical. As well, for grasshopper and locust pests, entomopoxviruses are not likely virulent enough to be considered for use as microbial insecticides for crop protection but may still have the potential in an inoculative strategy to suppress grasshopper populations over a number of seasons in rangeland habitats where plant damage is not as critical as it is for cereal crops. Significantly, EPA Tier I safety tests were completed on MSEV and no evidence of infectivity or toxicity was observed in the vertebrate animals tested (Streett et al., 1997).

Although not in the ICTV current listing of recognized entomopoxviruses, a virus isolate from the German cockroach, *Blattella germanica*, was described as an entomopoxvirus and its possible utility as a biological control agent was outlined (Radek and Fabel, 2000). However, like many entomopoxviruses, its virulence is likely too low to be realistically considered as a potential biopesticide for cockroach control.

3.3 Nudiviridae

Nudiviruses are a diverse group of rod-shaped viruses with large circular double-stranded DNA genomes (~97–232 Kbp) that had previously been considered to be non-occluded baculoviruses (Jehle, 2010). However, nudiviruses have been reclassified into a new family, *Nudiviridae*, that contains two recognized genera, *Alphanudivirus* and *Betanudivirus*, that contain two and one recognized species, respectively (Jehle et al., 2013). Genomic and phylogenetic studies have confirmed that this family of viruses is related to the baculoviruses and the two virus families share a set of 20 core genes (Jehle, 2010). *Alphanudivirus* has two species, *Oryctes rhinoceros nudivirus* (OrNV) and *Gryllus bimaculatus nudivirus* (GbNV), while *Betanudivirus* includes a single species, *Heliothis zea nudivirus*. There are also several unclassified nudiviruses from dipteran insects and a marine crustacean (Jehle et al., 2013).

The OrNV was first discovered in the 1960s during a search for biological control agents for adult rhinoceros beetle, *Oryctes rhinoceros*, which was causing severe damage to coconut palm plantations on South Pacific islands where the insect had been accidentally introduced, possibly from Sri Lanka (Marshall et al., 2017). The OrNV, originally described from rhinoceros beetle larvae in Malaysia, is infectious by ingestion and produces a systemic infection with significant mortality occurring within 1–4 weeks (reviewed in Huger, 2005). The discovery and subsequent utilization of this virus in an IPM system is credited with saving the palm industry in the South Pacific. The Oryctes virus is a key example of a virus used successfully in an inoculative strategy. The larvae do not cause significant economic damage to palms as they typically feed on decaying vegetation in palm plantations. Initial control attempts with OrNV targeted the larval stage by applying homogenized infected larvae mixed with sawdust to leaf litter in palm plantations. This strategy significantly reduced the build-up of adult beetle populations and feeding damage on palm shoots. In addition, infected adult beetles disseminated OrNV to neighbouring plantations. The early stage of infection in adult beetles is restricted to the midgut epithelium, thus, large quantities of OrNV are voided in the faecal material, resulting in very efficient horizontal transmission to other adult beetles as well as to larvae. Virus infection of adults also leads to significantly reduced fecundity. Thus, it became a common practice to release infected beetles into infested palm plantations as a means of introducing the virus into the system. Initial trials conducted in Samoa showed reductions in palm damage of up to 95% in treatment versus control plots (Huger, 2005). Subsequently, introductions of virus-infected adults were made on other South Pacific islands leading to significantly lower economic damage to palms. This utilization strategy for OrNV led to the long-term control of this important pest with only occasional re-introductions of infected adults being required. This is an example of a very successful inoculative release strategy. The success is largely due to the specific knowledge of the virus-pest interaction. The adult beetle produces large quantities of OrNV inoculum in faecal material leading to infection of other beetles and the gregarious nature of the pest also contributes to the efficacy by which the virus spreads naturally.

Recently, the changes in agriculture practices have led to resurgence in the rhinoceros beetle as an economic pest in Malaysia. This has spurred renewed interest in OrNV as a biological control agent and led to additional studies on the ecology and genetics of the virus (Ramle et al., 2005). OrNV strains of differing virulence have been identified in natural populations, thus, there are opportunities for selecting the most efficacious strains for particular beetle populations. Recently new *O. rhinoceros* invasions have occurred on Guam and other South Pacific Islands (Marshall et al., 2017). These new invasion populations have proven difficult to control using standard OrNV isolates and IPM strategies using OrNV inoculations. A thorough study of the population genetics of *O. rhinoceros* populations throughout the South Pacific region based on CO1 barcode sequencing has shown the occurrence of four separate clades of haplotypes. One of these clades includes two subtypes of a distinct CO1 haplotype and these populations show increased resistance to infection with standard OrNV isolates (Marshall et al., 2017). Thus the ongoing success of the use of OrNV for *O. rhinoceros* control will require more basic knowledge of population genetics of both OrNV and its host and a full understanding of the environmental factors that determine the virus prevalence and dispersal in the pest population.

3.4 Baculoviridae

Baculoviridae is the most studied group of insect-specific viruses due to their long-recognized potential as biological control agents (LaPointe et al., 2012) and their use as eukaryotic expression vector systems (van Oers, 2011). Baculoviruses are characterized by rod-shaped enveloped virions and a circular, dsDNA genome of 80–180 kbp. As with Cypoviruses and Entomopoxviruses, in the late stages of infection virions are occluded within a proteinaceous OB of 0.5–5 μm diameter which is easily detectable using phase contrast light microscopy. Baculoviruses are infectious orally and the occlusion-derived virions (ODV) infect midgut epithelial cells after ingestion and dissolution of the OB in the alkaline condition of the midgut. Typically baculovirus replication induced by ODV infection of midgut cells leads to the production of budded virions (BV) that egress from midgut cells and spread infection to tissues throughout the host. Thus, typically two virion phenotypes occur in baculovirus infections: ODV that are responsible for horizontal transmission between insect hosts and BV that are responsible for systemic spread of infection within a host.

The family *Baculoviridae* is organized into four genera, *Alphabaculovirus* (formerly known as *Nucleopolyhedrovirus* (NPV) that infect lepidopteran hosts), *Betabaculovirus* (formerly known as *Granulovirus* (GV)), *Gammabaculovirus* (hymenopteran specific NPV) and *Deltabaculovirus* (dipteran specific NPV) (Herniou et al., 2012; Harrison et al., 2018). To date, baculoviruses have been isolated from over 700 species of insects.

3.4.1 Attributes of baculovirus: host interactions

Baculoviruses are often associated with spectacular epizootics particularly among the lepidopteran and sawfly forest pests (Cory, 2003, 2010; Erlandson, 2008). At later stages of baculovirus infection, insects may climb to the top of host plants, 'tree-top disease', where they die and are thus quite obvious. Recently, there have been a number of molecular studies investigating the manipulation of insect behaviour in baculovirus-infected larvae and the role of specific virus gene products in this process (van Houte et al., 2014; Zhang et al., 2017). Infected cadavers serve as a source of inoculum as they degrade and release OB onto the plant canopy or into the soil. Although the virions within OB are quite susceptible to UV inactivation, when OB are protected within intact cadavers or in the upper layers of the soil, the virus can remain viable for extended periods. In this way, the OB can serve as a reservoir for infection of subsequent host generations, potentially for several years as the virus is returned to the plant canopy. This pattern of virus cycling is particularly evident for insect populations in more stable ecosystems such as forests. In addition to horizontal transmission of baculoviruses there has been increasing evidence for the role of covert infections and vertical (tansovum and transovarial) infection in the cycling of baculoviruses within populations (Cory, 2003; Williams et al., 2017).

The very obvious observable nature of baculovirus epizootics spurred research into their potential as biological control agents. Baculoviruses have a number of potentially advantageous attributes that are attractive for a successful microbial insecticide, particularly in the current context of increased interest in more environmentally sustainable pest control in an IPM system. Baculoviruses are typically quite virulent compared to other insect viruses and kill their hosts in 5–14 days depending on the specific virus-host interaction, virus dose, stage of insect development, food plant and abiotic factors notably temperature. Baculoviruses often replicate to high levels with 1.0×10^8–1.0×10^9 alphabaculovirus OBs produced in a single host and even higher OB yields, 5.0×10^9–1.0×10^{11}, for betabaculovirus (Grzywacz and Moore, 2017). This results in large quantities of inoculum for secondary infection beyond the initial virus application but also has important implications for commercial production of baculovirus-based insecticides. Baculoviruses are typically quite host-specific, infecting a few closely related species. Narrow host specificity limits the risk of negative non-target effects and makes baculoviruses attractive for use in IPM systems. However, there are exceptions, for example, Autographa californica MNPV (AcMNPV) and Mamestra brassicae MNPV (MbMNPV) infect upwards of 100 species in several lepidopteran families (Cory, 2003). The fact that baculoviruses are occluded in OB leads to increased environmental stability and enables baculovirus-based insecticides to be applied using existing spray application technology. Most current strategies for production of baculovirus-based insecticides rely on in vivo production in insects, allowing for relatively low-technology production systems which can be advantageous in the developing world (reviewed in Grzywacz and Moore, 2017). It is also generally considered that the cost of registration of baculoviruses, as low-risk pest control alternatives, may be far less than that for registration of chemical insecticides (LaPointe et al., 2012).

However, some of the perceived advantages can also be a liability. For example, the host specificity of a baculovirus may limit its market size to one or a few pest species, and yet significant development and production costs are incurred. In contrast, costs are spread over a broader spectrum of target species in the case of a chemical insecticide. Although relatively virulent, the baculoviruses need to be ingested by and replicate in the host before feeding stops and mortality occurs. Thus, the lag time between application and control is significantly longer than for contact chemical insecticides whose knock-down effect occurs very rapidly. Education of agricultural producers and/or pest control companies on strategies for the use of baculovirus insecticides will be essential for successful adoption of this technology. There is also a perceived high cost of production for baculoviruses whether done in vivo in cultured insects, largely due to labour costs, or in insect cell culture systems which can be capital intensive and operationally expensive due to media components required for their maintenance. However, one of the major limitations of baculoviruses is their limited field stability in aqueous spray suspensions due to UV inactivation. The half-life of a spray application of unformulated baculovirus may be as little as half-a-day under certain field conditions and novel approaches to virus formulation are required to overcome this drawback.

3.4.2 Baculovirus genomics and molecular biology

Major advances in nucleic acid sequencing technology have led to a rapid increase in the available number of whole genome sequences for baculovirus isolates. Currently, 84 baculovirus genomes are available on the NCBI virus genome site (https://www.ncbi.nlm.nih.gov/genome/viruses/) including 54 alphabaculoviruses, 26 betabaculoviruses, 3 gamabaculoviruses and 1 deltabaculovirus. These whole genome sequences have allowed for the identification of 38 core genes common to all baculoviruses and defining essential genes required for baculovirus replication in insects, including, genes required for oral infection, DNA replication, host cell cycle arrest, late gene transcription and virus particle assembly (reviewed in Rohrmann, 2019). These datasets have also allowed for detailed genetic comparisons between the closely related baculovirus species as well as among the different strains of the same baculovirus species. Although the definition of and the boundaries of virus 'species' is somewhat nebulous, the detailed genomic information available for baculoviruses has allowed for a better understanding and definition of species boundaries (Wenmann et al., 2018). It has been long recognized that there is a high degree of genetic variation among isolates of any particular baculovirus species found in the same host population or in geographic populations of the host (reviewed in Harrison and Hoover, 2012). Indeed, a number of studies have confirmed that even within a single infected larvae mixed genotype populations of baculoviruses occur (Cory et al., 2005; Simón et al., 2005) and that infection with mixed genotype stocks can have increased levels of infectivity and pathogenicity compared to single genotype stocks. Thus the existence and maintenance of genetic variation within isolates of baculovirus species can serve as a selective advantage for viruses infecting host populations with genetic variation including resistance to baculovirus infection or even for a baculovirus' ability to infect multiple host species (Harrison and Hoover, 2012). The existence of genetic diversity and molecular tools to characterize it in detail have important implications in the selection of baculovirus isolate(s) for development as viral insecticides, for characterization of baculovirus stocks for commercial products and indeed for the registration process (LaPointe et al., 2012).

Historically, much of our detailed knowledge of baculovirus molecular biology stems from detailed research on AcMNPV leading to the development of baculoviruses as eukaryotic expression vectors, vaccine production systems and as mammalian gene delivery platforms (reviewed in Rohrmann, 2019). Aside from defining the essential genes required for baculovirus replication in terms of DNA replication, host cell cycle arrest, virus particle assembly and gene transcription regulation, this body of knowledge identified baculovirus auxiliary genes and host range determinants that relate to infection and pathogenicity (reviewed in Rohrmann, 2019). As well, significant research effort, based on these advances in baculovirus molecular biology, has gone into the genetic manipulation of baculoviruses for enhanced efficacy or speed of kill. Two lines of investigation have been pursued, insertion and expression of exogenous genes, that impact negatively on normal insect function and manipulation or deletion of baculovirus genes that affect infectivity or virulence (reviewed in Inceoglu et al., 2006; Harrison and Hoover, 2012). Exogenous genes that have been expressed in baculoviruses include insect-specific neurotoxin genes from invertebrate sources, insect genes that encode peptide hormones and enzymes that affect hormones or hormonally regulated processes, and proteases. Recombinant baculoviruses expressing insect-specific neurotoxins from scorpion (e.g. AaIT) and mite sources using baculovirus late and very late gene promoters (*p6.9, p10* and *polyhedrin*) have produced good results, decreasing the time to kill host insects by 10–60% thus killing hosts in 48–72 h post-infection (Inceoglu et al., 2006). The expression of peptide hormones or enzymes-regulating hormone levels, for example, juvenile hormone esterase, has met with less success, possibly because of the complex nature of insect endocrine regulation. Significant increases in the speed of kill, up to 50%, have been observed for a recombinant AcMNPV virus expressing the flesh fly protease, cathepsin L, an enzyme known to digest proteins associated with the basal lamina of the insect midgut (Harrison and Bonning, 2001). It is postulated that protease degradation of the midgut basal lamina allows for more rapid spread of virus infection to the rest of the insect.

Manipulation of baculovirus genes for increased efficacy has for the most part focussed on deletion strategies, particularly deletion of the gene for *ecdysteroid UDP-glucosyl transferase* (*egt*). The *egt* gene encodes an enzyme that inactivates the insect hormone ecdysone which controls the timing of the moult. This gene is thought to confer an advantage on baculoviruses because delaying the moult, an energy- and resource-intensive process, leads to prolonged larval development and higher yields of virus. The *egt*-deletion viruses kill host insects up to 30% more quickly than the wild-type virus (O'Reily and Miller, 1991).

Among other baculovirus genes that increase efficacy is *enahncin*, a gene found only in GVs and a few NPVs. The *enhancin* gene encodes a metalloprotease that degrades the peritrophic matrix and its expression in baculoviruses which do not naturally contain the gene has been demonstrated to lower the dose required for infection of host larvae (reviewed in Erlandson et al., 2019).

A number of recombinant baculoviruses have been tested in small-scale field trials to evaluate efficacy, persistence, non-target insect impacts and safety. The most commonly used approach is a combination of the *egt*-deletion strategy and scorpion-toxin gene expression. AcMNPV and HaSNPV recombinants of this type have been field-tested in different crop systems and have demonstrated significant increases in efficacy over wild-type viruses, in some cases giving equivalent crop protection as standard chemical insecticides (Popham et al., 1997; Sun et al., 2009).

To date, no recombinant baculoviruses have been registered or commercialized despite their promise as more efficacious control agents. This is in large part due to the perceived added cost of any attempt to register a recombinant baculovirus, potential difficulties in production strategies and the general public perception of GM technology (Grzywacz, 2017).

3.4.3 Baculovirus biopesticide development

Currently, a substantial number of baculoviruses are registered as insect control products worldwide (Table 2). There are recent excellent reviews of the various facets of the development of commercial baculovirus active ingredients: including, laboratory and small-scale field assessments, production methods, formulation, registration issues and assessments of producer uptake (Harrison and Hoover, 2012; LaPointe et al., 2012; Lacey et al., 2015; Grzywacz, 2017; Grzywacz and Moore, 2017). Thus only a brief description of major success is provided here.

Table 2 Historic and currently registered/commercialized virus-based bioinsecticides

Virus	Jurisdiction	Target pest	Production/crop system
Baculoviruses			
Alphabacuvirus			
Lymantria dispar MNPV	NA	Gypsy moth (*L. dispar*)	Forestry
Orgyia pseudotsugata MNPV	NA	Douglas-fir tussock moth (*O. pseudotsugata)*	Forestry
Neodiprion sertifer NPV	NA	European spruce sawfly	Forestry
Neodiprion leconti NPV	NA	Redheaded pine sawfly	Forestry
Autographa californica MNPV	CH, LA, NA	Multiple pest targets	Horticulture, glasshouse and field crops
Anagrapha falcifera MNPV		Multiple pest targets	Horticulture, glasshouse and field crops
Anticarsia gemmetalis MNPV	LA	Velvet bean caterpillar Sugar cane borer	Soybean
Chrysodeixis includens NPV	LA	Soybean looper (*Chrysodeixis includens*)	Soybean, cotton
Ectropis oblique NPV	CH	Tea geometrid	Tea
Euproctis pseudoconspera NPV	CH	Tea caterpillar	Tea
Helicoverpa armigera NPV	CH, EU, ID, LA, SA	*Heliothis/Helicoverpa* complex	Cotton, tobacco, horticulture, field crops
Helicoverpa zea MNPV	LA	*Heliothis/Helicoverpa* complex	Cotton, horticulture
Mamestra brassicae MNPV	CH	Cabbage moth (*M. brassicae*)	Horticulture
Mamestra configurata MNPV-A		Bertha armyworm (*M. configurata*)	
Spodoptera albula NPV + AcMNPV	LA	Multiple pest targets	Horticulture
Spodoptera exigua MNPV	CH, EU, LA	Beet armyworm (*S. exigua*)	Horticulture, glasshouse and field crops
Spodoptera frugiperda MNPV	LA	Fall armyworm (*S. frugipera*)	
Spodoptera litura NPV	CH, ID	Cotton leaf worm (*S. littoralis*)	Horticulture

(Continued)

Table 2 (*Continued*)

Virus	Jurisdiction	Target pest	Production/crop system
Spodoptera littoralis NPV	EU	Fall armyworm (*S. frugipera*)	Horticulture
		Cotton leaf worm (*S. littoralis*)	Field crops
Spodoptera sunia NPV	LA	*Spodoptera* spp.	Horticulture
Betabaculovirus			
Adoxphyes orana GV	EU	Summer fruit tortrix Moth	Orchard
Adoxphyes honmai GV	JP	Smaller tea torttix (*Adoxphyes honmai*)	Tea
Cydia pomonella GV	EU, LA, NA, SA	Codling moth	Orchard
Cyrptophlebia leucotreta GV	SA	False codling moth	Orchard
Erinnyis ello GV	LA	*Erinnyis ello*	Cassava/rubber
Homona magnanima GV	JP	Oriental tea tortrix (*Homona magnanima*)	Tea
Pieris rapae GV	CH	Cabbage white butterfly	Horticulture
Plutella xylostella GV	CH	Diamondback moth	Horticulture
Phtorimaea operculella GV	LA	*Phtorimaea operculella*, *Tecia solanivora*	Potato
Densoviruses			
Pefudensovirus			
Periplaneta fuliginosa DNV	CH	Cockroach	Sanitation
Reovirus			
Cypovirus			
Dendrolimus spectabilis CPV	CH	Pine caterpillar	Forestry
Dendrolimus punctatus CPV	CH	Pine caterpillar	Forestry

China (CH), Europe (E), India (ID), Japan (JP), Latin America (LA), North America (NA), South Africa (SA).

Historically, many of the baculoviruses registered for control of lepidopteran and sawfly forest pests were developed and registered via governmental agencies, for example, Gypcheck and Neocheck-S were developed and produced by the US Forest Services and Abbietiv by Forestry Canada (Hajek and van Frankenhuyzen, 2017). These viruses have typically been used to treat pest populations before they reach their maximum outbreak phase, in essence creating an epizootic before it would naturally occur in a delayed density-dependent manner. This use pattern is an example of an inoculative release strategy. As many of these forest pest populations are cyclical in nature, the production and stockpiling of baculovirus-based insecticides has not been taken up by commercial ventures but rather by government agencies to be used when required.

A more extensive number of baculoviruses are used as microbial insecticides or inundative control agents in relatively large-scale applications. Gryzwacz has outlined a few factors that are predictive of successful baculovirus insect control products (Grzywacz, 2017). These include selection of highly pathogenic strains, cost point of mass production, ease of application to insect feeding sites in crop targets, crop and pest characteristics that foster high horizontal virus transmission post application and utilization in IPM systems where chemical insecticides are not acceptable. One additional factor could be added, that being a crop target that is of high-value and where a chemical insecticide-free product has an incremental value.

The Cydia pomonella granulovirus (CpGV) for control the codling moth in apple orchards has been successfully marketed for more than 20 years primarily in Europe and North America (Lacey et al., 2008). Although marketed under a number of product names, the majority of products have been based on the Mexican GV isolate (CpGV-M). The CpGV was originally touted as a possible solution to secondary pest outbreaks in orchards after extensive use of broad-spectrum chemical pesticides for codling moth control. However, CpGV products have become an attractive control alternative for producers to combat insecticide-resistant *C. pomonella* populations, where there is a pressure to reduce chemical pesticide applications in the orchard industry and specifically in organic orchards. Despite the fact that the codling moth larvae spend only a limited amount of time grazing on the surface of fruit before burrowing into the fruit, thus limiting the opportunity for virus dose uptake, and the occurrence of multiple generations of the pest per season, CpGV products have produced good control results based on multiple applications in weekly or bi-weekly spray programmes (Lacey et al., 2008). One of the key factors in the success of CpGV has been its implementation in IPM systems that include cultural controls, sterile male release programmes, crop residue clean-up, predator and parasitoid support or pheromone mating disruption (Lacey et al., 2008). The success of CpGV for *C. pomonella* control has also spurred the development of other GVs for control of other codling moth species, for example, the false codling moth in Africa (Hatting et al., 2018). However, the implementation of CpGV product utilization has not been without issues and in recent years at least three different genetic modes of resistance to CpGV infection have been discovered in *C. pomonella* populations (Sauer et al., 2017). This has necessitated the search for additional CpGV isolates that overcome these host-resistance mechanisms and the development as new CpGV products (Schmitt et al., 2013).

There are also a few examples of baculovirus products being used to control lepidopteran pests of field crops. These include: Helicoverpa zea NPV and Helicoverpa armigera SNPV (HaSNPV) for control of members of the *Helicoverpa/Heliothis* pest complex on cotton; Spodoptera exigua MNPV for control of beet armyworm, *S. exigua,* in greenhouse ornamental and vegetable crops, as well as field crops; and Anticarsia gemmatalis MNPV (AgMNPV) for control of velvetbean caterpillar, *Anticarsia gemmatalis*, in soybean. Although not used extensively Autographa californica MNPV and Anigrapha falcifera NPV have great potential due to their putative broader host range. Both HaSNPV and AgMNPV have been used extensively. Each year, HaSNPV is applied to more than 100 000 ha of cotton in China (Sun, 2015) and it has been used extensively in Australia (Buerger et al., 2007).

The development of AgMNPV was one of the most successful examples of a baculovirus-based insecticide, being applied annually to more than 2 million ha of soybean in Brazil (Moscardi, 2007). This virus was produced from infected field-collected and/or laboratory-reared caterpillars. A number of factors have been cited for the success of AgMNPV as a viral insecticide – it is relatively virulent and gives good control when applied at 1.5×10^{11} OB/ha, a rate 5–10 times lower than that for some other baculovirus products and formulated AgMNPV cost producers less than chemical insecticides. It is efficiently transmitted within the pest population following application. Thus one AgMNPV application gives control equivalent to that from 1.8 applications of conventional chemical insecticides. In addition, soybean can withstand substantial defoliation without yield reductions and an IPM programme and education strategy for producers was developed during its introduction in Brazilian agriculture. However, in recent years the use of AgMNPV has declined in Brazil owing to changes in agronomic practices in soybean production including no-till GM cropping using prescribed herbicides and broad-spectrum chemical insecticides (Grzywacz, 2017). In addition, soybean production in South America now faces a more complex lepidopteran pest situation with the emergence of looper species as a significant problem (Muraro et al., 2019).

3.4.4 Baculovirus production and formulation

Grzywacz and Moore (2017) provide a thorough review of baculovirus production systems and formulation approaches so only a brief overview will be provided here. Currently, virtually all commercial baculovirus products are produced with in vivo systems using host insect larvae reared on artificial diet in laboratory-type production facilities.

Despite the success of baculoviruses as a bioinsecticides, the delay between virus application and pest mortality and the rapid inactivation of virus by UV radiation are still considered major drawbacks. The latter problem is being addressed for baculoviruses in general by formulation research. Previous formulation efforts were based on approaches typically used with chemical pesticides in terms of wetting, sticker and emulsifying agents, as well as UV-blocking and absorption agents. Recently, more novel UV-protection agents have been investigated, including stilbene-derived optical brighteners. One unexpected outcome of this research was the demonstration of increased infectivity of baculoviruses in the presence of optical brighteners, possibly due to their interaction with the peritrophic matrix, a protective barrier that lines the insect midgut. This has led to increased research efforts to identify other synergistic compounds that increase efficacy when included in a formulation cocktail for baculovirus insecticides.

3.4.5 Development of host resistance to baculoviruses

Baculoviruses were initially thought to have a low potential for eliciting the development of resistance within the targeted host population due to their complex mode of action. However, there are, now, a number of recognized examples of resistance development both in laboratory-designed experiments to select for resistance development (Abot et al., 1996; Milks and Myers, 2003) and in field application situations, for example, codling moth CpGV resistance (Asser-Kaiser et al., 2010; Gebhardt et al., 2014; Sauer et al., 2017). This is an important issue that should be considered in the development of any baculovirus bioinsecticide and should be addressed in upstream studies. A useful example of attempts to obtain prior knowledge of the potential for resistance development is provided by the Chrysodeixis includens NPV - *Chyrsodeixis includens* study (Muraro et al., 2019). These authors conducted baseline susceptibility bioassay studies on geographic populations of *C. includens* in Brazil to determine the variability in susceptibility of different populations. These type of studies along with detailed population genetic studies (e.g. Palma et al., 2015) can give some fore knowledge of the potential for resistance development and allow for selection of alternate virus isolates with different infectivity and pathogenicity patterns.

4 Conclusion

Insect-specific viruses have good potential for the development as biological control agents of insect pests in the context of sustainable IPM systems. However, several drawbacks including the lag time between virus application and insect control, virus production technology and limited field stability will need to be addressed by further research and development. To date, baculoviruses have been the most successfully exploited insect viruses in terms of the number of species registered and commercially developed for use in specific markets. There is a potential for other virus groups including cypoviruses, tetraviruses and entomopoxviruses to be developed as more detailed information on their genetics and biology is forthcoming. For all these virus groups the possibility of genetic engineering holds promise for increased efficacy. However, it is quite likely that innovative approaches that diverge from the chemical insecticide model may be needed to fully exploit the potential of viruses as insect biological control agents.

5 Where to look for further information

There are a number of recent book chapters and special issues of peer reviewed journals that provide in-depth reviews of both the basic and applied science of insect viruses and their current and potential use as biologically based insect control agents.

Special Peer Review Journal Issues:

1 'Global status of microbial control programs and practices'. *Journal of Invertebrate Pathology* 165, 1–88. 2019. (Eds. S. P. Arthurs and S. K. Dara). This special issue contains 10 articles which provide an overview of microbial control programs, practices and products, including entomopathogenic viruses, from different regions of the world.
2 *Insect Viruses and Pest Management*, Viruses Special Issue 2019 (ISSN 1999-4915) (Guest Editor, M. López-Ferber). This issue presents 20 open access articles on the most current information on the use of insect viruses in pest control and emerging approaches to virus implementation for insect pest management strategies. (https://www.mdpi.com/journal/viruses/special_issues/insect_pest).

Reference Texts:

1 *Microbial Control of Insect and Mite Pests, From Theory to Practice*. 2017. (Ed. L. A. Lacey). Academic Press, London, UK. This book is a very comprehensive review of microbial control agents and

strategies. It includes two chapters on basic and applied aspects of the development, production and application of baculoviruses as insect control agents.

2 *Insect Virology*. 2010. (Eds. S. Asgari and K. N. Johnson). Caister Academic Press, Norfolk, UK. This book contains fundemental information about many of the insect virus groups discussed in this review chapter.
3 *Insect Molecular Virology: Advances and Emerging Trends*. 2019. (Ed. B. C. Bonning). Caister Academic Press, Norfolk, UK. This book includes numerous chapters with cutting edge information on the molecular biology and genomics of insect virus groups discussed in this review chapter.

6 References

Abot, A. R., Moscardi, F., Fuxa, J. R., Sosa-Gomez, D. R. and Richter, A. R. 1996. Development of resistance by *Anticarsia gemmatalis* from Brazil and the United States to a nuclear polyhedrosis virus under laboratory selection pressure. *Biol. Control* 7(1), 126–30. doi:10.1006/bcon.1996.0075.

Asser-Kaiser, S., Heckel, D. G. and Jehle, J. A. 2010. Sex linkage of CpGV resistance in a heterogeneous field strain of the codling moth *Cydia pomonella* (L.). *J. Invertebr. Pathol.* 103(1), 59–64. doi:10.1016/j.jip.2009.10.005.

Attoui, H., Mertens, P. P. C., Becnel, J., Belaganahalli, S., Bergoin, M., Brussaard, C. P., et al. 2012. Reoviridae. In: King, A. M. Q., Adams, M. J., Carstens, E. B. and Lefkowitz, E. J. (Eds), *Virus Taxonomy: Ninth Report of the International Committee on Taxonomy of Viruses*. Elsevier, Oxford, UK, pp. 541–637.

Belloncik, S. and Mori, H. 1998. Cypoviruses. In: Miller, L. K. and Ball, L. A. (Eds), *Insect Viruses*. Plenum Press, New York, NY, pp. 337–69.

Bergoin, M. and Tijssen, P. 1998. Biological and molecular properties of Densoviruses and their use in protein expression and biological control. In: Miller, L. K. and Ball, L. A. (Eds), *The Insect Viruses*. Plenum Press, New York, NY, pp. 141–69.

Buerger, P., Hauxwell, C. and Murray, D. 2007. Nucleopolyhedrovirus introductions in Australia. *Virol. Sin.* 22(2), 173–9. doi:10.1007/s12250-007-0019-y.

Chiu, E., Hijnen, M., Bunker, R. D., Boudes, M., Rajendran, C., Aizel, K., Oliéric, V., Schulze-Briese, C., Mitsuhashi, W., Young, V., Ward, V. K., Bergoin, M., Metcalf, P. and Coulibaly, F. 2015. Structural basis for the enhancement of virulence by viral spindles and their *in vivo* crystallization. *Proc. Natl. Acad. Sci. U. S. A.* 112(13), 3973–8. doi:10.1073/pnas.1418798112.

Christian, P. D., Murray, D., Powell, R., Hopkinson, J., Gibb, N. N. and Hanzlik, T. N. 2005. Effective control of a field population of *Helicoverpa armigera* by using the small RNA virus *Helicoverpa armigera* stunt virus (Tetraviridae: Omegatetravirus). *J. Econ. Entomol.* 98(6), 1839–47, doi:10.1093/jee/98.6.1839.

Cory, J. S. 2003. Ecological impacts of virus insecticides: host range and non-target organisms. In: Hokkanen, H. M. T. and Hajek, A. E. (Eds), *Environmental Impacts of Microbial Insecticides: Need and Methods for Risk Assessment*. Kluwer Academic Publishers, Dordrecht, the Netherlands, pp. 73–91.

Cory, J. S. 2010. The ecology of baculoviruses. In: Asgari, S. and Johnson, K. N. (Eds), *Insect Virology*. Caister Academic Press, Norfolk, UK, pp. 412-27. ISBN: 978-1-904455-71-4.

Cory, J. S., Green, B. M., Paul, R. K. and Hunter-Fujita, F. 2005. Genotypic and phenotypic diversity of a baculvoirus population within an individual insect host. *J. Invertebr. Pathol.* 89(2), 101–11. doi:10.1016/j.jip.2005.03.008.

Dorrington, R. A. and Short, J. R. 2010. Tetraviruses. In: Asgari, S. and Johnson, K. N. (Eds), *Insect Virology*. Caister Academic Press, pp. 285–306. ISBN: 978-1-904455-71-4.

Dorrington, R. A., Gorbalenya, A. E., Gordon, K. H. J., Lauber, C. and Ward, V. K. 2012. Tetraviridae. In: King, A. M. Q., Adams, M. J., Carstens, E. B. and Lefkowitz, E. J. (Eds), *Virus Taxonomy: Ninth Report of the International Committee on Taxonomy of Viruses*. Elsevier, Oxford, pp. 1091–102.

Erlandson, M. A. 2008. Insect pest control by viruses. In: *Encyclopedia of Virology* (3rd edn., vol. 3), pp. 125–33.

Erlandson, M. A. and Streett, D. A. 1997. Entomopoxviruses associated with grasshoppers and locusts: biochemical characterization. *Mem. Entomol. Soc. Can.* 129(171), 131–46. doi:10.4039/entm129171131-1.

Erlandson, M. A., Toprak, U. and Hegedus, D. D. 2019. Role of the peritrophic matrix in insect-pathogen interactions. *J. Insect Physiol.* 117, 103894. doi:10.1016/j.jinsphys.2019.103894.

Gebhardt, M. M., Eberle, K. E., Radtke, P. and Jehle, J. A. 2014. Baculovirus resistance in codling moth is virus-isolate dependent and consequence of a mutation in viral gene p38. *Proc. Natl. Acad. Sci. U. S. A.* 111(44), 15711–6. doi:10.1073/pnas.1411089111.

Gordon, K. H. J., Williams, M. R., Baker, J. S., Gibson, J. M., Bawden, A. L., Millgate, A. G., Larkin, P. J. and Hanzlik, T. N. 2001. Replication-independent assembly of an insect virus (Tetraviridae) in plant cells. *Virology* 288(1), 36–50. doi:10.1006/viro.2001.1049.

Grzywacz, D. 2017. Basic and applied research: baculovirus. In: Lacey, L. A. (Ed.), *Microbial Control of Insect and Mite Pests: From Theory to Practice*. Academic Press, London, pp. 27–36.

Grzywacz, D. and Moore, S. D. 2017. Production, formulation, and bioassay of baculoviruses for pest control. In: Lacey, L. A. (Ed.), *Microbial Control of Insect and Mite Pests: From Theory to Practice*. Academic Press, London, pp. 109–24.

Hajek, A. E. and van Frankenhuyzen, K. 2017. Use of entomopathogens against forest pests. In: Lacey, L. A. (Ed.), *Microbial Control of Insect and Mite Pests: From Theory to Practice*. Academic Press, London, pp. 313–30.

Harrison, R. L. and Bonning, B. C. 2001. Use of proteases to improve the insecticidal activity of baculoviruses. *Biol. Control* 20(3), 199–209. doi:10.1006/bcon.2000.0899.

Harrison, R. L. and Hoover, K. 2012. Baculoviruses and other occluded insect viruses. In: *Insect Pathology*. Elsevier, pp. 73–131. Chapter 4. doi:10.1061/B978-0-12-384984-7.00004-X.

Harrison, R. L., Herniou, E. A., Jehle, J. A., Theilmann, D. A., Burand, J. P., Becnel, J. J., Krell, P. J., van Oers, M. M., Mowery, J. D., Bauchan, G. R. and ICTV Report Consortium. 2018. ICTV virus taxonomy profile: *Baculoviridae*. *J. Gen. Virol.* 99(9), 1185–6. doi:10.1099/jgv.0.001107.

Hatting, J. L., Moore, S. D. and Malan, A. P. 2018. Microbial control of phytopagaous invertebrate pests in South Africa: current Status and future prospects. *J. Invertebr. Pathol.* 165, 54–66. doi:10.1016/j.jip.2018.02.004.

Herniou, E. A., Arif, B. M., Becnel, J. J., Blissard, G. W., Bonning, B., Harrison, R., Jehle, J. A., Theilmann, D. A. and Vlak, J. M. 2012. Baculoviridae. In: King, A. M. Q., Adams, M. J., Carstens, E. B. and Lefkowitz, E. J. (Eds), *Virus Taxonomy: Ninth Report of the International Committee on Taxonomy of Viruses*. Elsevier, Oxford, pp. 163-73.

Hu, Y. Y., Zhang, J. M., Lizuka, T. and Bando, H. 1994. A densovirus newly isolated from the smoky-brown cockroach, *Periplaneta fuliginosa*. *Arch. Virol.* 138, 365-72.

Huger, A. M. 2005. The *Oryctes* virus: its detection, identification, and implementation in biological control of coconut palm rhinoceros beetle, *Oryctes rhinoceros* (Coleoptera: Scarabaeidae). *J. Invertebr. Pathol.* 89(1), 78-84. doi:10.1016/j.jip.2005.02.010.

Hukuhara, T., Hayakawa, T. and Wijonarko, A. 1999. Increased baculovirus susceptibility of armyworm larvae feeding on transgenic rice plants expressingan entomopoxvirus gene. *Nat. Biotechnol.* 17(11), 1122-4. doi:10.1038/15110.

Inceoglu, A. B., Kamita, S. G. and Hammock, B. D. 2006. Genetically modified baculoviruses: a historical overview and future outlook. *Adv. Virus Res.* 68, 323-60. doi:10.1016/S0065-3527(06)68009-3.

James, C. 2014. Brief global status of commercialized biotech/GM crops: 2013. International Service for the Acquisition of Agri-Biotech Applications (ISAAA). Available at: http://www.isaaa.org/resources/publications/briefs/46/executivesummary/(accessed on 04 May 2019).

Jehle, J. A. 2010. Nudiviruses: their biology and genetics. In: Asgari, S. and Johnson, K. N. (Eds), *Insect Virology*. Caister Academic Press, Norfolk, UK, pp. 153-70. ISBN: 978-1-904455-71-4.

Jehle, J. A., Burand, J., Herniou, E., Harrison, R., Arif, B., Theilmann, D., van Oers, M. and Becnel, J. 2013. Creation of a new family Nudiviridae including two genera and three species. *ICTV*. Available at: https://talk.ictvonline.org/taxonomy/p/taxonomy-history?taxnode_id=20133589&src=NCBI&ictv_id=20133589.

King, L. A., Wilkinson, N., Miller, D. P. and Marlow, S. A. 1997. Entomopoxviruses. In: Miller, L. K. and Ball, L. A. (Eds), *The Insect Viruses*. Plenum Press, New York, NY, pp. 1-29.

Kunimi, Y. 2007. Current status and prospects on microbial control in Japan. *J. Invertebr. Pathol.* 95(3), 181-6. doi:10.1016/j.jip.2007.03.007.

Lacey, L. A., Thomson, D., Vincent, C. and Arthurs, S. P. 2008. Codling moth granulovirus: a comprehensive review. *Biocontrol Sci. Technol.* 18(7), 639-63. doi:10.1080/09583150802267046.

Lacey, L. A., Grzywacz, D., Shapiro-Ilan, D. I., Frutos, R., Brownbridge, M. and Goettel, M. S. 2015. Insect pathogens as biological control agents: back to the future. *J. Invertebr. Pathol.* 132, 1-41. doi:10.1016/j.jip.2015.07.009.

LaPointe, R., Thumbi, D. and Lucarotti, C. J. 2012. Recent advances in our knowledge of baculovirus molecular biology and its relevance for the registration of baculovirus-based products for insect pest population control. In: Larramendy, M. L. and Soloneski, S. (Eds), *Integrated Pest Management and Pest Control - Current and Future Tactics*. IntechOpen. Chapter 21. doi:10.5772/31389.

Lefkowitz, L. J., Davison, A. J., Sabanadzovic, S., Siddell, S. J. and Simmonds, P. (Eds). 2017. International Committee on Virus Taxonomy 10th Report (Online Version). Available at: https://talk.ictvonline.org/ictv-reports/ictv_online_report/.

MarketsAndMarkets. 2018. Biopesticides market by type (bioinsecticides, biofungicides, and bionematicides), source (microbials, plant extracts, and beneficial insects), mode of application, formulation, crop application, and region -global forecast to

2023. Available at: https://www.marketsandmarkets.com/Market-Reports/biopesticides-267.html.

Marshall, S. D. G., Moore, A., Vaqalo, M., Noble, A. and Jackson, T. A. 2017. A new haplotype of the coconut rhinoceros beetle, *Oryctes rhinoceros*, has escaped biological control by *Oryctes rhinoceros* nudivirus and is invading Pacific Islands. *J. Invertebr. Pathol.* 149, 127–34. doi:10.1016/j.jip.2017.07.006.

McGuire, M. R., Streett, D. A. and Shasha, B. S. 1991 Evaluation of starch encapsulation for formulation of grasshopper (Orthoptera: Acrididae) entomopoxviruses. *J. Econ. Entomol.* 84(6), 1652–6. doi:10.1093/jee/84.6.1652.

Milks, M. L. and Myers, J. H. 2003. Cabbage looper resistance to a nucleopolyhedrovirus confers cross-resistance to two granuloviruses. *Environ. Entomol.* 32(2), 286–9. doi:10.1603/0046-225X-32.2.286.

Mitsuhashi, W., Kawakita, H., Murakami, R., Takemoto, Y., Saiki, T., Miyamoto, K. and Wada, S. 2007. Spindles of an entomopoxvirus facilitate its infection of the host insect by disrupting the peritrophic membrane. *J. Virol.* 81(8), 4235–43. doi:10.1128/JVI.02300-06.

Mori, H. and Metcalf, P. 2010. Cypoviruses. In: Asgari, S. and Johnson, K. N. (Eds), *Insect Virology*. Caister Academic Press, Norfolk, UK, pp. 307–23. ISBN: 978-1-904455-71-4.

Moscardi, F. 2007. Development and use of the nucleopolyhedrovirus of the velvetbean caterpillar in soybeans. In: Vincent, C., Goettel, M. S. and Lazarovits, G. (Eds), *Biological Control: a Global Perspective*. CAB International, Wallingford, UK, pp. 344–53.

Muraro, D. S., Giacomelli, T., Stacke, R. F., Godoy, D. N., Marcon, P., Popham, H. J. R. and Bernardi, O. 2019. Baseline susceptibility of Brazilian populations of *Chrysodeixis includens* (Lepidoptera: Noctuidae) to *C. includens* nucleopolyhedrovirus and diagnostic concentrations for resistance management. *J. Econ. Entomol.* 112(1), 349–54. doi:10.1093/jee/toy361.

Olfert, O. O. and Erlandson, M. A. 1991. Wheat foliage consumption by grasshoppers (Orthoptera: Acrididae) infected with *Melanoplus sanguinipes* entomopoxvirus. *Environ. Entomol.* 20(6), 1720–4. doi:10.1093/ee/20.6.1720.

O'Reilly, D. R. and Miller, L. K. 1991. Improvement of a baculovirus pesticide by the deletion of the egt gene. *Biotechnology* 9(11), 1086–9. doi:10.1038/nbt1191-1086.

Palli, S. R., Ladd, T. R., Tomkins, W. L., Shu, S., Ramaswamy, S. B., Tanaka, Y., Arif, B. and Retnakaran, A. 2000. Choristoneura fumiferana entomopoxvirus prevents metamorphosis and modulates juvenile hormone and ecdysteroid titers. *Insect Biochem. Mol. Biol.* 30(8–9), 869–76. doi:10.1016/s0965-1748(00)00060-6.

Palma, J., Maebe, K., Guedes, J. V. C. and Smagghe, G. 2015. Molecular variability and genetic structure of *Chrysodeixis includens* (Lepidoptera: Noctuidae), an important soybean defoliator in Brazil. *PLoS ONE* 10(3), e0121260. doi:10.1371/journal.pone.0121260.

Perera, S., Li, Z., Pavlik, L. and Arif, B. 2010. Entomopoxviruses. In: Asgari, S. and Johnson, K. N. (Eds), *Insect Virology*. Caister Academic Press, Norfolk, UK, pp. 83–102. ISBN: 978-1-904455-71-4.

Popham, H. J. R., Li, Y. and Miller, L. K. 1997. Genetic improvement of *Helicoverpa zea* nuclear polyhedrosis virus as a biopesticide. *Biol. Control* 10(2), 83–91. doi:10.1006/bcon.1997.0552.

Radek, R. and Fabel, P. 2000. A new entomopoxvirus from a cockroach: light and electron microscopy. *J. Invertebr. Pathol.* 75(1), 19–27. doi:10.1006/jipa.1999.4896.

Ramle, M., Wahid, M. B., Norman, K., Glare, T. R. and Jackson, T. A. 2005. The incidence and use of Oryctes virus for control of rhinoceros beetle in oil palm plantations in Malaysia. *J. Invertebr. Pathol.* 89(1), 85–90. doi:10.1016/j.jip.2005.02.009.

Rohrmann, G. F. 2019. *Baculovirus Molecular Biology* (4th edn.). National Center for Biotechnology Information (USA), Bethesda, MD. Available at: https://www.ncbi.nlm.nih.gov/books/NBK543458/.

Sauer, A. J., Schulze-Bopp, S., Fritsch, E., Undorf-Spahn, K. and Jehle, J. A. 2017. A third type of resistance to *Cydia pomonella* granulovirus in codling moths shows a mixed Z-linked and autosomal inheritance pattern. *Appl. Environ. Microbiol.* 83(17), e01036-17. doi:10.1128/AEM.01036-17.

Schmitt, A., Bisutti, I. L., Ladurner, E., Benuzzi, M., Sauphanor, B., Kienzle, J., Zing, D., Undorf-Spahn, K., Fritsch, E., Hüber, J. and Jehle, J. A. 2013. The occurrence and distribution of resistance of codling moth to *Cydia pomonella* granulovirus in Europe. *J. Appl. Entomol.* 137(9), 641–9. doi:10.1111/jen.12046.

Simón, O., Williams, T., López-Ferber, M. and Caballero, P. 2005. Functional importance of deletion mutants genotypes in an insect baculovirus population. *Appl. Environ. Microbiol.* 71, 5579–88.

Skinner, M. A., Buller, R. M., Damon, I. K., Lefkowitz, E. J., McFadden, G., McInnes, C. J., Mercer, A. A., Moyer, R. W. and Upton, C. 2012. Poxviridae. In: King, A. M. Q., Adams, M. J., Carstens, E. B. and Lefkowitz, E. J. (Eds), *Virus Taxonomy: Ninth Report of the International Committee on Taxonomy of Viruses*. Elsevier, Oxford, UK, pp. 291–309.

Streett, D. A. and Woods, S. A. 1990. Grasshopper pathogen field evaluation: virus, pp. 210–17. In: *Cooperative Grasshopper Integrated Pest Management Project 1990 Annual Report*. USDA/APHIS, 282p.

Streett, D. A., Woods, S. A. and Erlandson, M. A. 1997. Entomopoxviruses of grasshoppers and locusts: biology and biological control potential. *Mem. Entomol. Soc. Can.* 129(171), 115–30. doi:10.4039/entm129171115-1.

Sun, X. 2015. History and current status of development and use of viral insecticides in China. *Viruses* 7(1), 306–19. doi:10.3390/v7010306.

Sun, X., Wu, D., Sun, X., Jin, L., Ma, Y., Bonning, B. C., Peng, H. and Hu, Z. 2009. Impact of *Helicoverpa armigera* nucleopolyhedroviruses expressing a cathepsin L-like protease on target and nontarget insect species on cotton. *Biol. Control* 49(1), 77–83. doi:10.1016/j.biocontrol.2008.12.011.

Takemoto, Y., Mitshuhashi, W., Murakami, R., Konishi, H. and Miyamoto, K. 2008. The N-termial region of an entomopoxvirus fusolin is essential for the enhancement of peroral infection, whereas the C-terminal region is eliminated in digestive juice. *J. Virol.* 82(24), 12406–15. doi:10.1128/JVI.01605-08.

Tijssen, P., Agbandje-McKenna, M., Almendral, J. M., Bergoin, M., Flegel, T. W., Hedman, K., Kleinschmidt, J., Li, Y., Pintel, D. J. and Tattersall, P. 2018. Parvovirdiae. In: Lefkowitz, L. J., Davison, A. J., Sabanadzovic, S., Siddell, S. J. and Simmonds, P. (Eds), *International Committee on Virus Taxonomy 10th Report (Online Version)*. Available at: https://talk.ictvonline.org/ictv-reports/ictv_online_report/.

van Houte, S., van Oers, M. M., Han, Y., Vlak, J. M. and Ros, V. I. D. 2014. Baculovirus infection triggers a positive phototactic response in caterpillars to induce 'tree-top' disease. *Biol. Lett.* 10(12), 20140680. doi:10.1098/rsbl.2014.0680.

van Oers, M. M. 2011. Opportunities and challenges for the baculovirus expression system. *J. Invertebr. Pathol.* 107(Suppl.), S3–15. doi:10.1016/j.jip.2011.05.001.

Wenmann, J. T., Keilwagen, J. and Jehle, J. A. 2018. Baculovirus Kimura two-parameter species demarcation criterion is confirmed by the distances of 38 core gene nucleotide sequences. *J. Gen. Virol.* 99(9), 1307–20. doi:10.1099/jgv.0.001100.

Williams, T., Virto, C., Murillo, R. and Caballero, P. 2017. Covert infection of insects by baculoviruses. *Front. Microbiol.* 8, 1337, doi:10.3389/fmicb.2017.01337.

Woods, S. A., Streett, D. A. and Henry, J. E. 1992. Temporal mortality from an entomopoxvirus and strategies for control of the migratory grasshopper *Melanoplus sanguinipes* (f.). *J. Invertebr. Pathol.* 60(1), 33–9. doi:10.1016/0022-2011(92)90150-3.

Zhang, S., An, S., Hoover, K., Li, Z., Li, X., Liu, X., Shen, Z., Ros, V. I. D., Zhang, Q. and Liu, X. 2017. Host miRNAs are involved in hormonal regulation of HaSNPV-triggered climbing behaviour in *Helicoverpa armigera*. *Mol. Ecol.* 27, 459–75.

Chapter 9

Advances in the use of entomopathogenic nematodes (EPNs) as biopesticides in suppressing crop insect pests

Albrecht M. Koppenhöfer, Rutgers University, USA; David I. Shapiro-Ilan, USDA-ARS, USA; and Ivan Hiltpold, University of Delaware, USA*

1 Introduction

The complex of entomopathogenic nematodes (EPNs) and their symbiotic bacteria has been commercialized as biopesticides since the 1980s, and at least 13 EPN species have been in commercial development to date (Table 1). Commercialization has been aided by their exclusion from pesticide registration requirements in many countries due to the high level of safety to humans, nontarget organisms, and the environment (Ehlers, 2005). Since the 1990s, research on the application, biology, and ecology of these biocontrol agents has seen exponential growth, and since the mid-2000s basic research on EPNs has further expanded due to the use of the nematodes and their symbiotic bacteria as tractable model systems for many basic biological and ecological questions. The resulting extensive body of literature has been reviewed comprehensively and extensively in the chapters of several seminal books (Gaugler and Kaya, 1990; Bedding et al., 1993; Gaugler, 2002; Grewal

* Current address: Agroscope, DEFR, Switzerland

http://dx.doi.org/10.19103/AS.2020.0073.10

Table 1 Currently or recently commercialized entomopathogenic nematode species

Heterorhabditis spp.	*Steinernema* spp.
H. bacteriophora	*S. carpocapsae*
H. indica	*S. feltiae*
H. marelata	*S. kraussei*
H. megidis	*S. kushidai*
H. zealandica	*S. longicaudum*
	S. riobrave
	S. scapterisci
	S. scarabaei

et al., 2005a; Campos-Herrera, 2015). Given the wide scope of this chapter, we will use these books and chapters therein as our primary references with additional references listed for important and relevant studies or more specific reviews that were not or only briefly described in these chapters.

While the emphasis of this chapter is on the use of EPNs as biopesticides (see Section 3), we will also provide overviews on EPN biology and ecology (see Section 2) and mass production and application technology and interactions with other management tools (see Section 4) as they are necessary for a better understanding of the strengths and limitations of EPNs as biopesticides.

2 Entomopathogenic nematode (EPN) biology and ecology

2.1 EPN biology

Nematodes and insects have evolved a wide range of relationships varying along a continuum spanning from phoretic nematodes (using insects to disperse quicker) having little or no effect on the arthropod host to obligate parasitic nematode eventually causing the death of the insect host. EPNs are the most important group of these lethal nematodes in terms of biocontrol and IPM (Dillman et al., 2012) and are mostly classified in the families Heterorhabditidae and Steinernematidae. While phylogenetically not closely related, the two families share a suite of common characteristics as a result of convergent evolution: (1) they are obligate parasites in nature; (2) they are only horizontally transmitted; (3) they possess an infective juvenile (IJ) stage that actively invades the insect host; and (4) they are always associated with symbiotic bacteria that participate in host infection. The biology of EPNs and their symbiotic bacteria has been extensively reviewed in Griffin et al. (2005), Lewis and Clarke (2012), Stock (2015), and Shapiro-Ilan et al. (2017, 2018).

The IJ is the only stage that can survive outside the host due to greater resistance to abiotic stressors compared to the parasitic stages. Its function is to locate, recognize, accept, and infect an insect host within which it will continue its development. Outside of a host, IJs do not feed nor develop and they rely on resources accumulated while in their previous host. IJs use host and environmental cues to locate, recognize, and accept insects as potential hosts. They penetrate their host through natural openings (mouth, anus, spiracles) or via direct penetration through thin parts of the cuticle (Burnell and Stock, 2000). Once established in the hemocoel, IJs molt and regurgitate or defecate their symbiotic bacteria. The bacteria suppress the host immune system and kill the insect within 24–48 h via toxemia and septicemia. *Heterorhabditis* spp. (sole genus within Heterorhabditidae) IJs develop into hermaphroditic adults in the first generation, but subsequent generations include males, females, and hermaphroditic individuals (Fig. 1). Conversely, *Steinernema* spp. (Steinernematidae includes the genera *Steinernema* and *Neosteinernema*) adults are always males and females (with the exception of *Steinernema hermaphroditum*). Over a period of 1–4 weeks, the nematodes go through one to three generations inside the host cadaver, feeding on bacteria cells and host tissues digested by the bacteria. When host resources are depleted, a new cohort of IJs with symbiotic bacteria cells in their nonfunctional digestive system leaves the host cadaver to forage for new insect hosts (Fig. 1).

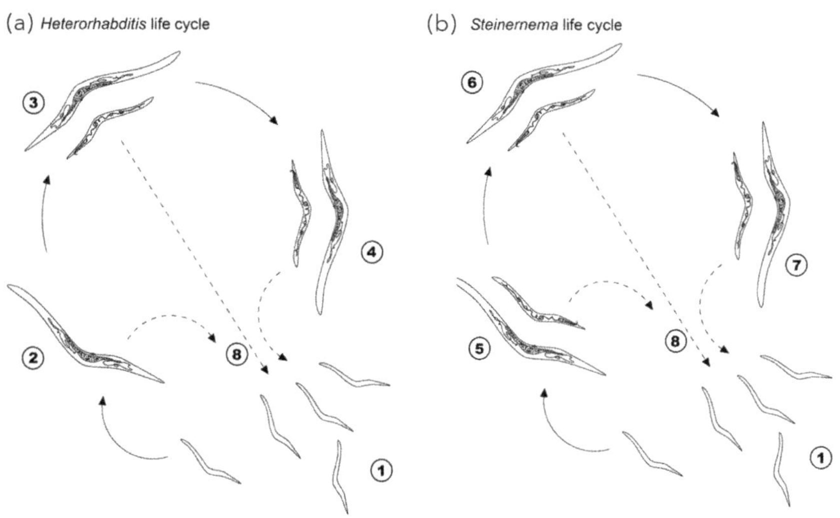

Figure 1 The life cycles of *Heterorhabditis* and *Steinernema* entomopathogenic nematodes. (1) Infective juveniles; (2) first generation of hermaphrodite females; (3) second-generation females (can either be amphimictic or hermaphrodite) and rare males; (4) third-generation hermaphrodite females and rare males; (5) first-generation amphimictic females and males; (6) second-generation amphimictic females and males; (7) third-generation amphimictic females and males. Dotted arrows imply some recruitment from an adult stage to the infective juvenile stage.

EPNs have evolved a characteristic mutualistic association with symbiotic bacteria in the genera *Photorhabdus* (*Heterorhabditis* spp.) and *Xenorhabdus* (*Steinernema* spp.). The bacteria rely on the nematode to vector them from insect to insect as they have no infective capabilities and cannot persist outside the nematodes or insect host. In return, the bacteria kill the insect host (though certain EPN species also contribute to killing the host), serve as food resources for the nematode, and maintain optimal developmental conditions within the host (production of antibiotics preventing secondary invasions). IJs of *Heterorhabditis* spp. carry the bacteria attached to their pre-intestinal valve whereas *Steinernema* spp. IJs harbor them in a specialized vesicle in their anterior gut. The association between the nematode and the bacterium is highly specific.

2.2 Host range, host immune response, and EPN virulence

Despite the economic importance of EPNs because of their potential in insect pest management and biological control (see Section 3), the host range of EPNs remains mostly unknown for most of the at least 96 *Steinernema* and 21 *Heterorhabditis* species described to date (Peters, 1996; Lewis and Clarke, 2012; Shapiro-Ilan et al., 2017, 2018). This is due to the fact that naturally infected insects are very difficult to retrieve from the field and that most field isolations of new species are conducted using *Galleria mellonella* larvae, an ideal bait insect because of its high susceptibility to EPN infections, but which provides no information about the natural host range of the isolated nematodes. While many EPN species may infect a wide range of insect species in laboratory assays (e.g. *S. carpocapsae* >200 insects across ten orders), after field applications and especially in nature, the host range is much narrower due to the ecology of the nematodes and its potential hosts as well as environmental factors. Some known EPN species are particularly well adapted to a narrow group of hosts species, with poor infection of other hosts even under laboratory conditions (i.e. *S. scapterisci* is adapted to Orthoptera and *S. scarabaei* to larvae of Scarabaeidae).

Host defenses and immune reactions in response to EPN infection have been studied in more detail, although only in a few EPN species-insect species combinations (reviewed in Lewis and Clarke, 2012; Shapiro-Ilan et al., 2017, 2018). Some insects have developed behavioral defenses against EPNs including intensive grooming behavior when in contact with IJs (scarabaeid white grubs) to prevent infection or even evading an area with high numbers of IJs (white grubs). Insects may also have evolved various physical barriers to prevent IJs from reaching their hemocoel. Access to the hemocoel via the mouth may be limited with forward projecting hairs in the preoral cavity (elaterid wireworms) or a thick peritrophic membrane protecting the midgut epithelium (white grubs). Access to the hemocoel via the tracheal system may be limited by fine sieve-like plates covering the spiracles (white grubs) or narrow, slit-like openings of the spiracles (wireworms).

If IJs can overcome insect behavioral and physical defenses, they still have to overcome the host's immune response (reviewed in Lewis and Clarke, 2012; Shapiro-Ilan et al., 2017, 2018). A typical strategy adopted by EPNs to evade the host immune response is the shedding of the second-stage-juvenile cuticle (sheath) that typically still covers the IJs and depositing *de-novo* produced or host-sequestered immune factors as a camouflage to avoid recognition by the host immune system. EPNs can also interfere with the host immune system by secreting putative proteins as observed in *S. carpocapsae*. *Steinernema* spp. are able to kill their insect host in the absence of the bacterial symbiont via the production and release of proteases but will usually not be able to complete their life cycle in the absence of their bacterial partners. Axenic *Heterorhabditis* spp. do not have the ability to kill their insect host. In both EPN genera, the symbionts play complimentary roles in overcoming the insect defenses and actively suppressing the immune response. *Xenorhabdus* and *Photorhabdus* species produce and release several toxins lethal to the insect host as well as antibiotics to prevent secondary infections of the cadaver by other pathogens or scavengers.

2.3 Dispersal and foraging strategies in EPNs

To locate a potential host, IJs employ various dispersal and foraging behaviors which are a key component in their ecology and their use in biological control (reviewed in Griffin et al., 2005; Lewis and Clarke, 2012; Griffin, 2015; Shapiro-Ilan et al., 2017, 2018). IJ dispersal impacts their spatial distribution in a given landscape (Bal et al., 2017). EPNs can be passively dispersed through phoresy via mites or earthworms for instance. Since hosts may take several days to die from EPN infection after initial IJ attachment, mobile host species can also contribute to dispersal. Active dispersal can be complex and will depend upon an EPN species' ecology and various biotic and abiotic factors but is generally limited with only up to 90 cm horizontally and vertically within 30 days even for the highly mobile *S. glaseri* (Kaya, 1990; Schroeder and Beavers, 1987).

EPN foraging strategies span across a continuum from ambushing to cruising behavior (Campbell and Lewis, 2002). Ambushers are generally less mobile and adopt the very particular behavior of nictating (i.e. standing on their tail) to increase their chances of contacting a host near the substrate surface; nictation may be concluded with jumping toward a host cue (Campbell and Gaugler, 1993). In contrast, cruisers actively seek out hosts throughout the soil profile and are usually highly mobile (Gaugler et al., 1997). However, many species fall between these two extremes. Interestingly, IJ populations of the ambusher *S. carpocapsae* contain a low proportion of individuals which will actively disperse from the cadaver they emerged from (i.e. 'sprinters'). EPN species have been observed to display group behavior and seem to disperse as packs (Shapiro-Ilan et al., 2014). Abiotic factors such as sand or organic matter content in soils have been reported to impact the behavior of IJs and some species may adopt an ambusher behavior in sandy soils but more actively disperse in more complex substrates (Hiltpold and Hibbard, 2018). However, the cruiser/ambusher paradigm is still widely used and sets the basis of our understanding of EPN foraging behavior.

2.4 EPN chemical ecology

IJs respond to a variety of cues and signals to navigate through the soil profile and find hosts (Shapiro-Ilan et al., 2018; Fig. 2). While vibrations or magnetic fields have been shown to attract them over some distance, the chemical ecology of

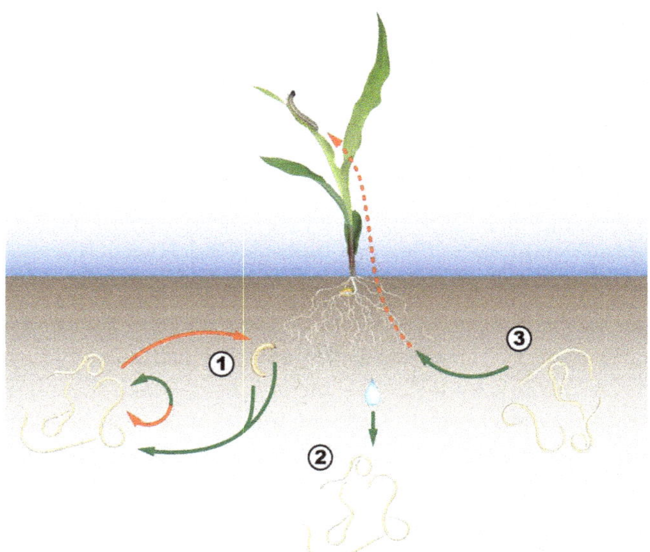

Figure 2 Chemically mediated reactions in entomopathogenic nematodes. (1) Entomopathogenic nematodes are responding to chemical cues emitted either directly by the insect host or by the roots of plants damaged by the insect hosts. Entomopathogenic nematodes are also able to evaluate the state of infection of insects by con- or heterospecific nematodes and to choose to either infect or not to infect this potential host. (2) Chemicals in exudates released in the soil by plant root tips have been shown to increase the entomopathogenic nematode activity in the rhizosphere and to improve their virulence. (3) Presence of entomopathogenic nematodes in the vicinity of roots negatively impacts insect feeding aboveground, adding to the complexity of the chemically mediated interaction between plants, insects, and entomopathogenic nematodes.

EPNs and how they exploit volatiles organic compounds has recently received a lot of attention. EPN respond to volatile cues from either their host (Baiocchi et al., 2017) or plants damaged by herbivorous insects and releasing alarm cues in the ground (Turlings et al., 2012). Because of the risks associated with triggering a host infection behavior, EPNs are able to detect and respond (positively or negatively) to host-specific volatiles emitted by various insects. Not only can IJs discriminate between potential hosts, they can also differentiate between already infected hosts and noninfected host. They will respond strongly to volatiles emitted by infected host and preferentially invade these insects. The invasion of a previously infected host may indeed offer certain advantages such as the presence of conspecific (necessary for species with amphimictic reproduction) or an already weakened immune system. However, research also indicates that positive chemotaxis to infected hosts decreases over time to even revert to repulsion from infected host (Grewal et al., 1996). The authors hypothesized that the IJs were avoiding such hosts with long infection history because of food depletion in the cadaver and a high level of conspecific competition.

EPNs respond to volatile organic compound in the soil emitted by plant roots after insect herbivory (Rasmann et al., 2005; Ali et al., 2010; Hiltpold et al., 2010). Recent contributions to the quickly evolving field highlight the importance of a good understanding of these finely tuned interactions when considering EPNs as a pest management strategy (e.g. Chiriboga et al., 2018; Stelinski et al., 2019). It has been suggested that IJs could be trained to better respond to particular volatiles and, therefore, increase their potential to manage insects (Willett et al., 2017). Using the appropriate combination of plant volatiles and IJs could also help mitigate resistance of certain insects to current pest management strategies (Hiltpold and Hibbard, 2018). Interestingly, cues emitted by IJs appear to negatively impact the performance of root insect herbivores (Helms et al., 2019) adding dual benefits of using EPNs in pest management strategies.

2.5 Factors affecting survival and efficacy

Of critical importance to efficacy is matching the best EPN species or strain to the target host; this includes consideration of innate virulence as well as suitability of environmental conditions. Also of great importance is the rate of EPN application with the general rule of thumb being 25 IJs per cm^2 of treated area. However, some target pests or cropping systems may require higher or (rarely) lower rates.

The persistence of EPN populations is determined by the longevity of individual IJs and the recycling of populations in host cadavers and numerous factors that can affect both mechanisms (reviewed in Griffin et al., 2005; Lewis and Clarke, 2012; Griffin, 2015; Stuart et al., 2015; Shapiro-Ilan et al., 2017,

2018). The innate longevity of IJs varies among EPN species between a few months to over 1 year and is affected by intrinsic (physiological, genetic, behavioral) and extrinsic factors (see in the following section). After soil applications, losses can reach 50% within hours until IJs have settled in the soil. Thereafter, losses may be in the range of 5–10% per day until often only around 1% of the original inoculum survives after 1–6 weeks. Nonetheless, in soil or similar substrates IJ populations generally remain high enough to provide effective control for 2–8 weeks. Some level of recycling commonly occurs after EPN application, but it is usually not sufficient to achieve multi-season control because IJ distribution becomes too patchy over time (but see Section 3.9).

UV radiation can inactivate and kill IJs within minutes, but the impact varies among EPN strains and species (Gaugler et al., 1992). Direct exposure to sunlight can be minimized if applications are made early in the morning or in the evening, by adding UV protectants to the IJ suspension, and by making soil applications in high carrier volumes combined with immediate rinsing in with sufficient amounts of water.

Soil moisture is the most critical factor affecting EPN survival and behavior. In order to move in the ground, EPNs require soil aggregates to be covered with a film of water thick enough to allow IJ movement. In saturated soils, anoxic conditions and low surface tension can be detrimental to movement and even survival. Dry conditions will limit IJ activity, but if moisture removal is gradual, IJs can persist longer due to reduced activity of the IJ. More desiccation-tolerant *S. carpocapsae* IJs can survive better in such conditions, whereas desiccation-intolerant *H. bacteriophora* IJs will actively seek out soil layers with higher moisture content. Moderate substrate moisture levels are essential for good IJ activity. IJ persistence in foliar applications may be extended by the addition of moisture-retaining adjuvants.

Temperature is an important factor for the survival, infection, and development of EPNs with optimal performance for most EPN species between 20–30°C (Grewal et al., 1994). Most species will become sluggish below 10–15°C and are inactivated above 30–40°C. However, various *Steinernema* spp. have been isolated from cold regions and other EPN species from hot semiarid or even arid regions. Such species may hold promise for use in environments with temperature extremes.

Soil texture also affects EPNs with movement and survival generally more restricted in fine textured soils than in sandy soils (Portillo-Aguilar et al., 1999). The latter may, however, be limiting as they can dry out more quickly. EPNs seem to fare well at a soil pH between 4 and 8; higher or lower pH reduces their efficacy. EPNs generally show little response to the application of synthetic fertilizers, are positively affected by composted manure whereas fresh manure or high rates of urea adversely impacts their persistence and pathogenicity.

EPN populations may be reduced by predators of IJs (e.g. mites, collembolans, tardigrades, predatory nematodes, nematophagous fungi), competition for hosts with other insect pathogens (e.g. entomopathogenic fungi, bacteria, or viruses), or scavengers feeding on the EPN-infected hosts.

3 Entomopathogenic nematode (EPN) use in different cropping systems and climatic zones

Given the extensive body of literature available on the subject, we mostly restricted our review to insect groups or species and crops or commodities in which EPN have been studied more intensely and where at least several field studies have shown good control potential by EPN (see also Table 2). Because this would still leave us with a very long list of references, we restricted the references mostly to relevant chapters in the comprehensive books by Lacey (2017), Grewal et al. (2005a), and Campos-Herrera (2015). Additional references were listed for important and relevant studies or more specific reviews that were not or only briefly described in these chapters. Additional, relevant chapters that may elaborate more on some of the older studies can be found in Bedding et al. (1993) and Gaugler and Kaya (1990).

3.1 EPN use in maize/corn

Research on the use of EPNs against insect pests of corn has been extensively summarized in Cabanillas et al. (2005) and Gassman and Clifton (2017) (see also references therein). Corn rootworms (CRW), *Diabrotica* spp., are among the most destructive pests of corn due to the root-feeding by the larvae. Earlier field studies conducted in North America on EPNs against larvae of western CRW, *D. virgifera virgifera*, and the northern CRW, *D. barberi*, produced variable results with failures often explained by the use of non-adapted EPN strains, soil aridity, or lack of alternative hosts (e.g. Jackson, 1996). More recent studies conducted in Europe after the introduction of the western CRW have shown the potential of EPNs for the management of the pest (Toepfer et al., 2014 and references therein). In plant-scale field experiments, EPNs were applied at 2.8–3.5 × 10^9 IJs/ha during sowing in a solid stream spray into the soil or as a narrow stream spray onto the soil along rows of young corn plants (Toepfer et al., 2008). Mean reduction of adults/mean reduction in root damage after April and June applications, respectively, were 81/80% and 75/72% for *H. bacteriophora*, 49/54% and 69/67% for *H. megidis*, and 36/73% and 32/91% for *S. feltiae*. Multi-year and multi-site field-scale trials using conventional farm equipment showed that *H. bacteriophora* provided control similar to that by soil insecticides with overall averages between 30% and 80% (Toepfer et al., 2010). Toepfer et al. (2014) concluded that the application as a stream into

Table 2 Major target pests on which entomopathogenic nematodes have been used commercially or have shown promise as targets[a,b]

Order	Family	Common name	Scientific name	Key crops targeted	Primary EPN species used
Coleoptera	Bupestridae	Flat-headed rootborer	*Capnodis tenebrionis*	Stone fruit	Sf
	Chrysomelidae	Corn rootworms	*Diabrotica* spp.	Vegetables	Hb, Sc
		Cranberry rootworm	*Rhadopterus picipes*	Cranberry	Hb
	Curculionidae	Annual bluegrass weevil	*Listronotus maculicollis*	Turf	Sc
		Banana weevil	*Cosmopolites sordidus*	Banana	Hb, Sc, Sf
		Billbugs	*Sphenophorus* spp.	Turf	Hb, Sc
		Black vine weevil	*Otiorhynchus sulcatus*	Berries, ornamentals	Hb, Hd, Hmeg, Hm, Sc, Sg, Sf
		Citrus root weevils	*Pachnaeus* spp.	Citrus, ornamentals	Hb, Sr
		Diaprepes root weevil	*Diaprepes abbreviatus*	Citrus, ornamentals	Hb, Hi, Sr
		Large pine weevil	*Hylobius abietis*	Forest plantings	Hd, Sc
		Pecan weevil	*Curculio caryae*	Pecan	Sc
		Plum curculio	*Conotrachelus nenuphar*	Fruit trees	Sr
		Red palm weevil	*Rhynchophorus ferrugineus*	Palms	Sc
		Strawberry root weevil	*Otiorhynchus ovatus*	Berries	Hm, Sc
		Sugarbeet weevil	*Temnorhinus mendicus*	Sugarbeet	Hb, Sc
		Sweet potato weevil	*Cylas formicarius*	Sweet potato	Hb, Sc, Sf
	Scarabaeidae	Scarab grubs	Many species	Turf, ornamentals, blueberries	Hb, Hz, Sg, Sk, Ssc
Diptera	Agromyzidae	Leafminers	*Liriomyza* spp.	Vegetables, ornamentals	Sc, Sf
	Ephydridae	Shore flies	*Scatella* spp.	Ornamentals	Sc, Sf
	Tephritidae	Fruit flies	various	Fruits various	Hi, Sc

	Sciaridae	Fungus gnats	various	Mushrooms, greenhouse	Sf, Hb, Hi
	Tipulidae	Crane flies	*Tipula* spp.	Turf	Sc
Lepidoptera	Crambidae	Cranberry girdler	*Chrysoteuchia topiaria*	Cranberries	Sc
	Gelichiidae	Tomato leafminer	*Tuta absoluta*	Tomato	Hb, Sc, Sf
	Noctuidae	Armyworms	Various	Vegetables, turf	Sc, Sf, Sr
		Black cutworm	*Agrotis ipsilon*	Turf, vegetables	Sc
		Corn earworm	*Helicoverpa zea*	Vegetables	Sc, Sf, Sr
		Turnip cutworm	*Agrotis segetum*	Vegetables	Sc, Sf
		Iris borer	*Macronoctua onusta*	Iris	Hb, Sc
	Pterophoridae	Artichoke plume moth	*Platyptilia carduidactyla*	Artichoke	Sc
	Pyralidae	Navel orangeworm	*Amylois transitella*	Nut and fruit trees	Sc
	Sesiidae	Clearwing borer moths	*Synanthedon* spp., others	Fruit trees, ornamentals	Hb, Sc, Sf
	Tineidae	Banana moth	*Opogona sacchari*	Ornamentals	Hb, Sc
	Tortricidae	Codling moth	*Cydia pomonella*	Pome fruit	Hz, Sc, Sf
		False codling moth	*Thaumatotibia leucotreta*	Citrus	Hb
Orthoptera	Gryllotalpidae	Mole crickets	*Scapteriscus* spp.	Turf, pasture	Ss
Thysanoptera	Thripidae	Western flower thrips	*Frankliniella occidentalis*	Greenhouse, flowers	Sc, Sf

[a] Modified after Shapiro-Ilan et al. (2018).

[b] At least one scientific paper reported ≥70% suppression of the pest in the field; table is not meant to be an exhaustive list.

Hb = *Heterorhabditis bacteriophora*, Hd = *H. downesi*, Hm = *H. marelata*, Hmeg = *H. megidis*, Hz = *H. zealandica*, Sc = *Steinernema carpocapsae*, Sf = *S. feltiae*, Sg = *S. glaseri*, Sk = *S. kushidai*, Sr = *S. riobrave*, Ss = *S. scapterisci*, Ssc = *S. scarabaei*.

the soil when sowing was the most promising approach. The use of capsules containing *H. bacteriophora* holds some promise as the IJs survived in the capsules, were able to escape them, and this formulation was more effective than water sprays in field trials (Hiltpold et al., 2012). Capsules coated with maize kairomones attracted CRW larvae in the laboratory; however, in the field the coating did not improve control efficacy compared to capsules with IJs only.

Corn earworm, *Helicoverpa zea*, larvae damage corn by feeding on whorls, silk, tassels, or kernels. Mature larvae drop to the ground and pupate in the soil at 5–10 cm depth making prepupae and pupae a good target for EPNs. *Steinernema riobrave*, a natural control agent of the pest in Texas, at a rate of 2×10^9 IJs/ha provided better control when applied at the time when 50% of the larvae were late instars (still on plant; 100% control) or when 10% of the larvae had dropped to the ground (95% control) than when 40% of the larvae were medium size (40% control); the same *S. riobrave* rate was more effective when applied via in-furrow irrigation (95%) than when applied before (56%) or after (84%) irrigation (Cabanillas and Raulston, 1995). In another study, *S. carpocapsae* did not cause any mortality of prepupae and pupae while *S. riobrave* provided 95% control, likely due to its greater tolerance of high soil temperatures (>38°C) in the system (Cabanillas and Raulston, 1996).

Black cutworm, *Agrotis ipsilon*, larvae cause damage to corn seedlings by feeding at the base of the plant, often completely cutting the plant off at the base. The efficacy of *S. carpocapsae* applications has ranged from 50% reduction in seedling damage at 5.4×10^9 IJs/ha (Capineira et al., 1988) to 76–83% at 1–10 days after treatment with 1.25×10^9 IJs/ha (Levine and Oloumi-Sadeghi, 1992).

3.2 EPN use in vegetable and tuber crops

Research on the use of EPNs against insect pests of vegetable and tuber crops has been extensively summarized in Bélair et al. (2005) and Georgis et al. (2006; see also references therein). Cucumber beetles/rootworms (*Diabrotica* spp., *Acalymma vittatum*) damage a variety of cucurbit crops and are susceptible to various EPNs species. For example, *S. riobrave* was capable of causing a 50% decrease in survival of *A. vittatum* larvae in both organic and conventional soil management systems under commercial field conditions.

Larvae of various cutworm species in different vegetables are amenable to control by EPNs. For example, turnip cutworm, *Agrotis segetum*, was effectively controlled with *S. feltiae* (as effective as standard insecticides at 2.5×10^9 IJs/ha in sandy soil and at 10^{10} IJs/ha in loamy soil) in lettuce fields (Lössbroek and Theunissen, 1985).

Sugarbeet weevil, *Temnorhinus mendicus*, larvae cause damage by feeding on the roots of sugarbeet. In field trials, all larval stages, pupae, and newly emerged adults were susceptible to EPNs, but the greatest efficacy was achieved with direct sprays on the crop at first larval hatch followed by irrigation or rainfall with 90–95% control at 2.5 × 10^9 IJs/ha; *S. carpocapsae* tended to be more effective than *H. bacteriophora*, and EPNs provided better control than standard insecticides (Boselli et al., 1997).

Several other pests of vegetables including carrot weevil (*Listronotus oregonensis*), cabbage maggot (*Delia radicum*), Colorado potato beetle (*Leptinotarsa decemlineata*), sweet potato weevil (*Cylas formicarius*), and some lepidopteran leaf feeders and dipteran leaf miners have been studied as targets of EPN, but under field conditions the EPNs did not provide adequate control at economical applications rates or were not competitive with cheaper or more effective insecticides.

3.3 EPN use in small fruit

Research on the use of EPNs against insect pests of small fruit has been extensively summarized in Cowles et al. (2005) and Dara (2017; see also references therein). Root weevil, *Otiorhynchus* spp., larvae cause severe damage to the roots of field-grown small fruit and are especially well suited for management with EPNs. Cowles et al. (2005) and Dara (2017) extensively discuss the importance of EPN species, soil temperature, weevil life stages, soil environment, applications technique, crop and cropping system, and interactions among them on the efficacy of EPNs for these pests. Different *Otiorhynchus* species have been successfully controlled with various EPN species (e.g. *H. bacteriophora*, *H. megidis*, *H. marelata*, *S. feltiae, S. glaseri*, *S. carpocapsae*) in strawberries, cranberries, raspberries, or black currants (e.g. Shanks and Agudelo-Silva, 1990; Booth et al., 2002). Field research with other pests has been much more limited.

Other insect pest of small fruit that have been effectively controlled include cranberry girdler, *Chrysoteuchia topiaria*, larvae in cranberries (44–92% control with *S. carpocapsae* at 4.9 × 10^9 IJs/ha) and cranberry rootworm, *Rhabdopterus picipes*, larvae in cranberries (54-79% control with *H. bacteriophora* at 5 × 10^9 IJs/ha). In highbush blueberries, oriental beetle, *Anomala orientalis*, larvae were not controlled by *H. bacteriophora*, but *S. scarabaei* provided 86–95% control within 21 days at rates of 0.75–1.5 × 10^9 IJs/ha (Polavarapu et al., 2007). Since *S. scarabaei* can provide long-lasting suppression of oriental beetle larvae in turfgrass (Koppenhöfer and Fuzy, 2009) due to effective recycling in the pest and it also persists very well in acidic blueberry soil (Koppenhöfer and Fuzy, 2006), this EPN species should have excellent potential for white grub control in blueberries.

3.4 EPN use in orchards

Research on the use of EPNs against insect pests of orchards has been extensively summarized in Shapiro-Ilan et al. (2005, 2017), Maniania et al. (2017), and Moore and Duncan (2017; see also references therein). Codling moth, *Cydia pomonella*, larvae are significant pests of pears, quince, walnuts, and especially apples worldwide. Overwintering fifth instars are difficult to control with chemical insecticides due to the cryptic location of their hibernacula (e.g. tree bark, prop piles, wooden fruit bins) but have been controlled effectively with *S. carpocapsae*, *S. feltiae*, and *H. zealandica* in apple and pear orchards (≥70% at 10^6 IJs/tree ~ 10^9 IJs/ha; e.g. Lacey et al., 2006; De Waal et al., 2011) and in fruit bins (up to 97% at 10–25 IJs/mL with adjuvant and humectant gel added; e.g. Lacey et al., 2005; De Waal et al., 2010).

Plum curculio, *Conotrachelus nenuphar*, is a key pest of pome and stone fruit in eastern North America. Larvae feed inside fruit which then falls to the ground. Mature larvae exit the fruit and burrow in the ground (1–8 cm depth) to pupate. While *H. bacteriophora* was ineffective and *S. feltiae* efficacy was variable (0–58%), *S. riobrave* consistently provided excellent control (80–100% at 4–40 × 10^9 IJs/ha) in peach, apple, and cherry orchards across the eastern United States (e.g. Shapiro-Ilan et al., 2013).

Clearwing moth larvae in the genus *Synanthedon* are generally very susceptible to EPNs. Hatched larvae of the peachtree borer, *S. exitiosa*, bore into the trunk of stone fruit trees near the soil surface and tunnel toward the roots to feed on crown and roots. *S. carpocapsae* (0.3–1.5 × 10^6 IJs/tree ~ 0.75–3.75 × 10^8 IJs/ha) applied preventively during the egg-laying period or curatively to existing infestations has provided control (78–100%) similar to chlorpyrifos (e.g. Shapiro-Ilan et al., 2015b, 2016a). Notably, the amount of nematodes needed per hectare is greatly reduced compared to application to most other target pests because in the case of *S. exitiosa* the EPNs only have to be applied around the base of the tree (as opposed to covering the entire acreage). Larvae of the lesser peach borer, *S. pictipes*, bore into the inner bark and cambium of the trunk and scaffolding limbs of *Prunus* spp. in the eastern United States. Application of *S. carpocapsae* (10^6 IJs/infested tree wound) combined with a diluted formulation of a fire gel (Barricade) to enhance IJ survival suppressed *S. pictipes* at the same level (70–100%) as chlorpyrifos (Shapiro-Ilan et al., 2010).

Pecan weevil, *Curculio caryae*, is a key pest of pecan in the southeastern United States. The larvae develop inside the nut before dropping to the ground and when mature they burrow to soil depths of 8–25 cm. The weevils spend most of their 2–3-year life cycle inactive in the soil as larvae, pupae, and adults, before emerging and feeding for a short period on developing nuts. Individual applications of *S. carpocapsae* (2.5–5.0 × 10^9 IJs/ha) targeting emerging adults provided insufficient (60%) and short-lived control. However, three seasonal applications of *S. carpocapsae* (50–100 IJs/cm^2 ~ 5–10 × 10^9 IJs/ha) over a 2-year period resulted in only 1% *C. caryae* survival (Shapiro-Ilan and Gardner, 2012).

Navel orangeworm, *Amyelois transitella*, larvae feed inside pistachio and almond nuts and are a key pest of these crops in California. Larvae overwinter in nuts left after harvest on the tree or ground and are a significant source for reinfestation in the following season. Applications of *S. carpocapsae* (10^9 IJs/ha) applied in a high spray volume (1870 L water/ha) followed by irrigation provided 75% control (Siegel et al., 2006).

Larvae of the red palm weevil, *Rhynchophorus ferrugineus*, tunnel in the trunk of various palm species and are a major pest of palms around the Persian Gulf and Mediterranean Sea. A combination of *S. carpocapsae* and chitosan (5 × 10^6 IJs/palm) applied to the top of the palm stipe was as effective as imidacloprid and provided 99% weevil control and improved palm survival from 17% to 74% (Dembilio et al., 2015).

The banana weevil, *Cosmopolites sordidus*, is the most important insect pest of bananas and plantains worldwide. Damage is caused by the larvae tunneling in the rhizome and the base of the pseudostem. *S. carpocapsae* applied with a water thickener into cuts or holes made in residual rhizomes has provided 43–68% control of larvae. A method involving cutting out several cones from residual rhizomes, adding 2.5 × 10^5 IJs/hole in a formulation with adjuvants, and replacing the cones killed nearly all adults that were attracted to the bait holes (Treverrow and Bedding, 1993).

The guava weevil, *Conotrachelus psidii*, is a major pest of guava in the Americas. Feeding of the larvae in the fruit causes fruit drop. The mature larvae exit the fruit and bury into the ground to pupate. Applications of *H. bacteriophora* provided 79–85% control and low rates of *H. indica* (1 and 10 IJs/cm^2) caused 33–50% mortality.

Various species of fruit flies are important pests of a variety of fruit crops around the world. Their mature larvae, prepupae, and pupae are susceptible to EPNs. For example, at high rates, *S. carpocapsae* provided 100% control of Mediterranean fruit fly, *Ceratitis capitata*, prepupae in a papaya field (500 IJs/cm^2), and 86% control of oriental fruit fly, *Bactrocera dorsalis* (300 IJs/cm^2) in a guava field. However, *H. indica* caused 66–93% mortality at much lower rates (1 and 10 IJs/cm^2).

EPN use has been particularly well studied in citrus orchards. In South Africa, the false codling moth, *Thaumatotibia leucotreta*, is an important pest of citrus. Its larvae develop in the fruit but fall to the ground to pupate at a very shallow depth. Single applications of *H bacteriophora* (10–20 IJs/cm^2) have reduced fruit infestation by up to 81%, but monthly applications at 5–10 IJs/cm^2 provided more consistent control. Citrus root weevils, especially the

Diaprepes root weevil, *Diaprepes abbreviatus*, are important pests in Florida, the Caribbean, and California with continuous and overlapping generations per year. Larval feeding on the roots, particularly girdling of large roots, causes severe damage. Since the 1990s, several EPN species have been commercialized for weevil control, including *S. carpocapsae*, *H. bacteriophora*, *H. indica*, and *S. riobrave* with the latter being most effective. Single applications of *S. riobrave* have provided up to 90% control in some studies, and twice-yearly applications through micro-sprinkler irrigation systems at 20–40 IJs/cm^2 under the tree canopies have reduced emerging adult weevil populations by about 50% (e.g. Duncan et al., 2007; Bender et al., 2014). After successful commercial use for about 15 years in the United States, *S. riobrave* products were discontinued in 2011 due to insufficient sales related to the advent of Asian citrus greening disease and the intensive chemical insecticide use for its vector. However, after it was recognized that weevil feeding on the roots is even more damaging to top tree health in the presence of the greening disease, *S. riobrave* production resumed in 2016.

Conservation and manipulation of natural EPN populations has been well studied in citrus orchards in Florida (Campos-Herrera et al., 2015). Surveys showed an unusually rich and abundant EPN fauna composed of 9 different species. In areas with drier soils, *S. diaprepesi*, *H. indica*, and *H. zealandica* were the dominant species and played an important role in significantly suppressing weevil populations. In areas with poorly drained soils (higher clay and organic matter content), *H. indica* and *Steinernema* sp. dominated but weevils were more abundant. Field trials in an orchard with the poorly draining soils showed that addition of coarse sand to imitate the more weevil-suppressive soil and augmentation with the four EPN species resulted in the same weevil-suppressive conditions as observed in orchards that naturally had those conditions (Duncan et al., 2013).

3.5 EPN use in turfgrass and pastures

Research on the use of EPNs against insect pests of turfgrass and pastures has been extensively summarized in Grewal et al. (2005b), Georgis et al. (2006), and Koppenhöfer et al. (2015, 2017; see also references therein). Three species of *Scapteriscus* mole crickets became important turfgrass pests in the southeastern United States after accidental introduction from South America and EPNs have been extensively studied for their control. In Florida, mole cricket populations were controlled with broadcast applications at 2.5×10^9 IJs/ha of *S. carpocapsae* (58%) and *S. scapterisci* and *S. riobrave* (both 75%). However, only the mole cricket-specific *S. scapterisci* can effectively reproduce in *Scapteriscus* spp. and provide long-term suppression of mole cricket populations. The introduction into the United States of *S. scapterisci*, isolated from mole crickets in Uruguay and Argentina, was the first successful use of an EPN in classical biological control (Frank and Walker, 2006). It is more effective against the southern mole cricket (*S. borellii*) than against the tawny mole cricket (*S. vicinus*) and least effective against the short-winged mole cricket (*S. abbreviatus*); it is most effective against adult mole crickets and is ineffective against small nymphs (Parkman and Frank, 1992). The use of *S. scapterisci* in more damage sensitive areas has been limited due to the competition from insecticides and by the need for application of nematicides for the control of plant-parasitic nematodes. In pastures, it has been applied using slit injectors in strips covering 12.5% of the area from which it then spreads over a period of several years (Frank and Walker, 2006) making it more cost-effective than chemical insecticides. Even though the *S. scapterisci*-based product Nematac-S appeared to become a success story, especially for uses in low maintenance turf and in pastures, it was discontinued in 2012 due to insufficient demand.

The larvae of numerous scarab beetles, referred to as white grubs, are important turfgrass pests around the world and have been extensively studied as targets of EPNs. White grub species differ in their susceptibility to EPNs and the relative virulence of different EPN species also varies among white grub species (e.g. Grewal et al., 2002; Koppenhöfer et al., 2004, 2006). Moreover, larval stages of white grubs may also differ in their susceptibility, but the effect varies with white grub and EPN species (e.g. Koppenhöfer and Fuzy, 2004; Power et al., 2009). Grewal et al. (2005a, b) provide an extensive summary of field studies on EPN efficacy against white grubs using as a qualifier for 'good' control at least 70% control (at 2.5×10^9 IJs/ha). Good control of Japanese beetle, *Popillia japonica*, has been achieved with *S. scarabaei* (100%), *H. bacteriophora* (34–97%), and *H. zealandica* (73–98%). *Steinernema scarabaei* is the only species that has provided high field control of the oriental beetle, *Anomala orientalis* (87–100%), the Asiatic garden beetle, *Maladera castanea* (71–86%), and the European chafer, *Rhizotrogus majalis* (89%). Good control of the northern masked chafer, *Cyclocephala borealis*, has been observed with *H. zealandica* (72–96%), *S. scarabaei* (84%), and *H. bacteriophora* (47–83%).

EPN persistence beyond a season following application against third-instar white grubs has been reported for *H. bacteriophora* and *S. scarabaei*, suggesting the potential impact of EPNs on multiple generations of white grubs (e.g. Klein and Georgis, 1992; Koppenhöfer and Fuzy, 2009). Due to the excellent adaptation of *S. scarabaei* to white grubs as hosts (high virulence and recycling in hosts) and outstanding persistence of individual IJs, a single application of this species provided not only high control of *A. orientalis* within 1 month of application at low application rates (77–100% control at $0.25–2.5 \times 10^9$ IJs/ha). It also provided additional control of overwintered larvae in the following spring (86–100% control at $0.1–2.5 \times 10^9$ IJs/ha) and persisted and suppressed *A. orientalis* for up to 4 years after application (Koppenhöfer and Fuzy, 2009). Thus, this species could be an excellent candidate for long-term suppression of white grubs with periodic reapplication at low rates.

The larvae of various billbug species can be controlled with EPNs. In the United States, applications at 2.5×10^9 IJs/ha targeting the larvae of the bluegrass billbug, *Sphenophorus parvulus*, in the soil, provided control by *S. carpocapsae* (average 78%) and *H. bacteriophora* (average 74%) similar to that by standard insecticides. On golf courses in Japan, *S. carpocapsae* (2.5×10^9 IJs/ha) was very effective for control of the hunting billbug, *S. venatus vestitus* (average 84%), due to favorable environmental conditions (temperature and rainfall) and the adoption of EPN-friendly application protocols. However, *S. carpocapsae* sales for billbug control in Japan significantly declined after the registration of imidacloprid for turfgrass uses.

Larvae of the annual bluegrass weevil, *Listronotus maculicollis*, are an important pest of golf course fairways, tees, and greens in eastern North America. In field trials with multiple species (*S. carpocapsae*, *S. feltiae*, *H. bacteriophora*), rates, and species combinations, control was variable (0–94% control at 2.5×10^9 IJs/ha) and IJ persistence was limited (McGraw et al., 2010). However, ongoing research indicates that split applications of *S. carpocapsae* (about 1 week apart) improve control and that simultaneous application of *S. carpocapsae* or *H. bacteriophora* with imidacloprid (applied for white grub control) has an additive effect on larval mortality (Koppenhöfer et al., unpublished data). Even though the destructive nature of *L. maculicollis* necessitates high levels of control, the dearth of effective insecticides for insecticide-resistant populations may increase opportunities for EPN use.

Larvae of various moth species are amenable to control by EPNs. Black cutworm larvae are a cosmopolitan pest of golf course greens and tees. Among commercial EPN products (2.5×10^9 IJs/ha) targeting fourth instar larvae, *S. carpocapsae* was the best performing species due to a combination of high (average 83%) and most consistent (70–90%) control rates (at 7 DAT) and fastest kill (average 68% at 4 DAT); *S. feltiae* (average 68%) and *H. bacteriophora* (average 62%) were less consistent, whereas *S. riobrave* was ineffective (average 32%; Ebssa and Koppenhöfer, 2011). Additional studies showed that (1) combinations of two EPN species at half rate each did not provide significantly better control than either of the two species alone at full rate, (2) syringing (i.e. twice daily small amount of irrigation) provided some limited improvement of *S. carpocapsae* efficacy under warm, sunny conditions, and (3) split application (two applications at half rate 3 days apart) improved *S. carpocapsae* efficacy by around 20% (Koppenhöfer et al., unpublished data). Despite this high efficacy, EPNs are not widely used for black cutworm control because of extremely low damage thresholds on golf course tees and especially greens. Field studies with other turfgrass caterpillars are more limited. Against fifth instars of the armyworm *Mythimna* (*Pseudaletia*) *separata*, Korean strains of *S. carpocapsae*,

S. longicaudum, *Heterorhabditis* sp., and *S. monticolum* provided 65%, 60%, 50%, and 45% control, respectively, at a rate of 10^9 IJs/ha.

Larvae of the European crane fly, *Tipula paludosa*, are important turfgrass pests in various regions of the United States and Canada. Larvae were controlled effectively (75–82%) at a rate of 5×10^9 IJs/ha when targeting first and second instars with *S. carpocapsae* but not with *S. feltiae*; both EPN species were ineffective against second to third instars, likely due to a combination of decreasing susceptibility of older larvae and decreasing soil temperatures (Oestergaard et al., 2006).

3.6 EPN use in landscape plants and nurseries

Research on the use of EPNs against insect pests of landscape and nursery plants has been extensively summarized in van Tol and Raupp (2005), Georgis et al. (2006), and Arthurs and Bruck (2017; see also references therein). Several wood-boring Lepidoptera including clearwing moths and the *Diaprepes* root weevil attack ornamental shade and fruit trees. The use of EPNs for their control has already been discussed in Section 3.4.

The black vine weevil, *Otiorhynchus sulcatus*, is an important pest of many different ornamental and landscape plants grown in pots and soil in nurseries and in the landscape. Damage is done primarily by the root-feeding larvae, although foliar feeding by the parthenogenetic females also causes damage. Overwintered females lay eggs around May and June, newly emerged females around July, resulting in a mix of first through third stage larvae in fall. The larvae finish their development in spring and pupate around May. EPNs use against black vine weevil in nursery production has been extensively summarized and analyzed by van Tol and Raupp (2005). Maximal control was generally achieved with about 0.5×10^6 IJs per pot and 10^6 IJs/m^2 in field trials. Soil temperature is an important factor that interacts with the weevil stages present. Summer applications, when temperatures are optimal, do not control eggs and freshly hatched larvae. In fall, low soil temperatures limit EPN activity. Best conditions tend to occur in late spring when soil temperatures are high enough and mature larvae and pupae are susceptible. But for many nursery plants, damage during fall and winter is not acceptable. For effective control, only a few hours of ≥12°C were sufficient in pot studies but longer periods of ≥12°C in the open field (at least 1 week of ≥12°C was recommended). Plant species also had a significant effect with control in conifers (*Thuja*, *Taxus*) higher than in *Rhododendron*, *Waldsteinia*, *Cornus*, or *Primula*.

The root-feeding larvae of white grub species (e.g. Japanese beetle, oriental beetle, European chafer) are pests of various plants species grown in pots or in the ground in nurseries. EPN efficacy against these pests in nurseries has been highly variable (0–100%; van Tol and Raupp, 2005), likely due to factors similar to how their control is affected in turfgrass (i.e. EPN species/strain thereof, white grub species/stage; see Section 3.5). However, research has been more limited than in turfgrass, and none of the most promising species or strains more recently tested in turfgrass (see Section 3.6) has been tested in nurseries.

3.7 EPN use in greenhouse production

Research on the use of EPNs against insect pests of greenhouses has been extensively summarized in Tomalak et al. (2005) and Wraight et al. (2017; see also references therein). Fungus gnats, *Bradysia* spp., are among the most important pests of many greenhouse-grown crops grown in soils with high organic matter content. The maggots feed primarily on fungi and organic matter but can also cause serious damage to roots and interfere with the callus development and root formation of fresh cuttings for propagation, and provide routes of entry for soil-borne pathogens. Due to the continuous and overlapping generations of fungus gnats in greenhouses, their control is very difficult. Among numerous EPNs tested, *S. feltiae* is the only one that is as effective as chemical insecticides giving 80–100% control at 2.5×10^6 IJs/m^2 (e.g. Harris et al., 1995; Jagdale et al., 2004). The second through fourth instars appear to be the most susceptible stages. While *S. feltiae* should be used at temperatures of 15–26°C, *H. indica* and *H. bacteriophora* may be a better choice if higher temperatures occur.

Thrips, in particular western flower thrips, *Frankliniella occidentalis*, is a major pest of greenhouse crops. Various studies have shown that the soil-dwelling stages (prepupae, pupae) can be controlled with various *Heterorhabditis* species causing higher mortality (24–60%) than various *Steinernema* species (3–54%), but the rates required for effective control (around 400 IJs/cm^2) were uneconomical (e.g. Ebssa et al., 2004). High-volume foliar applications of *S. feltiae* combined with adjuvants have been successful enough that the species is now commercially used in Europe and Canada.

3.8 EPN use in forestry

The large pine weevil, *Hylobius abietis*, is a major pest in replanted coniferous forests in northern Europe. Feeding of adults on young seedling trees causes around 50% mortality of newly planted trees in unprotected sites. However, even though the adults cause the damage and spend up to 4 years primarily on or in the soil, their low susceptibility to infection makes them poor targets for EPN applications. Rather, it is the larvae and pupae developing under the bark of stumps and roots of freshly cut trees that are the target of EPN applications. Torr et al. (2005) and Georgis et al. (2006) extensively summarized the earlier research on EPN use against this pest that showed that applications of *S. feltiae* and *S. carpocapsae* were more effective than *H. megidis* in reducing emergence of new adults from stumps. Large-scale applications of *S. carpocapsae* (3.5×10^6 IJs in 500 mL per stump = 7.5×10^9 IJs/ha) provided 60–75% reduction in adult emergence. Additional research showed that *H. downesi* was similarly effective as *S. carpocapsae* (Kapranas et al., 2017, and references therein). On both mineral and peat soil types, these EPNs suppressed adult emergence below the economic threshold in most treatments with average control rates of 62% (range 38–80%) for *H. downesi* and 55% (33–77%) for *S. carpocapsae*.

3.9 EPN use in mushroom production

Several species of flies in the families Sciaridae (*Lycoriella* and *Bradysia* spp.) and Phoridae (*Megasalia* spp.) are important pests of mushroom production. The larvae and pupae are found in the mushroom growth media. Sciarid larvae cause damage by feeding on mycelia and the developing sporophors and because their excretions render the substrate unsuitable for mycelial development. Phorids cause damage primarily as adults by vectoring fungal pathogens. Jess et al. (2005) give an extensive summary of EPN research and use for controlling these dipteran pests. In mushroom culture, effective sciarid control has been achieved by direct IJ application into or on to casing at casing time with *S. feltiae* providing control comparable to chemical insecticides which may be economically more advantageous than the chemicals (e.g. Scheepmaker et al., 1998). In contrast, EPN have only provided marginal control of phorids thus far.

3.10 Persistent native EPN strain for long-term pest suppression

The vast majority of EPN research has focused on using EPNs as biopesticides applied inundatively with little concern for long-term effects. By the time EPN strains are field tested, they generally have been reared under lab condition for several, if not many, years. Commercial strains have additionally been selected for ability to be mass-produced and formulated effectively. As a result, they may have lost some of their original ability to persist effectively in the environment (Wang and Grewal, 2002).

Inoculation of native EPNs strains for long-term persistence and long-term suppression of target pests has been suggested as an economically beneficial approach to applying EPNs (Shields, 2015; Shields and Testa, 2015; and reference therein). The native strains are hypothesized to be better adapted to local environmental conditions than non-native strains that may be introduced from commercial sources. Indeed, certain native strains have exhibited high levels of persistence in an alfalfa system when targeting alfalfa snout weevil, *Otiorhynchus ligustici* (Shields and Testa, 2017) or across crop rotations (Shields et al., 2018). In these studies, combinations of two EPN species (*S. carpocapsae* with *S. feltiae* or *H. bacteriophora*) were generally more effective at suppressing the pests than a single species because the two EPN species differed in their ability to attack different host stages and in their distribution in the soil profile. Especially the species combination significantly reduced the damage to the alfalfa crop and the population densities of *O. ligustici*. In addition to persistence of the native EPN strains, success with the approach likely also requires presence of susceptible pest stage and/or alternative hosts for the EPNs over an extended period of the year to allow effective recycling of the EPNs. Therefore, the crop should have a relatively high tolerance for the pest to allow EPN recycling and build-up of EPN densities over several weeks for additional pest suppression.

3.11 Slug-parasitic nematodes and their use

Slugs and snails in several mollusk families are worldwide pest of crops, particularly in cool moist climates. The adoption of more sustainable practices in agriculture has led to an increase in their significance as direct drilling, minimal tillage, and incorporation of crop residues create more favorable condition for slugs and snails. While many species of nematodes are known to be associated with slugs and snails, only the species *Phasmorhabditis hermaphrodita* has been developed as a biocontrol agent. Despite being successfully commercialized since 1994, the biology and ecology of this species is relatively poorly studied.

The life cycle of *P. hermaphrodita* is similar to that of EPNs (Wilson and Grewal, 2005; Wilson and Rae, 2015). It possesses an IJ stage that persists in the environment and infects the host. These juveniles enter the hosts through the dorsal integument pouch into the shell cavity where they recover and develop to adults. The infection eventually spreads to other body regions. Unlike in EPNs, the host is killed more slowly (4–21 days) and the nematode is a facultative parasite that can develop on many different substrates including slug feces, dead earthworms, dead insects, compost, and even leaf litter. While the nematode is commercially grown on the bacterium *Moraxella osloensis*, there is little evidence that it is specifically associated with any bacterium that it has been found developing on.

Most applications of *P. hermaphrodita* are made in greenhouse crops and field vegetables including greenhouse-grown flowers, lettuce, cabbage, Brussel sprouts, green asparagus, oilseed rape, wheat, and sugarbeet (Ester and Wilson, 2005; Wilson and Rae, 2015). The high value of these crops, the season-long threat of slug damage, and the limitations of the few available molluscides have promoted the use of the nematode. It is also particularly effective against one of the most import slug pest species, *Deroceras reticulatum*. The nematode can be applied at a single high rate (3×10^9/ha), but repeated applications at lower rates ($0.5–1.5 \times 10^9$/ha) are often more effective.

3.12 Nematode use in relation to cost:benefit and factors affecting commercial success

Shapiro-Ilan et al. (2002) examined the factors affecting the commercial success of EPNs using cotton, turfgrass, and citrus as model commodities and drew general conclusions from these systems. Their findings are generally still valid with slight variations among different commodities and pests. The two most important factors for successful EPNs commercialization are (1) a suitable nematode for the target pest and (2) favorable economics. Some other factors that can affect the success include the relative efficacy of the EPNs compared to other available control tactics, the reliable provision of a nematode product of high quality, and ease of use (shelf life, compatibility with typical application technology).

Nematode suitability is affected by various biological and ecological factors. First, the nematode has be able to withstand typical environmental extremes (e.g. temperature, moisture, UV exposure) in the target system including application procedures. Next, it as to be able to effectively locate hosts (adequate foraging behavior). Finally, it needs to be able to invade and kill the host effectively (i.e. high virulence). Limitations in virulence may be to some extent compensated for by high persistence in the system (see Section 3.9). Ultimately, host suitability has be proven by high and consistent performance under field conditions.

Factors that affect the economic viability of EPN use include the value of the commodity (i.e. profit margin), the cost of the EPNs relative to other control options, and the growers' perceived need to control the pest. The overall importance of the commodity in the agricultural market affects viability because small niche markets tend to have fewer control options due to registration costs for chemicals. EPN registration costs tend to be lower due to fewer safety requirements. In many cases, registration may not be required at all (e.g. the United States).

Management of *D. abbreviatus* in Florida has been a good example for successful EPN use at least until the advent of citrus greening disease. Several EPN species with high host suitability are available, the citrus system, particularly in the sandier central ridge region, is highly conducive for EPN activity, application can be concentrated under the trees where the IJs are protected from UV light and which effectively reduces the necessary IJ rate per total orchard hectarage by 3–10×, and chemical insecticides have not been available for soil applications. In contrast, use of EPNs for control of turfgrass pests in the United States has been limited despite some examples of good host suitability and favorable conditions at least in irrigated turfgrass areas. The major limitation has been the availability of many insecticides that are cheaper (by a factor of around 10× if off patent), easier to use (long shelf life, less limited by environmental extremes), and often even more effective.

4 New developments to improve efficacy

Despite extensive success in certain cropping systems (as indicated previously), EPNs are utilized in only a relatively small proportion of IPM programs. A number of approaches can be implemented to enhance EPN efficacy and thereby lead to wider usage. Areas that can be improved to increase efficacy include strain development, mass production, formulation and application technology, environmental manipulation, and combination with other management tools (Shapiro-Ilan et al., 2017).

4.1 Strain improvement and stabilization

Insect pest control efficacy can be improved through strain discovery, selection, hybridization, or transgenic manipulation (Gaugler, 1987; Shapiro-Ilan et al., 2017). Discovery consists of simply finding an EPN strain or species superior to those already in culture. In addition, genetic mechanisms can be employed to improve the nematodes. Directed selection has been used to improve EPNs for specific traits that are important to biocontrol such as host-finding, dispersal, and environmental tolerance (Shapiro-Ilan et al., 2017). One drawback to the approach can be inadvertent selection for one trait (reducing biocontrol potential) while directing selection toward another (trade-offs); additionally, reversion once selection pressure is removed is another potential concern. Hybridization, that is, controlled crosses, offers another avenue for strain improvement as in a kind of breeding program for superior nematodes. Selection and hybridization approaches can be combined (Mukuka et al., 2010).

A significant concern for both improved and wild-type EPNs is strain stability. Deterioration of biocontrol traits (such as virulence, reproductive capacity, and environmental tolerance) can result from genetic or nongenetic causes during repeated culturing in laboratory or commercial settings (Shapiro-Ilan et al., 2018). Trait deterioration can be avoided through improved cryopreservation methods or through creation of single or multiple purebred homozygous lines that are impervious to trait changes (Bai et al., 2005; Anbesse et al., 2013)

4.2 Mass production

EPNs are mass-produced commercially via *in-vivo* or *in-vitro* methods; the *in-vitro* approaches entail solid or liquid fermentation (Shapiro-Ilan et al., 2012). Generally, *in-vivo* production is based on a two-dimensional system that relies on a series of shelves and trays. Most *in-vivo* production is extrapolated from a White trap system, which is commonly used for laboratory-scale culturing. The most common insects for mass production include *G. mellonella* or *Tenebrio molitor* larvae, but other hosts can be used. The quantity and size of White traps can be expanded based on need (Shapiro-Ilan et al., 2012). *In-vivo* production is deemed to be of a low economy of scale as major costs are tied to labor and cost of insects. However, several approaches can be taken to improve the *in-vivo* process (Shapiro-Ilan et al., 2012, 2016b). Insect host production can be streamlined through mechanization (Morales-Ramos et al., 2011). Utilizing tri-trophic interactions, host diet can be optimized in terms of lipids, proteins, or other elements to increase EPN fitness (e.g. higher reproductive capacity and virulence). EPN inoculation of hosts as well as harvesting and packaging IJs can also be automated to reduce labor costs (Shapiro-Ilan et al., 2016b).

The bulk of EPNs are produced in liquid culture, which is deemed to have the most efficient economy of scale (Shapiro-Ilan et al., 2012). Solid fermentation falls between *in vivo* and liquid culture in terms of economy of scale, labor requirements, and capital outlay needed for a start-up. Production operations in some countries depending on labor and raw materials have found solid culture to be highly profitable. *In-vitro* culture can be improved through optimization of media and fermentation parameters (including physical factors of pH, aeration, etc.; Shapiro-Ilan et al., 2012; Leite et al., 2017).

4.3 Formulation and application technology

EPNs are applied effectively using diverse agricultural sprayers or irrigation systems (Shapiro-Ilan et al., 2017). Optimum application depends on a variety of factors including environmental conditions (moisture/humidity and avoiding UV radiation), appropriate application rates, nozzle and sprayer type, and so forth. Numerous formulations have been used for EPN application including activated charcoal, alginate and polyacrylamide gels, clay, diatomaceous earth, paste, peat, polyurethane sponge, vermiculite, and water dispersible granules (Shapiro-Ilan et al., 2012). User-friendly formulations that can simply be added to the tank and sprayed are desirable. Currently, vermiculite and diatomaceous earth are popular formulations.

Improved formulations can expand EPN utility in IPM systems. In particular, formulations that allow applications aboveground (thus protecting the nematodes from UV and desiccation) promise to increase EPN usage. As indicated previously, a sprayable gel, Barricade, has been used to control the wood-boring pest *S. pictipes*. Other formulations or adjuvants intended for aboveground EPN use have also been developed (Van Niekerk and Malan, 2015; Shapiro-Ilan et al., 2017).

Novel EPN application approaches can also enhance efficacy. For example, applying nematodes in their infected hosts (Dolinski et al., 2015), with phoretic hosts (Shapiro-Ilan and Brown, 2013), or in pre-infected hosts that reach cryptic habitats (Gumus et al., 2015) has resulted in superior control relative to traditional applications in aqueous sprays. As alluded to previously, certain plant volatiles may be used to enhance EPN efficacy by directing nematode behavior and enhancing infectivity (Willett et al., 2018). Moreover, adding infected host macerate or nematode pheromones (ascarosides) can boost EPN dispersal and has been shown to improve efficacy in a soil profile (Wu et al., 2018; Oliveira-Hofman et al., 2019).

4.4 Environmental manipulation

Practices that enhance biotic or abiotic factors to make the environment more conducive to EPN persistence, dispersal, or infectivity will lead to improved efficacy. For example, ground covers, mulches, or crop residues can be added to the cropping system to protect EPNs from UV radiation or desiccation (Shapiro-Ilan et al., 2015a, 2017). Optimization of basic soil parameters such as pH or texture can also be implemented. Duncan et al. (2013) improved the soil environment in a citrus orchard to be more conducive to EPNs by amending the soil texture; trees were planted in 'islands' of soil with appropriate sand content such that EPN efficacy was superior to endemic soils.

Conceivably, biotic agents that interact positively with EPNs could also be encouraged within the cropping system. For example, manure or other organic materials could be added to increase earthworm populations and thereby improve EPN dispersal via phoresy (Shapiro-Ilan et al., 2017). Other organisms that synergize EPN efficacy (see Section 4.5) might also be encouraged within the cropping system through environmental manipulations and thus lead to improved pest suppression. For example, entomopathogenic fungi that interact synergistically with EPNs might be enhanced by improving ground cover or increasing planting density.

4.5 Combinations of EPNs with other control agents

Combining EPNs with other control agents has the goal of providing effective pest control while reducing the use of more hazardous synthetic insecticides, increase consistency and level of pest control, and lower costs by being able to use reduced rates of chemicals and EPN. Much of the work on combining EPN with other control agents has been conducted in turfgrass particularly with different white grub species (Koppenhöfer and Grewal, 2005; Koppenhöfer et al., 2015; and references therein). Combination of two EPN species against third instars of several white grub species and against larvae of *A. ipsilon* and *L. maculicollis* (Koppenhöfer et al., unpublished data) have generally resulted in additive effects on mortality. Combinations of *Bacillus thuringiensis* subsp. *japonensis* (*Btj*) with EPNs have resulted in additive and synergistic effects on mortality of third instars of different white grub species (e.g. Koppenhöfer et al., 1999). The fungus *Metarhizium anisopliae* showed a strong synergistic interaction with *H. megidis* and *S. glaseri* in second- and third-instar *Hoplia philanthus* when the fungus was applied 3–4 weeks before the EPNs (e.g. Ansari et al., 2006). Combinations of *M. anisopliae* or *Beauveria bassiana* against third-instar southern masked chafer, *C. lurida* (in laboratory and greenhouse experiments only) generally resulted in additive mortality irrespective of application interval.

The organophosphate diazinon and *S. kushidai* caused additive mortality of third-instar *C. hirta* and *A. orientalis,* while the organophosphate chlorpyrifos and *Heterorhabditis* sp. Gyeongsan or *S. longicaudum* caused synergistic mortality of *A. orientalis*. Combinations of the anthranilic diamide chlorantraniliprole and *H. bacteriophora* resulted in mostly synergistic but also additive mortality of third-instar *A. orientalis*, *P. japonica*, and *C. borealis*. The neonicotinoid imidacloprid mostly interacted synergistically in combinations of several EPN species (*S. glaseri*, *H. bacteriophora*, *H. marelata*, *H. megidis*) in third instars of *Cyclocephala* spp., *P. japonica*, and *A. orientalis* (e.g. Koppenhöfer et al., 2002). However, two rather scarab-specific species, *S. kushidai* and *S. scarabaei*, generally did not interact with imidacloprid against several white grub species. Imidacloprid-*H. bacteriophora* combinations provide more consistent synergism against third instars but higher control rates against second instars and early third instars in laboratory and field trials using rates as low as 25% and 50% of the full rates for *H. bacteriophora* and imidacloprid, respectively (Koppenhöfer and Fuzy, 2008).

Only the interactions of nematodes with *Btj*, imidacloprid, and, to a more limited extent, chlorantraniliprole have been sufficiently studied to allow for recommendations to be made. Against medium to large third-instar white grubs, all three combinations required relatively high rates of both agents, limiting their economic feasibility. All three combinations should be more effective or equally effective at lower rates when applied against younger larvae, that is, young third instars or second instars, and species that are more susceptible to at least one of the control agents.

5 Conclusion and future trends

Extensive research efforts since the 1990s have led to the commercial use of EPNs in many agricultural and other commodities with varying degrees of success as affected by various factors (Section 3). Most importantly, the availability of EPN strains suitable for the target pest in the target system and favorable economics has increased, but the relative efficacy of the EPNs compared to other available control tactics and their ease of use has also improved. However, EPN use is generally still very limited compared to that of synthetic insecticides which are generally less expensive and easier to use and often more effective. While concerns about health risks and environmental hazards of pesticides have led to pesticide legislation in many countries that have resulted in the loss of many insecticides, they have been replaced by new active ingredients from several new insecticide classes that are considered low-risk insecticides (e.g. neonicotinoids, oxadiazines, anthranilic diamides). Nonetheless, public concerns about insecticide use persist.

To take advantage of the growing interest in alternatives to synthetic insecticides and in organic agriculture and land care, EPNs will need to be further improved with respect to efficacy (more virulent strains and species from field populations or through biotechnology, improved application technologies), costs (better production technologies), and especially ease of use (formulations with extended shelf life and tolerance to temperature extremes). Improvements in most of these areas will likely be only gradual, but, at least where pesticide regulations, local ordinances, and public opinion already impinge on the use of synthetic insecticides, they may suffice to significantly increase EPN use.

Another avenue to increase EPN use could be through better exploitation of their ability to recycle in hosts to provide long-term pest suppression. This may be achieved using persistent native and more generalist strains as discussed previously (Section 3.10) or species that are highly adapted to a narrow range of hosts such as *S. scapterisci* or *S. scarabaei* (Section 3.5). Depending on the characteristics of the system into which they would be applied and the biology and ecology of the target pest(s), either approach could be achieved with inoculative releases of the EPN or periodic augmentative releases. For maintaining persistent strains as well as other improved strains, advanced methods are needed to ensure stabilization such as through creation of homozygous purebred lines or other techniques.

Education and legislations will also play important roles in the furtherance of EPN use, whether as biopesticides or inoculative agents. Major changes in insecticides use pattern will have to be encouraged through legislative incentives, regulations, and restrictions in a similar fashion as they have been happening in Canada since the late 1990s for turfgrass uses (Bélair et al., 2010).

6 Where to look for further information

Excellent introductions into the field of entomopathogenic nematodes in general and their application in pest management can be found in previously cited books and chapters therein (Gaugler and Kaya, 1990; Bedding et al., 1993; Gaugler, 2002; Grewal et al., 2005a; Campos-Herrera, 2015; Lacey, 2017). Important societies for the field include the Society for Invertebrate Pathology (http://sipweb.org/), particularly the Division of Nematodes thereof (http://sipweb.org/nematodes/div_nematodes.html) and the Society of Nematologists (https://nematologists.org/), both of which hold annual meetings. Important scientific journal that regularly publish articles related to the field include the *Journal of Invertebrate Pathology* (https://www.journals.elsevier.com/journal-of-invertebrate-pathology/), the *Journal of Nematology* (https://www.exeley.com/journal/journal_of_nematology), *Biological Control* (https://www.journals.elsevier.com/biological-control/), and *Biocontrol Science and Technology* (https://www.tandfonline.com/loi/cbst20), but numerous other entomology- nematology- or pest management-related journals also publish articles related to the use of entomopathogenic in pest management.

7 References

Ali, J. G., Alborn, H. T. and Stelinski, L. L. 2010. Subterranean herbivore-induced volatiles released by citrus roots upon feeding by *Diaprepes abbreviatus* recruit entomopathogenic nematodes. *J. Chem. Ecol.* 36(4), 361–8. doi:10.1007/s10886-010-9773-7.

Anbesse, S., Sumaya, N. H., Dörfler, A. V., Strauch, O. and Ehlers, R. U. 2013. Stabilization of heat tolerance traits in *Heterorhabditis bacteriophora* through selective breeding and creation of inbred lines in liquid culture. *BioControl* 58(1), 85–93. doi:10.1007/s10526-012-9467-x.

Ansari, M. A., Shah, F. A., Tirry, L. and Moens, M. 2006. Field trials against *Hoplia philanthus* (Coleoptera: Scarabaeidae) with a combination of an entomopathogenic nematode and the fungus *Metarhizium anisopliae* CLO 53. *Biol. Control* 39(3), 453–9. doi:10.1016/j.biocontrol.2006.07.004.

Arthurs, S. P. and Bruck, D. J. 2017. Microbial control of nursery ornamental and landscape plant pests. In: Lacey, L. A. (Ed.), *Microbial Control of Insect and Mite Pests*. Elsevier, Amsterdam, The Netherlands, pp. 355–66.

Bai, C., Shapiro-Ilan, D. I., Gaugler, R. and Hopper, K. R. 2005. Stabilization of beneficial traits in *Heterorhabditis bacteriophora* through creation of inbred lines. *Biol. Control* 32(2), 220–7. doi:10.1016/j.biocontrol.2004.09.011.

Baiocchi, T., Lee, G., Choe, D. H. and Dillman, A. R. 2017. Host seeking parasitic nematodes use specific odors to assess host resources. *Sci. Rep.* 7(1), 6270. doi:10.1038/s41598-017-06620-2.

Bal, H. K., Acosta, N., Cheng, Z., Grewal, P. S. and Hoy, C. W. 2017. Effect of habitat and soil management on dispersal and distribution patterns of entomopathogenic nematodes. *Appl. Soil Ecol.* 121, 48–59. doi:10.1016/j.apsoil.2017.08.018.

Bedding, R., Akhurst, R. and Kaya, H. 1993. *Nematodes as the Biological Control of Insect Pests*. CSIRO Publishing, East Melbourne, Australia.

Bélair, G., Wright, D. J. and Curto, G. 2005. Vegetable and tuber crop applications. In: Grewal, P. S., Ehlers, R.-U. and Shapiro-Ilan, D. I. (Eds), *Nematodes as Biocontrol Agents*. CABI Publishing, Wallingford, UK, pp. 231–54.

Bélair, G., Koppenhöfer, A. M., Dionne, J. and Simard, L. 2010. Current and potential use of pathogens in the management of turfgrass insects as affected by new pesticide regulations in North America. *Int. J. Pest Manag.* 56(1), 51–60. doi:10.1080/09670870903076012.

Bender, G. S., Bates, L. M., Bethke, J. A., Lewis, E., Tanizaki, G., Morse, J. G. and Godfrey, K. E. 2014. Evaluation of insecticides, entomopathogenic nematodes, and physical barriers for control of *Diaprepes abbreviatus* (Coleoptera: Curculionidae) in citrus. *J. Econ. Entomol.* 107(6), 2137–46. doi:10.1603/EC14150.

Booth, S. R., Tanigoshi, L. K. and Shanks Jr., C. H. 2002. Evaluation of entomopathogenic nematodes to manage root weevil larvae in Washington State cranberry, strawberry, and red raspberry. *Environ. Entomol.* 31(5), 895–902. doi:10.1603/0046-225X-31.5.895.

Boselli, M., Curto, G. M. and Tacconi, R. 1997. Field efficacy of entomopathogenic nematodes against the sugar-beet weevil *Temnorhinus* (=*Conorrhynchus*) *mendicus* Gyll. (Coleoptera: Curculionidae). *Biocontrol Sci. Technol.* 7(2), 231–8. doi:10.1080/09583159730929.

Burnell, A. M. and Stock, S. P. 2000. *Heterorhabditis*, *Steinernema* and their bacterial symbionts – lethal pathogens of insects. *Nematology* 2(1), 31–42. doi:10.1163/156854100508872.

Cabanillas, H. E. and Raulston, J. R. 1995. Impact of *Steinernema riobravis* (Rhabditida: Steinernematidae) on the control of *Helicoverpa zea* (Lepidoptera: Noctuidae) in corn. *J. Econ. Entomol.* 88(1), 58–64. doi:10.1093/jee/88.1.58.

Cabanillas, H. E. and Raulston, J. R. 1996. Evaluation of *Steinernema riobravis, S. carpocapsae*, and irrigation timing for the control of corn earworm, *Helicoverpa zea*. *J. Nematol.* 28(1), 75–82.

Cabanillas, H. E., Wright, R. J. and Vyas, R. V. 2005. Cereal, fibre, oilseed and medicinal crop applications. In: Grewal, P. S., Ehlers, R.-U. and Shapiro-Ilan, D. I. (Eds), *Nematodes as Biocontrol Agents*. CABI Publishing, Wallingford, UK, pp. 265–79.

Campbell, J. F. and Gaugler, R. 1993. Nictation behavior and its ecological implications in the host search strategies of entomopathogenic nematodes (Heterorhabditidae and Steinernematidae). *Behaviour* 126(3–4), 155–69. doi:10.1163/156853993X00092.

Campbell, J. F. and Lewis, E. E. 2002. Entomopathogenic nematode host-search strategies. In: Lewis, E. E., Campbell, J. F. and Sukhdeo, M. V. K. (Eds), *The Behavioural Ecology of Parasites*. CABI Publishing, Wallingford, UK, pp. 13–38.

Campos-Herrera, R. 2015. *Nematode Pathogenesis of Insect and Other Pests*. Springer International Publishing, Cham, Switzerland and Berlin, Germany, 531pp.

Campos-Herrera, R., El-Borai, F. E. and Duncan, L. W. 2015. It takes a village: entomopathogenic nematode community structure and conservation biological control in Florida (U.S.) orchards. In: Campos-Herrera, R. (Ed.), *Nematode Pathogenesis of Insect and Other Pests*. Springer International Publishing, Cham, Switzerland and Berlin, Germany, pp. 329–51.

Capineira, J. L., Pelissier, L. D., Menout, G. S. and Epsky, N. D. 1988. Control of black cutworm, *Agrotis ipsilon*, with entomogenous nematodes (Nematoda: Steinernematidae, Heterorhabditidae). *J. Invertebr. Pathol.* 52, 427–35.

Chiriboga, X., Guo, H., Campos-Herrera, R., Röder, G., Imperiali, N., Keel, C., Maurhofer, M. and Turlings, T. C. J. 2018. Root-colonizing bacteria enhance the levels of (E)-β-caryophyllene produced by maize roots in response to rootworm feeding. *Oecologia* 187(2), 459–68. doi:10.1007/s00442-017-4055-5.

Cowles, R. S., Polavarapu, S., Williams, R. N., Thies, A. and Ehlers, R.-U. 2005. Soft fruit applications. In: Grewal, P. S., Ehlers, R.-U. and Shapiro-Ilan, D. I. (Eds), *Nematodes as Biocontrol Agents*. CABI Publishing, Wallingford, UK, pp. 191–213.

Dara, S. K. 2017. Microbial control of arthropod pests in small fruit and vegetables in temperate climate. In: Lacey, L. A. (Ed.), *Microbial Control of Insect and Mite Pests*. Elsevier, Amsterdam, The Netherlands, pp. 209–21.

Dembilio, O., Llácer, E., del Mar Martínez de Altube, M. and Jacas, J. A. 2015. Field efficacy of imidacloprid and *Steinernema carpocapsae* in a chitosan formulation against the red palm weevil *Rhynchophorus ferrugineus* (Coleoptera: Curculionidae) in *Phoenix canariensis*. *Pest Manag. Sci.* 66, 365–70.

De Waal, J. Y., Malan, A. P., Levings, J. and Addison, M. F. 2010. Key elements in the successful control of diapausing codling moth, *Cydia pomonella* (Lepidoptera: Tortricidae) in wooden fruit bins with as South African isolate of *Heterorhabditis zealandica* (Rhabditida: Heterorhabditidae). *Biocontrol Sci. Technol.* 20, 1161–76.

De Waal, J. Y., Malan, A. P. and Addison, M. F. 2011. Efficacy of entomopathogenic nematodes (Rhabditida: Heterorhabditidae and Steinernematidae) against codling moth, *Cydia pomonella* (Lepidoptera: Tortricidae) in temperate regions. *Biocontrol Sci. Technol.* 21(10), 1161–76. doi:10.1080/09583157.2011.607922.

Dillman, A. R., Chaston, J. M., Adams, B. J., Ciche, T. A., Goodrich-Blair, H., Stock, S. P. and Sternberg, P. W. 2012. An entomopathogenic nematode by any other name. *PLoS Pathog.* 8(3), e1002527. doi:10.1371/journal.ppat.1002527.

Dolinski, C., Shapiro-Ilan, D. I. and Lewis, E. E. 2015. Insect cadaver applications: pros and cons. In: Campos-Herrera, R. (Ed.), *Nematode Pathogenesis of Insect and Other Pests*. Springer International Publishing, Cham, Switzerland and Berlin, Germany, pp. 207–30.

Duncan, L. W., Graham, J. H., Zellers, J., Bright,, D., Dunn, D. C., El-Borai, F. E. and Porazinska, D. L. 2007. Food web responses to augmentation biological control using entomopathogenic nematodes in bare and composted-manure amended soil. *J. Nematol.* 39(2), 176–89.

Duncan, L. W., Stuart, R. J., El-Borai, F. E., Campos-Herrera, R., Pathak, E., Giurcanu, M. and Graham, J. H. 2013. Modifying orchard planting sites conserves entomopathogenic nematodes, reduces weevil herbivory and increases citrus tree growth, survival and fruit yield. *Biol. Control* 64(1), 26–36. doi:10.1016/j.biocontrol.2012.09.006.

Ebssa, L. and Koppenhöfer, A. M. 2011. Efficacy and persistence of entomopathogenic nematodes for black cutworm control in turfgrass. *Biocontrol Sci. Technol.* 21(7), 779–96. doi:10.1080/09583157.2011.584610.

Ebssa, L., Borgemeister, C. and Poehling, H. M. 2004. Effectiveness of different species/strains of entomopathogenic nematodes for control of western flower thrips (*Frankliniella occidentalis*) at various concentrations, host densities and temperatures. *Biol. Control* 29(1), 145–54. doi:10.1016/S1049-9644(03)00132-4.

Ehlers, R.-U. 2005. 'Forum on safety and regulation'. In: Grewal, P. S., Ehlers, R.-U. and Shapiro-Ilan, D. I. (Eds), *Nematodes as Biocontrol Agents*. CABI Publishing, Wallingford, UK, pp. 107–14.

Ester, A. and Wilson, M. J. 2005. Application of slug-parasitic nematodes. In: Grewal, P. S., Ehlers, R.-U. and Shapiro-Ilan, D. I. (Eds), *Nematodes as Biocontrol Agents*. CABI Publishing, Wallingford, UK, pp. 431–44.

Frank, J. H. and Walker, T. J. 2006. Permanent control of pest mole crickets (Orthoptera: Gryllotalpidae) in Florida. *Am. Entomol.* 52, 138–44.

Gassman, A. J. and Clifton, E. H. 2017. Current and potential applications of biopesticides to manage insect pests of maize. In: Lacey, L. A. (Ed.), *Microbial Control of Insect and Mite Pests*. Elsevier, Amsterdam, The Netherlands, pp. 173–84.

Gaugler, R. 1987. Entomogenous nematodes and their prospects for genetic improvement. In: Maramorosch, K. (Ed.), *Biotechnology in Invertebrate Pathology and Cell Culture*. Academic Press, San Diego, CA, pp. 457–84.

Gaugler, R. 2002. *Entomopathogenic Nematology*. CABI Publishing, Wallingford, UK, 388pp.
Gaugler, R. and Kaya, H. K. 1990. *Entomopathogenic Nematodes in Biological Control*. CRC Press, Boca Raton, FL, 365pp.
Gaugler, R., Bednarek, A. and Campbell, J. F. 1992. Ultraviolet inactivation of Heterorhabditid and Steinernematid nematodes. *J. Invertebr. Pathol.* 59(2), 155–60. doi:10.1016/0022-2011(92)90026-Z.
Gaugler, R., Lewis, E. E. and Stuart, R. J. 1997. Ecology in the service of biological control: the case of entomopathogenic nematodes. *Oecologia* 109(4), 483–9. doi:10.1007/s004420050108.
Georgis, R., Koppenhöfer, A. M., Lacey, L. A., Belair, G., Duncan, L. W., Grewal, P. S., Samish, M., Torr, P. and van Tol, R. W. H. M. 2006. Successes and failures of entomopathogenic nematodes. *Biol. Control* 38, 103–23.
Grewal, P. S., Selvan, S. and Gaugler, R. 1994. Thermal adaptation of entomopathogenic nematodes-niche breadth for infection, establishment and reproduction. *J. Therm. Biol.* 19(4), 245–53. doi:10.1016/0306-4565(94)90047-7.
Grewal, P. S., Lewis, E. E. and Gaugler, R. 1996. Response of infective juvenile stage parasites (Rhabditida: Steinernematidae) to volatile cues from infected hosts. *J. Chem. Ecol.* 23, 503–15.
Grewal, P. S., Grewal, S. K., Malik, V. S. and Klein, M. G. 2002. Differences in the susceptibility of introduced and native white grub species to entomopathogenic nematodes from various geographic localities. *Biol. Control* 24(3), 230–7. doi:10.1016/S1049-9644(02)00025-7.
Grewal, P. S., Ehlers, R.-U. and Shapiro-Ilan, D. I. 2005a. *Nematodes as Biocontrol Agents*. CABI Publishing, Wallingford, UK, 505pp.
Grewal, P. S., Koppenhöfer, A. M. and Choo, H. Y. 2005b. Lawn, turfgrass and pasture applications. In: Grewal, P. S., Ehlers, R.-U. and Shapiro-Ilan, D. I. (Eds), *Nematodes as Biocontrol Agents*. CABI Publishing, Wallingford, UK, pp. 115–46.
Griffin, C. T. 2015. Behaviour and population dynamics of entomopathogenic nematodes following application. In: Campos-Herrera, R. (Ed), *Nematode Pathogenesis of Insect and Other Pests*. Springer International Publishing, Cham, Switzerland and Berlin, Germany, pp. 57–95.
Griffin, C. T., Boemare, N. E. and Lewis, E. E. 2005. Biology and behavior. In: Grewal, P. S., Ehlers, R.-U. and Shapiro-Ilan, D. I. (Eds), *Nematodes as Biocontrol Agents*. CABI Publishing, Wallingford, UK, pp. 47–64.
Gumus, A., Karagoz, M., Shapiro-Ilan, D. I. and Hazir, S. 2015. A novel approach to biocontrol: release of live insect hosts pre-infected with entomopathogenic nematodes. *J. Invertebr. Pathol.* 130, 56–60. doi:10.1016/j.jip.2015.07.002.
Harris, M. A., Oetting, R. D. and Gardner, W. A. 1995. Use of entomopathogenic nematodes and a new monitoring technique for control of fungus gnats, *Bradysia coprophila* (Diptera: Sciaridae), in floriculture. *Biol. Control* 5(3), 412–8. doi:10.1006/bcon.1995.1049.
Helms, A. M., Ray, S., Matulis, N. L., Kuzemchak, M. C., Grisales, W., Tooker, J. F. and Ali, J. G. 2019. Chemical cues linked to risk: cues from below-ground natural enemies enhance plant defences and influence herbivore behaviour and performance. *Funct. Ecol.* 33(5), 798–808. doi:10.1111/1365-2435.13297.
Hiltpold, I. and Hibbard, B. E. 2018. Indirect root defenses cause induced fitness costs in Bt-resistant western corn rootworm. *J. Econ. Entomol.* 111(5), 2349–58. doi:10.1093/jee/toy220.

Hiltpold, I., Toepfer, S., Kuhlmann, U. and Turlings, T. C. J. 2010. How maize root volatiles affect the efficacy of entomopathogenic nematodes in controlling the western corn rootworm? *Chemoecology* 20(2), 155–62. doi:10.1007/s00049-009-0034-6.

Hiltpold, I., Hibbard, B. E., Wade French, B. W. and Turlings, T. C. J. 2012. Capsules containing entomopathogenic nematodes as a Trojan horse approach to control the western corn rootworm. *Plant Soil* 358(1–2), 11–25. doi:10.1007/s11104-012-1253-0.

Jackson, J. J. 1996. Field performance of entomopathogenic nematodes for suppression of western corn rootworm (Coleoptera: Chrysomelidae). *J. Econ. Entomol.* 89(2), 366–72. doi:10.1093/jee/89.2.366.

Jagdale, G. B., Casey, M. L., Grewal, P. S. and Lindquist, R. K. 2004. Effect of application rate and timing, potting medium, and host plant on efficacy of *Steinernema feltiae* against fungus gnat, *Bradysia coprophila*, in floriculture. *Biol. Control* 29(2), 296–305. doi:10.1016/S1049-9644(03)00164-6.

Jess, S., Schweizer, H. and Kilpatrick, M. 2005. Mushroom applications. In: Grewal, P. S., Ehlers, R.-U. and Shapiro-Ilan, D. I. (Eds), *Nematodes as Biocontrol Agents*. CABI Publishing, Wallingford, UK, pp. 191–213.

Kapranas, A., Malone, B., Quinn, S., Mc Namara, L., Williams, C. D., O'Tuama, P., Peters, A. and Griffin, C. T. 2017. Efficacy of entomopathogenic nematodes for control of large pine weevil, *Hylobius abietis*: effects of soil type, pest density and spatial distribution. *J. Pest Sci.* 90(2), 495–505. doi:10.1007/s10340-016-0823-y.

Kaya, H. K. 1990. Soil ecology. In: Gaugler, R. and Kaya, H. K. (Eds), *Entomopathogenic Nematodes in Biological Control*. CRC Press, Boca Raton, FL, pp. 93–115.

Klein, M. G. and Georgis, R. 1992. Persistence of control of Japanese beetle (Coleoptera: Scarabaeidae) larvae with steinernematid and heterorhabditid nematodes. *J. Econ. Entomol.* 85(3), 727–30. doi:10.1093/jee/85.3.727.

Koppenhöfer, A. M. and Fuzy, E. M. 2004. Effect of white grub developmental stage on susceptibility to entomopathogenic nematodes. *J. Econ. Entomol.* 97(6), 1842–9. doi:10.1603/0022-0493-97.6.1842.

Koppenhöfer, A. M. and Fuzy, E. M. 2006. Effect of soil type on infectivity and persistence of the entomopathogenic nematodes *Steinernema scarabaei*, *Steinernema glaseri*, *Heterorhabditis zealandica*, and *Heterorhabditis bacteriophora*. *J. Invertebr. Pathol.* 92(1), 11–22. doi:10.1016/j.jip.2006.02.003.

Koppenhöfer, A. M. and Fuzy, E. M. 2008. Early timing and new combinations to increase the efficacy of neonicotinoid–entomopathogenic nematodes (Rhabditida: Heterorhabditidae) combinations against white grubs (Coleoptera: Scarabaeidae). *Pest Manag. Sci.* 64(7), 725–35. doi:10.1002/ps.1550.

Koppenhöfer, A. M. and Fuzy, E. M. 2009. Long-term effects and persistence of *Steinernema scarabaei* applied for suppression of *Anomala orientalis* (Coleoptera: Scarabaeidae). *Biol. Control* 48(1), 63–72. doi:10.1016/j.biocontrol.2008.09.005.

Koppenhöfer, A. M. and Grewal, P. S. 2005. Compatibility and interactions with agrochemicals and other biocontrol agents. In: Grewal, P. S., Ehlers, R.-U. and Shapiro-Ilan, D. I. (Eds), *Nematodes as Biocontrol Agents*. CABI Publishing, Wallingford, UK, pp. 363–81.

Koppenhöfer, A. M. and Wu, S. 2017. Microbial control of insect pests of turfgrass. In: Lacey, L. A. (Ed.), *Microbial Control of Insect and Mite Pest: From Theory to Practice*. Elsevier, Amsterdam, The Netherlands, pp. 331–41.

Koppenhöfer, A. M., Choo, H. Y., Kaya, H. K., Lee, D. W. and Gelernter, W. D. 1999. Increased field and greenhouse efficacy with a combination of entomopathogenic nematode and *Bacillus thuringiensis* against scarab grubs. *Biol. Control* 14(1), 37–44. doi:10.1006/bcon.1998.0663.

Koppenhöfer, A. M., Cowles, R. S., Cowles, E. A., Fuzy, E. M. and Baumgartner, L. 2002. Comparison of neonicotinoid insecticides as synergists for entomopathogenic nematodes. *Biol. Control* 24(1), 90–7. doi:10.1016/S1049-9644(02)00008-7.

Koppenhöfer, A. M., Fuzy, E. M., Crocker, R. L., Gelernter, W. D. and Polavarapu, S. 2004. Pathogenicity of *Steinernema scarabaei*, *Heterorhabditis bacteriophora* and *S. glaseri* to twelve white grub species. *Biocontrol Sci. Technol.* 14(1), 87–92. doi:10.1080/0958315031000151701.

Koppenhöfer, A. M., Grewal, P. S. and Fuzy, E. M. 2006. Virulence of the entomopathogenic nematodes *Heterorhabditis bacteriophora*, *H. zealandica*, and *Steinernema scarabaei* against five white grub species (Coleoptera: Scarabaeidae) of economic importance in turfgrass in North America. *Biol. Control* 38(3), 397–404. doi:10.1016/j.biocontrol.2005.12.013.

Koppenhöfer, A. M., Kostromytska, O. S., McGraw, B. A. and Ebssa, L. 2015. Entomopathogenic nematodes in turfgrass: ecology and management of important insect pests in North America. In: Campos-Herrera, R. (Ed.), *Nematode Pathogenesis of Insect and Other Pests*. Springer International Publishing, Cham, Switzerland and Berlin, Germany, pp. 309–27.

Lacey, L. A. 2017. *Microbial Control of Insect and Mite Pests: From Theory to Practice*. Elsevier, Amsterdam, The Netherlands, 461pp.

Lacey, L. A., Neven, L. G., Headrick, H. L. and Fritts Jr., R. 2005. Factors affecting entomopathogenic nematodes (Steinernematidae) for the control of overwintering codling moth (Lepidoptera: Tortricidae) in fruit bins. *J. Econ. Entomol.* 98(6), 1863–9. doi:10.1603/0022-0493-98.6.1863.

Lacey, L. A., Arthurs, S. P., Unruh, T. R., Headrick, H. L. and Fritts Jr., R. 2006. Entomopathogenic nematodes for the control of codling moth (Lepidoptera: Tortricidae) in apple and pear orchards: effect of nematode species and seasonal temperatures, adjuvants, application equipment and post-application irrigation. *Biol. Control* 37(2), 214–23. doi:10.1016/j.biocontrol.2005.09.015.

Leite, L. G., Shapiro-Ilan, D. I., Hazir, S. and Jackson, M. A. 2017. Effect of inoculum age and physical parameters on *in vitro* culture of the entomopathogenic nematode *Steinernema feltiae*. *J. Helminthol.* 91(6), 686–95. doi:10.1017/S0022149X16000821.

Levine, E. and Oloumi-Sadeghi, H. 1992. Field evaluation of *Steinernema carpocapsae* (Rhabditida: Steinernematidae) against the black cutworm in field corn. *J. Entomol. Sci.* 27(4), 427–35. doi:10.18474/0749-8004-27.4.427.

Lewis, E. E. and Clarke, D. J. 2012. Nematodes parasites and entomopathogens. In: Vega, E. F. and Kaya, H. K. (Eds), *Insect Pathology*. Elsevier, Amsterdam, The Netherlands, pp. 395–424.

Lössbroek, T. G. and Theunissen, J. 1985. The entomogenous nematode *Neoaplectana bibionis* as a biological control agent of *Agrotis segetum* in lettuce. *Entomol. Exp. Appl.* 39(3), 261–4. doi:10.1111/j.1570-7458.1985.tb00468.x.

Maniania, N. K., Ekesi, S. and Dolinski, C. 2017. Entomopathogens routinely used in pest control strategies: orchards in tropical climate. In: Lacey, L. A. (Ed.), *Microbial Control of Insect and Mite Pests*. Elsevier, Amsterdam, The Netherlands, pp. 269–82.

McGraw, B. A., Vittum, P. J., Cowles, R. S. and Koppenhöfer, A. M. 2010. Field evaluation of entomopathogenic nematodes for the biological control of the annual bluegrass weevil, *Listronotus maculicollis* (Coleoptera: Curculionidae), in golf course turfgrass. *Biocontrol Sci. Technol.* 20(2), 149–63. doi:10.1080/09583150903440658.

Moore, S. D. and Duncan, L. W. 2017. Microbial control of insect and mite pests of citrus. In: Lacey, L. A. (Ed.), *Microbial Control of Insect and Mite Pests*. Elsevier, Amsterdam, The Netherlands, pp. 283–98.

Morales-Ramos, J. A., Rojas, M. G., Shapiro-Ilan, D. I. and Tedders, W. L. 2011. Automated insect separation system. US Patent 8,025,027 B1.

Mukuka, J., Strauch, O., Hoppe, C. and Ehlers, R.-U. 2010. Improvement of heat and desiccation tolerance in *Heterorhabditis bacteriophora* through cross-breeding of tolerant strains and successive genetic selection. *BioControl* 55(4), 511–21. doi:10.1007/s10526-010-9271-4.

Oestergaard, J., Belau, C., Strauch, O., Ester, A., van Rozen, K. and Ehlers, R.-U. 2006. Biological control of *Tipula paludosa* (Diptera: Nematocera) using entomopathogenic nematodes (*Steinernema* spp.) and *Bacillus thuringiensis* subsp. *israelensis*. *Biol. Control* 39(3), 525–31. doi:10.1016/j.biocontrol.2006.07.003.

Oliveira-Hofman, C., Kaplan, F., Stevens, G., Lewis, E., Wu, S., Alborn, H. T., Perret-Gentil, A. and Shapiro-Ilan, D. I. 2019. Pheromone extracts act as boosters for entomopathogenic nematodes efficacy. *J. Invertebr. Pathol.* 164, 38–42. doi:10.1016/j.jip.2019.04.008.

Parkman, J. P. and Frank, J. H. 1992. Infection of sound-trapped mole crickets, *Scapteriscus* spp., by *Steinernema scapterisci*. *Fla Entomol.* 75(1), 163–5. doi:10.2307/3495500.

Peters, A. 1996. The natural host range of *Steinernema* and *Heterorhabditis* spp. and their impact on insect populations. *Biocontrol Sci. Technol.* 6(3), 389–402. doi:10.1080/09583159631361.

Polavarapu, S., Koppenhöfer, A. M., Barry, J. D., Holdcraft, R. J. and Fuzy, E. M. 2007. Entomopathogenic nematodes and neonicotinoids for remedial control of oriental beetle, *Anomala orientalis* (Coleoptera: Scarabaeidae), in highbush blueberry. *Crop Prot.* 26(8), 1266–71. doi:10.1016/j.cropro.2006.10.026.

Portillo-Aguilar, C., Villani, M. G., Tauber, M. J., Tauber, C. A. and Nyrop, J. P. 1999. Entomopathogenic nematode (Rhabditida: Heterorhabditidae and Steinernematidae) response to soil texture and bulk density. *Environ. Entomol.* 28(6), 1021–35. doi:10.1093/ee/28.6.1021.

Power, K. T., An, R. and Grewal, P. S. 2009. Effectiveness of *Heterorhabditis bacteriophora* strain GPS11 applications targeted against different instars of the Japanese beetle *Popillia japonica*. *Biol. Control* 48(3), 232–6. doi:10.1016/j.biocontrol.2008.10.014.

Rasmann, S., Köllner, T. G., Degenhardt, J., Hiltpold, I., Toepfer, S., Kuhlmann, U., Gershenzon, J. and Turlings, T. C. J. 2005. Recruitment of entomopathogenic nematodes by insect-damaged maize roots. *Nature* 434(7034), 732–7. doi:10.1038/nature03451.

Scheepmaker, J. W. A., Geels, F. P., Smits, P. H. and van Griensven, L. J. L. D. 1998. Influence of *Steinernema feltiae* and diflubenzuron on yield and economics of the cultivated mushroom *Agaricus bisporus* in Dutch mushroom culture. *Biocontrol Sci. Technol.* 8(2), 269–75. doi:10.1080/09583159830333.

Schroeder, W. J. and Beavers, J. B. 1987. Movement of the entomogenous nematodes of the families Heterorhabditidae and Steinernematidae in soil. *J. Nematol.* 19(2), 257–9.

Shanks Jr., C. H. and Agudelo-Silva, F. 1990. Field pathogenicity and persistence of heterorhabditid and steinernematid nematodes (Nematoda) infecting black vine weevil larvae (Coleoptera: Curculionidae) in cranberry bogs. *J. Econ. Entomol.* 83(1), 107–10. doi:10.1093/jee/83.1.107.

Shapiro-Ilan, D. I. and Brown, I. 2013. Earthworms as phoretic hosts for *Steinernema carpocapsae* and *Beauveria bassiana*: implications for enhanced biological control. *Biol. Control* 66(1), 41–8. doi:10.1016/j.biocontrol.2013.03.005.
Shapiro-Ilan, D. I. and Gardner, W. A. 2012. Improved control of *Curuclio caryae* (Coleoptera: Curculionidae) through multi-stage pre-emergence applications of *Steinernema carpocapsae*. *J. Entomol. Sci.* 47(1), 27–34. doi:10.18474/0749-8004-47.1.27.
Shapiro-Ilan, D. I., Gouge, D. and Koppenhöfer, A. M. 2002. Factors affecting field efficacy: analysis of case studies in cotton, turf, and citrus. In: Gaugler, R. (Ed.), *Entomopathogenic Nematology*. CABI Publishing, Wallingford, UK, pp. 333-56.
Shapiro-Ilan, D. I., Duncan, L. W., Lacey, L. A. and Han, R. 2005. Orchard applications. In: Grewal, P. S., Ehlers, R.-U. and Shapiro-Ilan, D. I. (Eds), *Nematodes as Biocontrol Agents*. CABI Publishing, Wallingford, UK, pp. 215–29.
Shapiro-Ilan, D. I., Cottrell, T. E., Mizell, R. F., Horton, D. L., Behle, R. W. and Dunlap, C. A. 2010. Efficacy of *Steinernema carpocapsae* for control of the lesser peachtree borer, *Synanthedon pictipes*: improved aboveground suppression with a novel gel application. *Biol. Control* 54(1), 23–8. doi:10.1016/j.biocontrol.2009.11.009.
Shapiro-Ilan, D. I., Han, R. and Dolinski, C. 2012. Entomopathogenic nematode production and application technology. *J. Nematol.* 44(2), 206–17.
Shapiro-Ilan, D. I., Wright, S. E., Tuttle, A. F., Cooley, D. R. and Leskey, T. C. 2013. Using entomopathogenic nematodes for biological control of plum curculio, *Conotrachelus nenuphar*: effects of irrigation and species in apple orchards. *Biol. Control* 67(2), 123–9. doi:10.1016/j.biocontrol.2013.07.020.
Shapiro-Ilan, D. I., Lewis, E. E. and Schliekelman, P. 2014. Aggregative group behavior in insect parasitic nematode dispersal. *Int. J. Parasitol.* 44(1), 49–54. doi:10.1016/j.ijpara.2013.10.002.
Shapiro-Ilan, D. I., Hazir, S. and Leite, L. 2015a. Viability and virulence of entomopathogenic nematodes exposed to ultraviolet radiation. *J. Nematol.* 47(3), 184–9.
Shapiro-Ilan, D. I., Cottrell, T. E., Mizell III, R. F., Horton, D. L. and Zaid, A. 2015b. Field suppression of the peachtree borer, *Synanthedon exitiosa*, using *Steinernema carpocapsae*: effects of irrigation, a sprayable gel and application method. *Biol. Control* 82, 7–12. doi:10.1016/j.biocontrol.2014.12.003.
Shapiro-Ilan, D. I., Cottrell, T. E., Mizell III, R. F. and Horton, D. L. 2016a. Curative control of the peachtree borer using entomopathogenic nematodes. *J. Nematol.* 48(3), 170–6. doi:10.21307/jofnem-2017-024.
Shapiro-Ilan, D. I., Morales-Ramos, J. A. and Rojas, M. G. 2016b. *In vivo* production of entomopathogenic nematodes. In: Glare, T. and Moran-Diez, M. (Eds), *Microbial-Based Biopesticides - Methods and Protocols*. Human Press, New York, pp. 137–58.
Shapiro-Ilan, D., Arthurs, S. P. and Lacey, L. A. 2017. Microbial control of arthropod pests of orchards in temperate climates. In: Lacey, L. A. (Ed.), *Microbial Control of Insect and Mite Pests: From Theory to Practice*. Elsevier, Amsterdam, The Netherlands, pp. 253–67.
Shapiro-Ilan, D. I., Hiltpold, I. and Lewis, E. E. 2018. Ecology of invertebrate pathogens: nematodes. In: Hajek, A. E. and Shapiro-Ilan, D. I. (Eds), *Ecology of Invertebrate Diseases*. John Wiley & Sons, Ltd., Hoboken, NJ, pp. 415–40.
Shields, E. J. 2015. Utilizing persistent entomopathogenic nematodes in a more classical biological control approach. In: Campos-Herrera, R. (Ed.), *Nematode Pathogenesis*

of Insect and Other Pests. Springer International Publishing, Cham, Switzerland and Berlin, Germany, pp. 165–84.

Shields, E. J. and Testa, A. M. 2015. New York case study: biological control of *Otiorhynchus ligustici* with native persistent entomopathogenic nematodes using a more classical approach. In: Campos-Herrera, R. (Ed.), *Nematode Pathogenesis of Insect and Other Pests*. Springer International Publishing, Cham, Switzerland and Berlin, Germany, pp. 285–307.

Shields, E. J. and Testa, A. M. 2017. Biological control of alfalfa snout beetle with persistent entomopathogenic nematodes: expanding a single farm's success to an area-wide biological control program. *Am. Entomol.* 63(4), 216–23. doi:10.1093/ae/tmx046.

Shields, E. J., Testa, A. M. and O'Neil, W. J. 2018. Long-term persistence of native New York entomopathogenic nematode isolates across crop rotation. *J. Econ. Entomol.* 111(6), 2592–8, doi:10.1093/jee/toy258.

Siegel, J. P., Lacey, L. A., Higbee, B. S., Noble, P. and Fritts Jr., R. 2006. Effect of application rates and abiotic factors on *Steinernema carpocapsae* for control of overwintering navel orangeworm (Lepidoptera: Pyralidae, *Amyelois transitella*) in fallen pistachios. *Biol. Control* 36(3), 324–30. doi:10.1016/j.biocontrol.2005.09.012.

Stelinski, L. L., Willett, D., Rivera, M. J. and Ali, J. G. 2019. 'Tuning' communication among four trophic levels of the root biome to facilitate biological control. *Biol. Control* 131, 49–53. doi:10.1016/j.biocontrol.2019.01.006.

Stock, S. P. 2015. Diversity, biology and evolutionary relationships. In: Campos-Herrera, R. (Ed.), *Nematode Pathogenesis of Insect and Other Pests*. Springer International Publishing, Cham, Switzerland and Berlin, Germany, pp. 3–27.

Stuart, R. J., Barbercheck, M. E. and Grewal, P. S. 2015. Entomopathogenic nematode in the soil environment: distributions, interactions and the influence of biotic and abiotic factors. In: Campos-Herrera, R. (Ed.), *Nematode Pathogenesis of Insect and Other Pests*. Springer International Publishing, Cham, Switzerland and Berlin, Germany, pp. 97–137.

Toepfer, S., Peters, A., Ehlers, R. U. and Kuhlmann, U. 2008. Comparative assessment of the efficacy of entomopathogenic nematode species at reducing western corn rootworm larvae and root damage in maize. *J. Appl. Entom.* 132(5), 337–48. doi:10.1111/j.1439-0418.2008.01274.x.

Toepfer, S., Hatala-Zseller, I., Ehlers, R. U., Peters, A. and Kuhlmann, U. 2010. The effect of application techniques on field-scale efficacy: can the use of entomopathogenic nematodes reduce damage by western corn rootworm larvae? *Agric. Forest Entom.* 12(4), 389–402. doi:10.1111/j.1461-9563.2010.00487.x.

Toepfer, S., Knuth, P., Glas, M. and Kuhlmann, U. 2014. Successful application of entomopathogenic nematodes for the biological control of western corn rootworm larvae in Europe – a min review. *Julius Kühn-Archiv* 444, 59–66.

Tomalak, M., Piggott, S. and Jagdale, G. B. 2005. Glasshouse applications. In: Grewal, P. S., Ehlers, R.-U. and Shapiro-Ilan, D. I. (Eds), *Nematodes as Biocontrol Agents*. CABI Publishing, Wallingford, UK, pp. 147–66.

Torr, P., Wilson, M. J. and Heritage, S. 2005. Forestry applications. In: Grewal, P. S., Ehlers, R.-U. and Shapiro-Ilan, D. I. (Eds), *Nematodes as Biocontrol Agents*. CABI Publishing, Wallingford, UK, pp. 281–93.

Treverrow, N. and Bedding, R. A. 1993. Development of a system for the control of the banana weevil borer, *Cosmopolites sordidus*, with entomopathogenic nematodes.

In: Bedding, R. A., Akhurst, R. and Kaya, H. K. (Eds), *Nematodes and the Biological Control of Insect Pests*. CSIRO Publishing, Melbourne, Australia, pp. 41–7.

Turlings, T. C. J., Hiltpold, I. and Rasmann, S. 2012. The importance of root-produced volatiles as foraging cues for entomopathogenic nematodes. *Plant Soil* 358(1–2), 51–60. doi:10.1007/s11104-012-1295-3.

van Niekerk, S. and Malan, A. P. 2015. Adjuvants to improve aerial control of the citrus mealybug *Planococcus citri* (Hemiptera: Pseudococcidae) using entomopathogenic nematodes. *J. Helminthol.* 89(2), 189–95. doi:10.1017/S0022149X13000771.

van Tol, R. W. H. M. and Raupp, M. J. N. 2005. Nursery and tree applications. In: Grewal, P. S., Ehlers, R.-U. and Shapiro-Ilan, D. I. (Eds), *Nematodes as Biocontrol Agents*. CABI Publishing, Wallingford, UK, pp. 167–90.

Wang, X. and Grewal, P. S. 2002. Rapid deterioration of environmental tolerance and reproductive potential of an entomopathogenic nematode during laboratory maintenance. *Biol. Control* 23(1), 71–8. doi:10.1006/bcon.2001.0986.

Willett, D. S., Alborn, H. T. and Stelinski, L. L. 2017. Multitrophic effects of belowground parasitoid learning. *Sci. Rep.* 7(1), 2067. doi:10.1038/s41598-017-02193-2.

Willett, D. S., Alborn, H. T., Stelinski, L. L. and Shapiro-Ilan, D. I. 2018. Risk taking of educated nematodes. *PLoS ONE* 13(10), e0205804. doi:10.1371/journal.pone.0205804.

Wilson, M. J. and Grewal, P. S. 2005. Biology, production and formulation of slug-parasitic nematodes. In: Grewal, P. S., Ehlers, R.-U. and Shapiro-Ilan, D. I. (Eds), *Nematodes as Biocontrol Agents*. CABI Publishing, Wallingford, UK, pp. 421–9.

Wilson, M. and Rae, R. 2015. *Phasmorhabditis hermaphrodita* as a control agent for slugs. In: Campos-Herrera, R. (Ed.), *Nematode Pathogenesis of Insect and Other Pests*. Springer International Publishing, Cham, Switzerland and Berlin, Germany, pp. 509–21.

Wraight, S. P., Lopes, R. B. and Faria, M. 2017. Microbial control of mite and insect pests of greenhouse crops. In: Lacey, L. A. (Ed.), *Microbial Control of Insect and Mite Pests*. Elsevier, Amsterdam, The Netherlands, pp. 237–52.

Wu, S., Kaplan, F., Lewis, E., Alborn, H. T. and Shapiro-Ilan, D. I. 2018. Infected host macerate enhances entomopathogenic nematode movement towards hosts and infectivity in a soil profile. *J. Invertebr. Pathol.* 159, 141–4. doi:10.1016/j.jip.2018.10.007.

Chapter 10

Advances in the use of entomopathogenic oomycetes as biopesticides in suppressing crop insect pests

Aurélien Tartar, Nova Southeastern University, USA

1 Introduction

Oomycetes are heterotrophic, eukaryotic microbes that are phylogenetically related to diatoms and brown algae, and grouped with these photosynthetic relatives in the phylum Heterokonta (Kamoun et al., 2015). For most microbiologists and agricultural scientists, oomycetes have been traditionally associated with two fundamental and immutable biological characteristics – first, oomycetes are primarily known as devastating agents of plant diseases, especially in agrosystems, and second, oomycetes, and by extension oomycete-caused diseases, are typically compared to fungi and fungal-caused diseases, since oomycetes are morphologically similar to fungi and have erroneously been classified as fungi based on these similarities (hence the use of the suffix -mycete to describe these microbes). The impact of oomycetes as plant pathogens in agrosystems is exemplified by one the best-characterized oomycete species: *Phytophthora infestans*. This organism is notorious for having caused the potato blight disease that triggered the Irish Potato Famine in the late 1840s. Since then, it has consistently ranked as one of the most serious pathogens of crop plants worldwide, and even

http://dx.doi.org/10.19103/AS.2020.0073.11

today is estimated to cause around $6 billion of damage to crops each year, mostly on potato and tomato plants (Nowicki et al., 2012). In addition to *P. infestans*, other oomycetes currently affecting crops and causing significant losses include *P. sojae* and *P.capsici*, as well as *Pythium* spp., which are known as agents responsible for root rots. A list of the most agriculturally important plant pathogenic oomycetes has recently been compiled and published (Kamoun et al., 2015), and reviews focused on the threats posed by current emerging infectious diseases have also routinely included oomycetes (Derevnina et al., 2016; Fisher et al., 2012; Anderson et al., 2004). Because of their impact on plants, the majority of the work directed at understanding the biology of oomycetes is aimed at controlling or eliminating these organisms from anthropogenic agroecosystems. Part of these efforts have included oomycete genome analyses that have contributed to detailing the molecular basis of oomycete-plant interactions, and led to the detection of several classes of novel molecules (effectors) that have been hypothesized to mediate recognition and infection of plant tissues by oomycetes (Bozkurt et al., 2012). Although comparisons to fungi remain unavoidable (Latijnhouwers et al., 2003), the discovery of an oomycete-specific molecular arsenal used during plant infections has contributed to further distinguish these microbes from fungi, and prompted a renewed interest in exploring fundamental oomycete biology. In this context, research on animal pathogenic oomycetes is also increasing, due in part to the need experienced by plant pathologists to discriminate oomycete genome signature(s) that may truly be associated with the ability to infect plant tissue from the more mundane genome content that is conserved in all oomycete species and their phylogenetic relatives, but also because several important emerging animal diseases have been linked to oomycetes (Phillips et al., 2008). Such diseases include saprolegniasis (caused by *Saprolegnia parasitica* and associated species) that impacts the fish farming industry (Derevnina et al., 2016), and pythiosis (caused by *Pythium insidiosum*), that is recognized as an emerging disease affecting vertebrates, including dogs, horses and in some cases humans (Gaastra et al., 2010). Based on the recurrent ability demonstrated by oomycetes to cause devastating and lethal diseases in diverse hosts, entomopathogenic oomycetes warrant special interest from entomologists and integrated pest management specialists. Although a few species of oomycetes that infect and kill insects are described, they have typically been regrouped with fungi in most insect pathology textbooks and have been overshadowed by the more mainstream entomopathogenic fungal species such as *Beauveria bassiana* and *Metarhizium anisopliae* (Araújo and Hughes, 2016). In this chapter, recent research progress on entomopathogenic oomycetes is summarized, and their potential as biopesticides or as sources or novel bioactive compounds against insect pests is assessed.

2 Oomycetes as invertebrate pathogens

Despite being primarily known as plant pathogens, emerging phylogenomic evidence inferred from oomycete genomes has suggested that phytopathogenicity is a derived trait that has evolved multiple times independently in several oomycetes subgroups. Phylogeny reconstructions have depicted several invertebrate pathogens as basal oomycetes (in relation to diatoms and brown algae), suggesting that oomycetes may have evolved primarily as marine animal pathogens, with sea nematodes and arthropods representing potential early hosts (McCarthy and Fitzpatrick, 2017). The extent of the oomycete diversity known today still includes isolates that are primarily known as lower invertebrate and insect pathogens, and insect pathologists and entomologists have routinely listed oomycetes as effective insect pathogens in review efforts aimed at describing the taxonomic diversity associated with entomopathogenicity. As stated earlier, mentions of oomycetes have typically been incorporated in larger sections focused on fungi, and have included several obscure species that remain difficult to track through scientific literature searches, and even more difficult to locate in microbial culture collections (Araújo and Hughes, 2016). For the purpose of clarity, Table 1 lists the best-characterized oomycete entomopathogens for which recent scientific literature is available, and that currently represent the main source of information for agricultural scientists interested in the potential of oomycetes as biocontrol agents against insect crop pests. The list includes the two most common and studied entomopathogenic oomycetes, *Lagenidium giganteum* and *Leptolegnia chapmanii* (Boucias and Pendland, 2012), as well as a recently described species, *Pythium guiyangense*, that may rapidly be developed as a biopesticide (Shen et al., 2019). The presence of these three species (Table 1) underscores a general trend that has long been associated with entomopathogenic oomycetes: these pathogens have been almost exclusively

Table 1 Main species of entomopathogenic oomycetes

Oomycete species	Insect hosts	Recent publications
Lagenidium giganteum	*Aedes* spp., *Culex* spp. Other mosquitoes, other insects	Quiroz Velasquez et al. (2014)
Leptolegnia chapmanii	*Aedes* spp. Other mosquitoes	Gutierrez et al. (2017)
Pythium guiyangense	*Aedes* spp., Other mosquitoes	Shen et al. (2019)
Saprolegnia parasitica	*Galleria mellonella*[a] Other insects[b]	Wuensch et al. (2018)

[a] heterologous host.
[b] *Saprolegnia* spp. infect other insects (Sarowar et al., 2013).

found infecting mosquito larvae, in the aquatic environments that are used as breeding grounds by mosquitoes. Although they have been highly regarded for their potential as biopesticides against pest and vector mosquitoes (Scholte et al., 2004), the best-studied entomopathogenic oomycetes arguably hold little value for agricultural scientists and entomologists focused on controlling crop insect pests. To date, there has been no description of oomycetes infecting major pests of crop plants, and only marginal instances of oomycetes infecting insects other than mosquitoes have been reported. These occurrences have remained localized in aquatic environments reminiscent of mosquito breeding grounds, and have involved other groups of aquatic, dipteran insect hosts, such as midge and caddisfly eggs and larvae (De Souza et al., 2014; Sarowar et al., 2013). Despite the lack of apparent connections between oomycetes and insect pests of crop plants, reports of well-studied mosquito-oomycete pathosystems highlight two major points that may be of significance for agricultural scientists: (i) species of oomycetes have evolved to become highly efficient pathogens of insects, capable of causing epizootics in natural populations, and (ii) these oomycetes have been demonstrated to use bioactive compounds to recognize and kill insects and therefore can be considered as sources of novel bioinsecticides. The recent detections of novel oomycete entomopathogens such as *P. guiyangense* (Shen et al., 2019) or *Saprolegnia* spp. (Sarowar et al., 2013) suggest that additional oomycete species infecting insects may have yet to be discovered. Given the ubiquitous presence of plant-associated oomycetes in crop agrosystems, it is possible that oomycetes infecting insects have evolved in these environments, but have remained undetected. Interestingly, oomycetes have long been recognized as ecologically similar to fungi, (i.e. occupying similar niches), and a recent study suggested that, like entomopathogenic fungi, entomopathogenic oomycetes may engage in tripartite interactions with insects and plants, and may have the ability to establish symbiotic, or at least asymptomatic, relationships with plant tissues (Leoro-Garzon et al., 2019). In this regard, integrated management specialists should pay special consideration to the expanding field of plant and agricultural soil microbiome surveys, which aim to inventory the microbial species inhabiting these specific environments. Pilot surveys focused on oomycete diversity have revealed that *Lagenidium* spp., which are traditionally associated with pathogenicity to invertebrates (Table 1), could be detected in low abundance in forest soil samples from Estonia (Riit et al., 2016), and in pea field soils from Canada (Esmaeili Taheri et al., 2017). Additional soil metagenomic studies could potentially provide evidence for the presence of novel oomycete isolates that negatively impact insect pests in crop systems, and that may be incorporated with other pest control management tools. An alternative avenue for the development for novel insecticides could assess known oomycetes against heterologous insect hosts. As illustrated in Table 1, an intriguing and

Figure 1 Interactions between oomycetes and insects. Oomycetes are also known as water molds, and they appear morphologically similar to fungi, with filamentous, branching hyphae forming a mycelium. Entomopathogenic oomycetes can infect and kill insects, using the nutrient-rich hemocoel as a source of energy for massive vegetative growth. The oomycete *Lagenidium giganteum* (pictured) naturally infect mosquito larvae and can be grown in the lab in artificial, defined media. Vegetative growth can also be sustained in minimal media, with heterologous host insects (the cricket *Gryllus assimilis*) as the only source of proteins.

promising study using *Saprolegnia parasitica* has highlighted the untapped potential of oomycetes as sources of novel compounds with demonstrated biological activities against insects (Wuensch et al., 2018). In agreement with this study, recent observations have demonstrated that *Lagenidium giganteum* can recognize and degrade insect substrates corresponding to heterologous hosts and potential pest targets (Fig. 1).

3 Case study: *Lagenidium giganteum* as a biocontrol agent against insects

Although the most studied entomopathogenic oomycetes are principally known as mosquito larvae pathogens (Table 1) and therefore may be seen as irrelevant to pest insect control in crop fields, discussions on oomycetes as potential biopesticides would not be complete without mentioning the fact that *L. giganteum* was registered as a microbial control agent by the US Environmental Protection Agency in 1995 (Registration number 69592-2), and developed as a commercial product under the trade name Laginex (Scholte et al., 2004). The approved formulations have since been deregistered and/or not renewed by the manufacturer, and are no longer used or available, but this attempt at commercialization of an oomycete-based biopesticide for the control of pest insects has provided manufacturing and legislative precedents that remain valuable for the potential development of future products. Formulation developments were accompanied by a large body of research work that was focused on the mass production of *L. giganteum* spores at industrial scales (May et al., 2006), and the efficiency of various Laginex concentrations in field trials (Hallmon et al., 2000). Earlier surveys demonstrated that *L. giganteum* can create massive epizootic in host populations (Kerwin and Washino, 1988), can persist and overwinter in the field (Fetter-Lasko and Washino, 1983), and was compatible with the use of other biological control agents such as the bacterium *Bacillus thuringiensis* (Orduz and Axtell, 1991). The release of commercial products was also preceded by numerous safety studies that concluded that prolonged exposure to high concentrations of various *L. giganteum* developmental stages had no impact of non-target organisms, including mammals (Kerwin et al., 1990). However these safety assessments have since been challenged by several reports of *L. giganteum* infections in dogs, which contributed to drastically decrease the potential of this oomycete as a biocontrol tool, and may be associated with the Laginex deregistration request (Vilela et al., 2015). Taxonomic characterizations are currently aimed at understanding the relationship between the mosquito-infecting and mammal-infecting isolates, and have suggested that thermotolerance may have enabled specific *Lagenidium* strains to affect warm-bodied animals (Vilela et al., 2015). Additional work is sorely needed to more clearly characterize these different strains, and all the many *Lagenidium* novel species that were recently proposed (Vilela et al., 2019). Aside from possible effects on non-target organisms, the viability of Laginex as a oomycete-based biopesticide may have also been compromised by other factors, such as inconsistent efficacy in the field, and especially, very poor shelf life and storage survival (Vandergheynst et al., 2007). The stability challenges encountered during Laginex storage have prompted additional studies aimed at defining more robust formulations for *L. giganteum* (Maldonado-Blanco et al., 2011). Advances in oomycete mass production have also contributed to the commercial release of another oomycete-based product (a biofungicide known under the trade name Polyversum) based on the mycoparasite *Pythium oligandrum* (Marjańska-Cichoń et al., 2009). Products such as Polyversum suggest that developing oomycetes as biopesticides remains a realistic endeavor, and that some of shelf life and stability drawbacks reported for Laginex may be overcome. In addition, the development, and subsequent deregistration, of Laginex has created both an industrial opportunity and a market void that may be exploited. The oomycetes *Leptolegnia chapmanii* and *Pythium guiyangense* (Table 1) are under continued, serious consideration for the development of a novel generation of biopesticides against insects (Liu et al., 2018). Significant progress on these species, as well as a more precise characterization of mosquito-infecting *Lagenidium* spp. strains, may re-establish oomycete-based products similar to Laginex as safe and reliable biocontrol agents against pest insects. Alternatively, strategies designed to avoid the Laginex drawbacks have shifted away from mass production and environmental release of infectious stages to promote molecular investigations and the identification of oomycete-produced molecules showing biological activity against insect pests. The detection of novel metabolites has been made possible by a large and growing body of research work detailing oomycete-insect interactions, and the pathogenicity process associated with oomycete infections in insects.

4 Mode of infection by entomopathogenic oomycetes

Detailed microscopic observations of the oomycete-mosquito pathosystems listed in Table 1 have shed lights on the pathogenicity process and infectious strategies used by oomycetes to kill insects. These observations were performed by insect pathologists and included descriptions of filamentous, mycelia-like hyphae (Fig. 1) growing out of susceptible insect hosts, and contributed to reinforce the parallels between oomycetes and fungi. Generally, the mode of infection is similar to entomopathogenic fungi: the oomycete infectious propagules (zoospores) have been showed to attach to the insect host cuticle and initiate a germinating process where a combination of mechanical and chemical forces is used to disrupt and penetrate the host exoskeleton. This allows the growing germinating structures to access the host hemocoel, where massive vegetative mycelial growth simultaneously overwhelms the host defense system and deplete the host nutrients (Gutierrez et al., 2017; Boucias and Pendland, 2012). Microbial growth within the hemocoel ultimately leads to starvation, tissue destruction and death. The presence of toxins has also been reported in *L. giganteum* (Vyas et al., 2007), possibly accelerating the disease process and allowing the pathogen to switch from biotrophy to necrotrophy. Following insect death and utilization of all nutrient sources, vegetative growth stops, and production of new infectious zoospores terminates the cycle. This infection strategy, based on cuticle disruption and access to hemocoel, appears conserved in all oomycete entomopathogens (Shen et al., 2019) and suggests that *L. giganteum*, *L. chapmanii* and *P. guiyangense* metabolites include molecules that recognize and digest insect cuticle substrates, and that may potentially be translated as broad-range, contact insecticides. Similarities with entomopathogenic fungi also offer the opportunity to develop comparative analyses aimed at identifying conserved molecules that are used to disrupt the insect cuticle and gain access to the hemocoel. Pioneering work on the *L. chapmanii* life cycle indicated that a second infectious strategy might consist of colonization of the gut system and disruption of the peritrophic membrane (Zattau and McInnis Jr., 1987). Interestingly, similar observations were recently reported for *P. guiyangense* (Shen et al., 2019), suggesting that a *per os* infectious pathway, potentially similar to what is known for entomopathogenic eukaryotes such as the alga *Helicosporidium parasiticum* (Tartar, 2013), may also be conserved among all entomopathogenic oomycetes. An oral infectious route has yet to be mentioned for *L. giganteum*, but may provide additional, untapped avenues for the development of oomycete-based insecticides. Regardless on the infection route, successful infections by entomopathogenic oomycetes terminate with access to the hemolymph, where defense systems such as melanization are triggered but ultimately evaded, and vegetative propagation leads to the host death. This stage may also be linked with the production of biologically active molecules with insecticidal potential. In particular, the use of *Galleria mellonella* as a heterologous host suggested that production of such molecules may be species-specific and restricted to animal pathogenic oomycetes, since *Saprolegnia parasitica*, but not *Phytophthora infestans*, is able to successfully colonize the host hemolymph (Wuensch et al., 2018). Overall, detailed descriptions of oomycete infections in susceptible insect hosts contributed to distinguish three major stages that can be further studied for the presence of pathogenicity determinants with activity against insects: (i) recognition, binding, disruption and penetration of the insect cuticle, (ii) interaction and ingress through the peritrophic membrane and (iii) debilitation of the host defense system and digestion of the host biomass. Future use of oomycetes as biopesticides may include identifying the active metabolites involved in any of these stages.

5 Oomycetes as sources of novel biopesticidal metabolites

Combining the absence of known oomycetes infecting crop pest insects, the technical difficulties associated with production and storage of previous oomycete-based biocontrol products, and the potential effects on non-target mammalian populations related with large scale environmental release of infectious microbes, it is understandable that the most recent and promising advances in using oomycetes for insect pest management have been directed at identifying novel metabolites. Current efforts in genome sequencing have especially been aimed at detecting genes responsible for the entomopathogenicity displayed by specific oomycetes (Quiroz Velasquez et al., 2014; Shen et al., 2019). Given the similarity between the fungal and oomycete infectious processes in susceptible insect hosts, an obvious, if not understated, objective is that oomycete gene products that can be associated with virulence to insects can almost immediately be tested to generate transgenic, hypervirulent strains of pest-specific fungi. Strategies that included producing fungi with increased cuticle degrading abilities, recombinant toxin (from spider/scorpion) production or enhanced nutrient depletion capabilities have all been previously tested and reported as effective means to design more efficient and safe mycopesticides, and been the subject of recent reviews (Zhao et al., 2016; Lovett and St. Leger, 2018). One of the first molecules demonstrated to be involved in the oomycete-insect interactions corresponded to a trehalase enzyme that was shown to be more effective in the lysis of trehalose than the host enzyme, allowing the pathogen to outcompete the host for access to trehalose (the major source of sugar in the insect hemolymph), and contributing to insect starvation (McInnis and Domnas, 1973). Interestingly, overexpression of a fungal trehalase through genetic transformation was recently used to produce strains with enhanced virulence (Peng et al., 2015), illustrating the potential associated with the identification of additional oomycete virulence factors. Oomycete gene products may offer the double benefit to be completely compatible with the fungal infectious process, yet represent molecular mechanisms that evolved independently in a phylogenetically distant lineage. The identification of novel, oomycete virulence factors with activity against insect has been recently boosted by two genomic and transcriptomic efforts, from *Pythium guiyangense* (Shen et al., 2019) and *Lagenidium giganteum* (Quiroz Velasquez et al., 2014). Genome sequences are also available for *Saprolegnia parasitica* (Jiang et al., 2013). Gene discovery in these genomes may be accelerated by taking advantage of the entomopathogenic oomycetes' unique position for dual comparative analysis (Fig. 2). Entomopathogenic oomycete genomes can be compared to available sequences from entomopathogenic fungi, and from phytopathogenic oomycetes, in an effort to, respectively, detect genes that may be conserved in filamentous entomopathogens, and/or genes than may be unique to entomopathogenic oomycetes and not present in the better characterized plant pathogens. Pioneering comparisons have led to the identification of several candidate molecules. Using comparative analyses to plant pathogenic oomycetes, genes orthologous to the oomycete-specific effectors crinkler proteins have been detected in both *P. guiyangense* and *L. giganteum* (Quiroz Velasquez et al., 2014; Shen et al., 2019). Crinkler (also known as CRN, for crinkling and necrosis) molecules were initially named for the characteristic crinkling phenotype that they induce during plant assays, and have since recurrently been associated with cell death-inducing activities in susceptible hosts (Stassen and Van Den Ackerveken, 2011). Their roles in plant infection by oomycetes such as *Phytophthora infestans* has been well established (Haas et al., 2009). However, it remains unclear if entomopathogenic oomycetes use CRN protein during insect infections. Gene expression analyses showed that *L. giganteum* CRN genes are not overexpressed in the presence of insect hosts (Olivera et al., 2016), but cell toxicity assays demonstrated that some *P. guiyangense* CRN proteins display cytotoxicity toward insect cells (Shen et al., 2019). Because CRN orthologs have been reported in other, unrelated animal pathogens (Sun et al., 2011), these proteins remain strong candidates

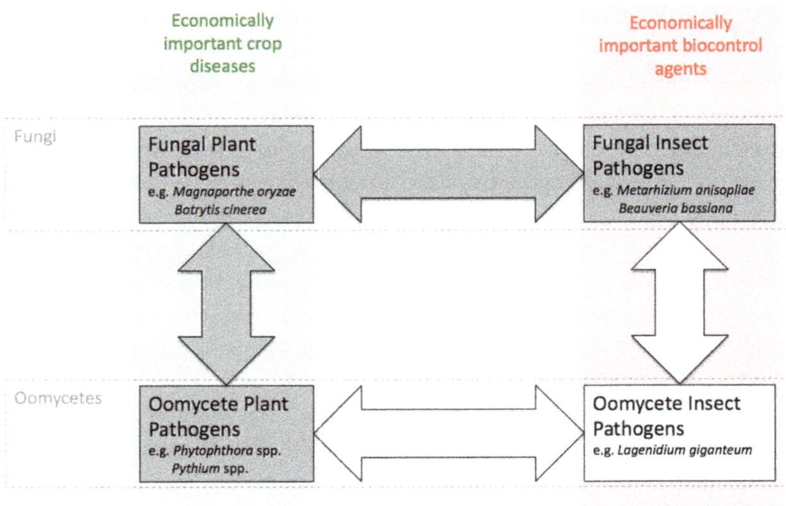

Figure 2 A model for future investigations on entomopathogenic oomycetes with relevance to agriculture. Molecular data generated from oomycete-insect pathosystems can be compared to both oomycetes that create diseases in crop plants, and entomopathogenic fungi that are currently considered and/or commercialized as biological control agents of crop pest insects (white block arrows). These comparative analyses may mirror more common comparisons between fungi or between plant pathogens (grey arrows), and allow for the identification of novel metabolites specifically used by entomopathogenic oomycetes against their insect targets.

as virulence factors associated with the later stages of the infectious cycle, where their cytotoxic activities may contribute to host tissue degradation or debilitation of the cell-based host defense system. CRN proteins were also recently associated with DNA damage in host cells (Ramirez-Garcés et al., 2016). Another potentially toxic molecule identified from *L. giganteum* is a glycoside hydrolase known as GH20 (Glycoside Hydrolase family 20), and predicted to function as a hexosaminidase (Olivera et al., 2016). In contrast to CRN effectors, GH20 proteins appeared restricted to animal pathogenic oomycetes, and they also have demonstrated activity against pest insects, as corn plants transformed to express GH20 proteins caused mortality and reduced growth in the cigarette beetle, *Lasioderma serricorne*, and in the fall armyworm, *Spodoptera frugiperda* (Dowd et al., 2007). Genomic analyses, demonstrating that GH20 genes are present in the *Saprolegnia parasitica* genome, but not in the *Phytophthora infestans* genome, provide additional support for the hypothesis that GH20 activity may explain the difference of pathogenic abilities between *S. parasitica* and *P. infestans* (Wuensch et al., 2018), suggesting that GH20 may contribute to oomycete infections in insects. Further work is needed to more precisely define the GH20 substrate and its potential role during the various stages of the infectious cycle. Finally, and similar to GH20, another glycoside hydrolase gene has been identified as part of a *L. giganteum* transcriptome survey. This gene, annotated as GH5_27 (family 5, subfamily 27), has been predicted to code for an endoglycosylceramidase (Quiroz Velasquez et al., 2014). More importantly, comparative genomics analyses have suggested that GH5_27 are preferentially present in entomopathogenic oomycete and fungi, and predominantly absent in non-entomopathogenic genomes, suggesting that they may play a key role in the infectious process. To date, these proteins have tentatively been linked to cuticle disruption, although a full characterization of their biochemical activity, and their impact on insects, has yet to be performed. The identification of GH5_27 has illustrated a final application linked with the study of entomopathogenic oomycetes: detailing oomycete-insect interactions may reveal pathogenicity factors that are shared between fungi and oomycetes, and therefore indirectly assist in the discovery of novel fungal compounds (from well-known biocontrol agents such as *Beauveria* or *Metarhizium* spp.) that may be more directly used against insect pest of crop plants. Overall, genome sequence analyses of oomycetes that cause diseases in insects have been initiated only very recently but led to the detection of several genes that warrant additional attention. Additional genes of interest have been listed for *P. guiyangense* (Shen et al., 2019) and this list is likely to grow bigger in the near future, setting up the possibility of seeing novel biopesticides derived from the growing understanding of the molecular basis of insect infections.

6 Conclusion and future trends

Oomycetes are effective and recognized pathogens of insects, and they represent a potential option for the development of novel insect control strategy. Early realization of the oomycete potential as biopesticides against insects has included the approval, mass production and commercialization of oomycete-based products. However, oomycetes are primarily mosquito pathogens and they have yet to be directly associated with pests of crop plants, which limits their potential for immediate use in agrosystems. In addition, early, fruitful and well-documented efforts in the development of oomycete-based biopesticides have since been mitigated, at least in part, by technical difficulties in production and storage, and reports of infections in non-target populations (including mammals). Alternative oomycete species remain available for the development of biopesticides mainly directed at controlling mosquitoes, and it is possible that modern microbial detection techniques will lead to the identification of novel isolates that directly impact pest insects of crop plants. A more viable strategy for the near future appears to favor oomycetes as sources of novel metabolites with activity against insects. Supported by preliminary microscopic observations and infection assays, several genome sequencing projects have recently advanced the study of entomopathogenic oomycetes, and initiated a search for oomycete genes that may be translated in safe and reliable insecticides (in the form of hypervirulent fungi for example). Gene products such as CRN, GH5_27 or GH20 illustrate the potential of these screening methodologies, and thorough characterizations of these diverse proteins may define their roles in the pathogenicity process and their impact on insects. Future work aimed at using oomycetes as biopesticides will combine these biochemical characterizations with the search for additional pathogenicity factors from sequenced genomes, as well as further genomic explorations in less-studied isolates. Long-term objectives will consist of turning this growing body of laboratory-based work in real-world applications, and the development of novel insect control strategies that will contribute in increasing the environmental and financial sustainability of crop agriculture systems.

7 Where to look for further information

7.1 Further reading

A quick guide on oomycetes has recently been made available:

- Thines, M. 2018. Oomycetes. *Current Biology* 28(15), R812–13.

Other reviews of interest include:

- Araújo, J. P. and Hughes, D. P. 2016. Diversity of entomopathogenic fungi: which groups conquered the insect body? *Advances in Genetics* 94, 1–39.
- Phillips, A. J., Anderson, V. L., Robertson, E. J., Secombes, C. J. and Van West, P. 2008. New insights into animal pathogenic oomycetes. *Trends in Microbiology* 16(1), 13–19.

7.2 Key conferences

The Oomycete Molecular Genetic Network (OMGN) is an active group that includes the international leaders in oomycete research as well as agriculture specialists that are deeply involved in monitoring crop diseases in farm systems, and similar extension efforts. The group holds annual meetings (www.omgn.org), sometimes in conjunction with the Fungal Genetics Conference.

8 References

Anderson, P. K., Cunningham, A. A., Patel, N. G., Morales, F. J., Epstein, P. R. and Daszak, P. 2004. Emerging infectious diseases of plants: pathogen pollution, climate change and agrotechnology drivers. *Trends in Ecology and Evolution* 19(10), 535–44. doi:10.1016/j.tree.2004.07.021.

Araújo, J. P. and Hughes, D. P. 2016. Diversity of entomopathogenic fungi: which groups conquered the insect body? *Advances in Genetics* 94, 1-39.
Boucias, D. G. and Pendland, J. C. 2012. *Principles of Insect Pathology*. Springer Science & Business Media.
Bozkurt, T. O., Schornack, S., Banfield, M. J. and Kamoun, S. 2012. Oomycetes, effectors, and all that jazz. *Current Opinion in Plant Biology* 15(4), 483-92. doi:10.1016/j.pbi.2012.03.008.
Derevnina, L., Petre, B., Kellner, R., Dagdas, Y. F., Sarowar, M. N., Giannakopoulou, A., De la Concepcion, J. C., Chaparro-Garcia, A., Pennington, H. G., van West, P. and Kamoun, S. 2016. Emerging oomycete threats to plants and animals. *Philosophical Transactions of the Royal Society of London. Series B, Biological Sciences* 371(1709). doi:10.1098/rstb.2015.0459.
De Souza, J. I., Gleason, F. H., Ansari, M. A., Lastra, C. C. L., Garcia, J. J., Pires-Zottarelli, C. L. A. and Marano, A. V. 2014. Fungal and oomycete parasites of Chironomidae, Ceratopogonidae and Simuliidae (Culicomorpha, Diptera). *Fungal Biology Reviews* 28(1), 13-23. doi:10.1016/j.fbr.2014.02.002.
Dowd, P. F., Johnson, E. T. and Pinkerton, T. S. 2007. Oral toxicity of β-N-acetyl hexosaminidase to insects. *Journal of Agricultural and Food Chemistry* 55(9), 3421-8. doi:10.1021/jf063562w.
Esmaeili Taheri, A., Chatterton, S., Gossen, B. D. and Mclaren, D. L. 2017. Metagenomic analysis of oomycete communities from the rhizosphere of field pea on the Canadian prairies. *Canadian Journal of Microbiology* 63(9), 758-68. doi:10.1139/cjm-2017-0099.
Fetter-Lasko, J. L. and Washino, R. K. 1983. In situ studies on seasonality and recycling pattern in California of *Lagenidium giganteum* Couch, an aquatic fungal pathogen of mosquitoes. *Environmental Entomology* 12(3), 635-40. doi:10.1093/ee/12.3.635.
Fisher, M. C., Henk, D. A., Briggs, C. J., Brownstein, J. S., Madoff, L. C., McCraw, S. L. and Gurr, S. J. 2012. Emerging fungal threats to animal, plant and ecosystem health. *Nature* 484(7393), 186-94. doi:10.1038/nature10947.
Gaastra, W., Lipman, L. J., De Cock, A. W., Exel, T. K., Pegge, R. B., Scheurwater, J., Vilela, R. and Mendoza, L. 2010. *Pythium insidiosum*: an overview. *Veterinary Microbiology* 146(1-2), 1-16. doi:10.1016/j.vetmic.2010.07.019.
Gutierrez, A. C., Rueda Páramo, M. E., Falvo, M. L., López Lastra, C. C. and García, J. J. 2017. *Leptolegnia chapmanii* (Straminipila: Peronosporomycetes) as a future biorational tool for the control of *Aedes aegypti* (L.). *Acta tropica* 169, 112-8. doi:10.1016/j.actatropica.2017.01.021.
Haas, B. J., Kamoun, S., Zody, M. C., Jiang, R. H., Handsaker, R. E., Cano, L. M., Grabherr, M., Kodira, C. D., Raffaele, S., Torto-Alalibo, T., Bozkurt, T. O., Ah-Fong, A. M., Alvarado, L., Anderson, V. L., Armstrong, M. R., Avrova, A., Baxter, L., Beynon, J., Boevink, P. C., Bollmann, S. R., Bos, J. I., Bulone, V., Cai, G., Cakir, C., Carrington, J. C., Chawner, M., Conti, L., Costanzo, S., Ewan, R., Fahlgren, N., Fischbach, M. A., Fugelstad, J., Gilroy, E. M., Gnerre, S., Green, P. J., Grenville-Briggs, L. J., Griffith, J., Grünwald, N. J., Horn, K., Horner, N. R., Hu, C. H., Huitema, E., Jeong, D. H., Jones, A. M., Jones, J. D., Jones, R. W., Karlsson, E. K., Kunjeti, S. G., Lamour, K., Liu, Z., Ma, L., Maclean, D., Chibucos, M. C., McDonald, H., McWalters, J., Meijer, H. J., Morgan, W., Morris, P. F., Munro, C. A., O'Neill, K., Ospina-Giraldo, M., Pinzón, A., Pritchard, L., Ramsahoye, B., Ren, Q., Restrepo, S., Roy, S., Sadanandom, A.,

Savidor, A., Schornack, S., Schwartz, D. C., Schumann, U. D., Schwessinger, B., Seyer, L., Sharpe, T., Silvar, C., Song, J., Studholme, D. J., Sykes, S., Thines, M., van de Vondervoort, P. J., Phuntumart, V., Wawra, S., Weide, R., Win, J., Young, C., Zhou, S., Fry, W., Meyers, B. C., van West, P., Ristaino, J., Govers, F., Birch, P. R., Whisson, S. C., Judelson, H. S. and Nusbaum, C. 2009. Genome sequence and analysis of the Irish potato famine pathogen *Phytophthora infestans*. *Nature* 461(7262), 393–8. doi:10.1038/nature08358.

Hallmon, C. F., Schreiber, E. T., Vo, T. and Bloomquist, A. 2000. Field trials of three concentrations of Laginex as biological larvicide compared to Vectobac-12AS as a biocontrol agent for *Culex quinquefasciatus*. *Journal of the American Mosquito Control Association* 16(1), 5–8.

Jiang, R. H., De Bruijn, I., Haas, B. J., Belmonte, R., Lobach, L., Christie, J., Van Den Ackerveken, G., Bottin, A., Bulone, V., Diaz-Moreno, S. M., Dumas, B., Fan, L., Gaulin, E., Govers, F., Grenville-Briggs, L. J., Horner, N. R., Levin, J. Z., Mammella, M., Meijer, H. J., Morris, P., Nusbaum, C., Oome, S., Phillips, A. J., Van Rooyen, D., Rzeszutek, E., Saraiva, M., Secombes, C. J., Seidl, M. F., Snel, B., Stassen, J. H., Sykes, S., Tripathy, S., Van Den Berg, H., Vega-Arreguin, J. C., Wawra, S., Young, S. K., Zeng, Q., Dieguez-Uribeondo, J., Russ, C., Tyler, B. M. and Van West, P. 2013. Distinctive expansion of potential virulence genes in the genome of the oomycete fish pathogen *Saprolegnia parasitica*. *PLoS Genetics* 9(6), e1003272. doi:10.1371/journal.pgen.1003272.

Kamoun, S., Furzer, O., Jones, J. D., Judelson, H. S., Ali, G. S., Dalio, R. J., Roy, S. G., Schena, L., Zambounis, A., Panabières, F., Cahill, D., Ruocco, M., Figueiredo, A., Chen, X. R., Hulvey, J., Stam, R., Lamour, K., Gijzen, M., Tyler, B. M., Grünwald, N. J., Mukhtar, M. S., Tomé, D. F., Tör, M., Van Den Ackerveken, G., McDowell, J., Daayf, F., Fry, W. E., Lindqvist-Kreuze, H., Meijer, H. J., Petre, B., Ristaino, J., Yoshida, K., Birch, P. R. and Govers, F. 2015. The Top 10 oomycete pathogens in molecular plant pathology. *Molecular Plant Pathology* 16(4), 413–34. doi:10.1111/mpp.12190.

Kerwin, J. L. and Washino, R. K. 1988. Field evaluation of *Lagenidium giganteum* (Oömycetes: Lagenidiales) and description of a natural epizoötic involving a new isolate of the fungus. *Journal of Medical Entomology* 25(6), 452–60. doi:10.1093/jmedent/25.6.452.

Kerwin, J. L., Dritz, D. A. and Washino, R. K. 1990. Confirmation of the safety of *Lagenidium giganteum* (Oomycetes: Lagenidiales) to mammals. *Journal of Economic Entomology* 83(2), 374–6. doi:10.1093/jee/83.2.374.

Latijnhouwers, M., De Wit, P. J. and Govers, F. 2003. Oomycetes and fungi: similar weaponry to attack plants. *Trends in Microbiology* 11(10), 462–9. doi:10.1016/j.tim.2003.08.002.

Leoro-Garzon, P., Gonedes, A. J., Olivera, I. E. and Tartar, A. 2019. Oomycete metabarcoding reveals the presence of *Lagenidium* spp. in phytotelmata. *PeerJ* 7, e7903. doi:10.7717/peerj.7903.

Liu, P., Yang, X., Wang, Y. and Su, X. 2018. A preliminary study on the sustainability of mosquito-killing effect of *Pythium guiyangense* Su. *Journal of Mosquito Research* 8, 14–7. doi:10.5376/jmr.2018.08.0002.

Lovett, B. and St. Leger, R. J. 2018. Genetically engineering better fungal biopesticides. *Pest Management Science* 74(4), 781–9. doi:10.1002/ps.4734.

Maldonado-Blanco, M. G., Leal-López, E. Y., Ochoa-Salazar, O. A., Elías-Santos, M., Galán-Wong, L. J. and Quiroz-Martínez, H. 2011. Effects of culture medium and formulation on the larvicidal activity of the mosquito pathogen *Lagenidium giganteum*

(Oomycetes: Lagenidiales) against *Aedes aegypti*. *Acta tropica* 117(2), 114–8. doi:10.1016/j.actatropica.2010.10.010.

Marjańska-Cichoń, B., Sapieha-Waszkiewicz, A. and Miętkiewski, R. 2009. Effectiveness of biofungicides Bioczos S and Polyversum used to soak strawberry cuttings in *Verticillium* wilt disease control. *Journal of Plant Protection Research* 49(4), 395–8. doi:10.2478/v10045-009-0062-9.

May, B. A., Vandergheynst, J. S. and Rumsey, T. 2006. The kinetics of *Lagenidium giganteum* growth in liquid and solid cultures. *Journal of Applied Microbiology* 101(4), 807–14. doi:10.1111/j.1365-2672.2006.02967.x.

McCarthy, C. G. P. and Fitzpatrick, D. A. 2017. Phylogenomic reconstruction of the oomycete phylogeny derived from 37 genomes. *mSphere* 2(2), e00095-17. doi:10.1128/mSphere.00095-17.

McInnis, T. and Domnas, A. 1973. The properties of trehalase from the mosquito-parasitizing water mold, *Lagenidium* sp. *Journal of Invertebrate Pathology* 22(3), 313–20. doi:10.1016/0022-2011(73)90157-2.

Nowicki, M., Foolad, M. R., Nowakowska, M. and Kozik, E. U. 2012. Potato and tomato late blight caused by *Phytophthora infestans*: an overview of pathology and resistance breeding. *Plant Disease* 96(1), 4–17. doi:10.1094/PDIS-05-11-0458.

Olivera, I. E., Fins, K. C., Rodriguez, S. A., Abiff, S. K., Tartar, J. L. and Tartar, A. 2016. Glycoside hydrolases family 20 (GH20) represent putative virulence factors that are shared by animal pathogenic oomycetes, but are absent in phytopathogens. *BMC Microbiology* 16(1), 232. doi:10.1186/s12866-016-0856-7.

Orduz, S. and Axtell, R. C. 1991. Compatibility of *Bacillus thuringiensis* var. *israelensis* and *Bacillus sphaericus* with the fungal pathogen *Lagenidium giganteum* (Oomycetes: Lagenidiales). *Journal of the American Mosquito Control Association* 7(2), 188–93.

Peng, G., Jin, K., Liu, Y. and Xia, Y. 2015. Enhancing the utilization of host trehalose by fungal trehalase improves the virulence of fungal insecticide. *Applied Microbiology and Biotechnology* 99(20), 8611–8. doi:10.1007/s00253-015-6767-y.

Phillips, A. J., Anderson, V. L., Robertson, E. J., Secombes, C. J. and Van West, P. 2008. New insights into animal pathogenic oomycetes. *Trends in Microbiology* 16(1), 13–9. doi:10.1016/j.tim.2007.10.013.

Quiroz Velasquez, P. F., Abiff, S. K., Fins, K. C., Conway, Q. B., Salazar, N. C., Delgado, A. P., Dawes, J. K., Douma, L. G. and Tartar, A. 2014. Transcriptome analysis of the entomopathogenic oomycete *Lagenidium giganteum* reveals putative virulence factors. *Applied and Environmental Microbiology* 80, 6427–36.

Ramirez-Garcés, D., Camborde, L., Pel, M. J., Jauneau, A., Martinez, Y., Néant, I., Leclerc, C., Moreau, M., Dumas, B. and Gaulin, E. 2016. CRN13 candidate effectors from plant and animal eukaryotic pathogens are DNA-binding proteins which trigger host DNA damage response. *The New Phytologist* 210(2), 602–17. doi:10.1111/nph.13774.

Riit, T., Tedersoo, L., Drenkhan, R., Runno-Paurson, E., Kokko, H. and Anslan, S. 2016. Oomycete-specific ITS primers for identification and metabarcoding. *MycoKeys* 14, 17. doi:10.3897/mycokeys.14.9244.

Sarowar, M. N., Van Den Berg, A. H., McLaggan, D., Young, M. R. and Van West, P. 2013. *Saprolegnia* strains isolated from river insects and amphipods are broad spectrum pathogens. *Fungal Biology* 117(11–12), 752–63. doi:10.1016/j.funbio.2013.09.002.

Scholte, E. J., Knols, B. G., Samson, R. A. and Takken, W. 2004. Entomopathogenic fungi for mosquito control: a review. *Journal of Insect Science* 4, 19. doi:10.1093/jis/4.1.19.

Shen, D., Tang, Z., Wang, C., Wang, J., Dong, Y., Chen, Y., Wei, Y., Cheng, B., Zhang, M., Grenville-Briggs, L. J., Tyler, B. M., Dou, D. and Xia, A. 2019. Infection mechanisms and putative effector repertoire of the mosquito pathogenic oomycete *Pythium guiyangense* uncovered by genomic analysis. *PLoS Genetics* 15(4), e1008116. doi:10.1371/journal.pgen.1008116.

Stassen, J. H. and Van Den Ackerveken, G. 2011. How do oomycete effectors interfere with plant life? *Current Opinion in Plant Biology* 14(4), 407–14. doi:10.1016/j.pbi.2011.05.002.

Sun, G., Yang, Z., Kosch, T., Summers, K. and Huang, J. 2011. Evidence for acquisition of virulence effectors in pathogenic chytrids. *BMC Evolutionary Biology* 11, 195. doi:10.1186/1471-2148-11-195.

Tartar, A. 2013. The non-photosynthetic algae *Helicosporidium* spp.: emergence of a novel group of insect pathogens. *Insects* 4(3), 375–91. doi:10.3390/insects4030375.

Vandergheynst, J., Scher, H., Guo, H.-Y. and Schultz, D. 2007. Water-in-oil emulsions that improve the storage and delivery of the biolarvacide *Lagenidium giganteum*. *BioControl* 52(2), 207–29. doi:10.1007/s10526-006-9021-9.

Vilela, R., Taylor, J. W., Walker, E. D. and Mendoza, L. 2015. *Lagenidium giganteum* pathogenicity in mammals. *Emerging Infectious Diseases* 21(2), 290–7. doi:10.3201/eid2102.141091.

Vilela, R., Humber, R. A., Taylor, J. W. and Mendoza, L. 2019. Phylogenetic and physiological traits of oomycetes originally identified as *Lagenidium giganteum* from fly and mosquito larvae. *Mycologia* 111(3), 408–22. doi:10.1080/00275514.2019.1589316.

Vyas, N., Dua, K. K. and Prakash, S. 2007. Efficacy of *Lagenidium giganteum* metabolites on mosquito larvae with reference to nontarget organisms. *Parasitology Research* 101(2), 385–90. doi:10.1007/s00436-007-0496-9.

Wuensch, A., Trusch, F., Iberahim, N. A. and Van West, P. 2018. *Galleria melonella* as an experimental in vivo host model for the fish-pathogenic oomycete *Saprolegnia parasitica*. *Fungal Biology* 122(2–3), 182–9. doi:10.1016/j.funbio.2017.12.011.

Zattau, W. C. and McInnis Jr., T. 1987. Life cycle and mode of infection of *Leptolegnia chapmanii* (oomycetes) parasitizing *Aedes aegypti*. *Journal of Invertebrate Pathology* 50(2), 134–45. doi:10.1016/0022-2011(87)90113-3.

Zhao, H., Lovett, B. and Fang, W. 2016. Genetically engineering entomopathogenic fungi. *Advances in Genetics* 94, 137–63. doi:10.1016/bs.adgen.2015.11.001.

Part 3

Natural substance-based biopesticides

Chapter 11

Advances in the use of semiochemicals in integrated pest management: pheromones

Gadi V. P. Reddy, USDA-ARS, Southern Insect Management Research Unit, USA; Anamika Sharma, Montana State University, USA; and Angel Guerrero, Institute of Advanced Chemistry of Catalonia-CSIC, Spain

1 Introduction

Chemical communication in insects is an integral part of insect survival along with tactile, visual and acoustic communication (Ali and Morgan, 1990). Semiochemicals are organic compounds that create chemical signals and enable intra- and inter-specific chemical communication (Landolt and Phillips, 1997). Pheromones are the semiochemicals released by an individual and received by conspecifics in which they elicit specific reactions (intraspecific chemical communication) (Karlson and Lüscher, 1959). They are species-specific, non-toxic and active at extremely low doses.[1] In insects, pheromones can stimulate and control various essential behaviors (finding host for food and oviposition, mating, trail marking, alarm, aggregation) and physiological processes (reproduction, diapause, morphosis) (Yew and Chung, 2015). The semiochemicals involved in interspecific chemical communication, known as

1 The synthetic (10E,12Z)-bombykol (Fig. 1) resulted to be as potent as the natural pheromone at concentrations as low as 10^{-12} µg.cm^{3} in petroleum ether.

http://dx.doi.org/10.19103/AS.2020.0073.12
Published by Burleigh Dodds Science Publishing Limited, 2020.

allelochemicals, are produced by one species and induce a behavioral response in a different species. Examples of allelochemicals are kairomones, allomones, synomones and apneumones and have also been implicated in IPM programs (Weatherston and Minks, 1995; Reddy and Guerrero, 2004; Smart et al., 2014).

In insects, the first pheromone was identified from hundreds of thousands of female gland extracts of the silkworm moth, *Bombyx mori* L. (Lepidoptera: Bombycidae) (Butenandt, 1959) (Fig. 1). In the 1960s, pheromone researchers began to use gas chromatography, mass spectrometry and nuclear magnetic resonance (NMR) along with electroantennography to identify insect pheromones. In the 1970s, the first pheromone was registered and farmers started using pheromones for monitoring pests. In the 1980s, pheromones were used to capture more than four billion beetles, curbing an epidemic of bark beetles in the forests of Norway and Sweden (Bakke, 1991). In the 1990s, pheromones were used for the first time to cause mating disruption and demonstrated to help curb insect damage in stone-pitted fruit orchards, tomato, rice, cotton and grape fields (National Academy of Science, 2003). In the last 50 years, research on pheromones has diversified and directed deep into the knowledge of the insect olfactory system, biosynthetic production of pheromones and biomolecular processes related to pheromone reception by pheromone receptor neurons (Reddy and Guerrero, 2004; Baker, 2008; de Bruyne and Baker, 2008; El-Sayed, 2008; Trematerra, 2012; Yew and Chung, 2015).

Classification of insect pheromones was initially done as 'olfactorily acting' pheromones where sex pheromones, alarm and marking pheromones were

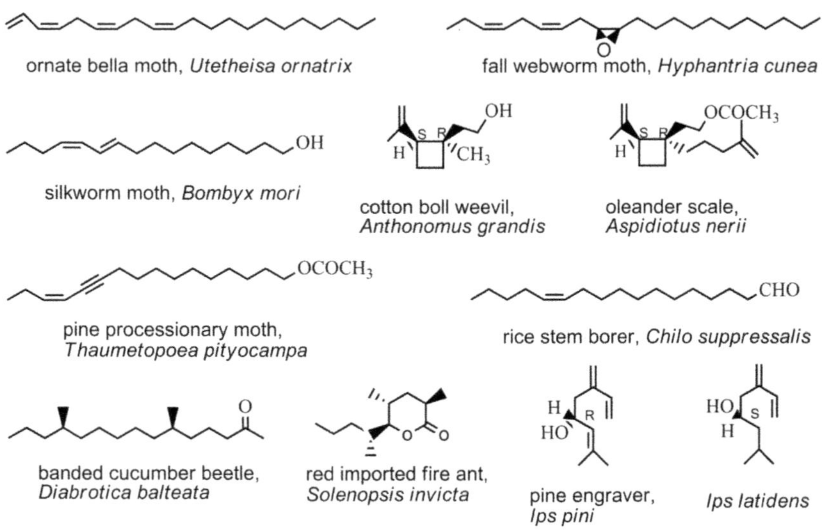

Figure 1 Structural diversity of pheromones.

Published by Burleigh Dodds Science Publishing Limited, 2020.

included (Karlson and Lüscher, 1959). Later on, Wilson (1963) classified them as 'releaser' and 'primer' pheromones, where releaser pheromones included sex attractants that are produced immediately and induce reversible changes in the behavior of the recipient, and primer pheromones that are produced to accelerate or retard sexual maturation in other members of the community to trigger physiological changes in the receiver. Butler (1967) classified pheromones according to their biological functions. Thus, sex pheromones serve as attractants to indicate the presence and reproductive status of potential mating partners (Witzgall et al., 2010a; Yew and Chung, 2015). Aggregation pheromones attract both sexes and aggregate them to locate potential food sources, mate finding and protection from dangers (Prado and Tjallingii, 1997; Gamberale and Tullberg, 1998; Riipi et al., 2001; Wertheim et al., 2005; Durisko et al., 2014). Alarm pheromones stimulate dispersal and aggression to warn or alert individuals of the same species of an impending danger (Wilson and Regnier, 1971).

Pheromones can play an important role in IPM to tackle structural, landscape, agricultural, stored-products and forest insect problems (Shea, 1995; Skillen et al., 1997; Trematerra, 2012). At present, pheromones are used for monitoring, mass trapping, mating disruption, attract and kill and in push-pull strategies (see below) (Baker, 2008; Witzgall et al., 2010b; Trematerra, 2012). Today, the database of insect pheromones and related attractants contains thousands of chemicals (El-Sayed, 2008), and their chemical composition can be assessed from *Pherobase* and *Pherolist* websites.

Pheromones have been used for a variety of guilds of insects, such as phytophagous insects, household insects, stored pests, predators and parasitoids. The major variation in the usage of pheromones for various guilds refers to the methods of delivery and deployment of traps (Burkholder, 1990; Landolt and Heath 1990; Shea, 1995; Phillips, 1997; Skillen et al., 1997; Trematerra, 2012; Evenden and Silk, 2016). The pheromone delivery systems used are passive pheromone dispensers, retrievable polymeric pheromone dispensers, spray method of microencapsulated pheromones, hollow fibers and high-emission dispensers (USDA, 2012). Other methods, such as aerial or ground application of pheromone-impregnated flakes, or polymer bags filled with large doses of pheromone (Biddinger and Krawczyk, 2009), have also been used. ISCA, an agricultural biotech company in California, developed the Specialized Pheromone and Lure Application Technology (SPLAT®) to apply semiochemicals with or without pesticides, in which biologically inert materials are used to control the release of the semiochemicals. The physical properties of this formulation (amorphous, flowable and controlled-release emulsion) allow a better modulation of its flexibility at the moment of application. The formulation provides a continuous controlled-release of semiochemicals for about 6 months. SPLAT products, which are applied as a spray, are available for the management of a variety of Lepidoptera, such as the oriental fruit moth, *Grapholita molesta* (Busck) (Lepidoptera: Tortricidae); the pink bollworm, *Pectinophora gossypiella* (Saunders) (Lepidoptera: Gelechiidae); the codling moth, *Cydia pomonella* (L.) (Lepidoptera: Tortricidae); the gypsy moth, *Lymantria dispar dispar* (L.) (Lepidoptera: Erebidae); the light brown apple moth, *Epiphyas postvittana* (Walker) (Lepidoptera: Tortricidae); the carob moth, *Ectomyelois ceratoniae* (Zeller) (Lepidoptera: Pyralidae) and the citrus leafminer, *Phyllocnistis citrella* (Stainton) (Lepidoptera: Gracillariidae) (ISCA Technologies, 2010; USDA, 2012; Mafra-Neto et al., 2014).

2 Structural diversity and specificity of pheromones

Pheromone compounds are composed of diverse chemical functional groups such as hydrocarbons, epoxides, alcohols, esters, aldehydes, ketones, lactones, acids and isoprenoids (Yew and Chung, 2015). A representative list of the pheromones structural diversity is shown in Fig. 1. Hydrocarbons (saturated or unsaturated, linear or branched) are quite common as pheromones of insects (Ginzel and Blomquist, 2016). In some cases, composition of the pheromone may vary geographically within populations and even from year to year. This is the case of the ornate bella moth, *Utetheisa ornatrix* L. (Lepidoptera: Erebidae), whose female pheromone is a mixture of the rather unstable (Z,Z,Z)-1,3,6,9-heneicosatetraene (Fig. 1), and the corresponding (Z,Z,Z)-3,6,9-triene and (Z,Z)-6,9-diene (Jain et al., 1983). In many cases, the Z geometry of the double bond(s) is crucial for activity. Epoxides, which are generated by epoxidation of the corresponding alkenes, are also common structures among pheromones. The epoxide pheromone of the fall webworm moth, *Hyphantria cunea* Drury (Lepidoptera: Arctiidae) (Fig. 1), a polyphagous defoliator that causes high levels of damage to forests, urban ornamental trees and agricultural crops in China, is highly useful in monitoring populations of this pest (Su et al., 2008). Alcohols and esters are likely the main constituents of most pheromone structures, particularly among Lepidoptera. As representative examples of the enormous structural variety shown in the *Pherolist* website, we have included the pheromones of the cotton boll weevil, *Anthonomus grandis* Boheman (Coleoptera: Curculionidae) (Hardee and Harris, 2003); the oleander scale, *Aspidiotus nerii* Bouché (Hemiptera: Diaspididae) (Einhorn et al., 1998) and the pine processionary moth, *Thaumetopoea pityocampa* Denis and Schiffermüller (Lepidoptera: Thaumetopoeidae) (Quero et al., 1997). Aldehydes and ketones are also common structures in insect pheromones. Thus, for instance, the major compound of the 3-aldehyde pheromone components of the rice stem borer, *Chilo suppressalis* (Lepidoptera: Crambidae), a widespread pest of rice in Asia, is (Z)-11-hexadecenal (Fig. 1). This chemical has been successfully used in combination with the other two aldehydes (Z)-13-octadecenal and

(Z)-9-hexadecenal in the ratio of 48:6:5 to control the pest via mass trapping and mating disruption (Chen et al., 2014). The females of the banded cucumber beetle, *Diabrotica balteata* Leconte (Coleoptera: Chrysomelidae), an economically important pest of vegetable and field crops in the United States and other countries, produce the chiral methyl-branched ketone (R,R)-6,12-dimethylpentadecan-2-one as pheromone (Fig. 1). Methyl-branched pheromones make up one of the most interesting chemical groups, because they are used for the species-specific communication of beetles, moths, ants, cockroaches, scale insects and so on. The absolute configurations of the stereogenic centers have been determined by enantioselective synthesis of all diastereoisomers/enantiomers and/or by application of special analytical techniques equipped with enantioselective GC and HPLC columns (Ando and Yamakawa, 2015). Lactones are also key functions in the structures of pheromones. Illustrative examples are the queen recognition pheromone of the red imported fire ant *Solenopsis invicta* Buren (Hymenoptera: Formicidae) (Fig. 1); the pheromones of the brown-banded cockroach, *Supella longipalpa* Fabricius (Dictyoptera: Blattellidae) and the Japanese beetle, *Popillia japonica* Newman (Coleoptera: Scarabaeidae); and the oviposition pheromone of the southern house mosquito, *Culex pipiens fatigans* L. (Diptera: Culicidae) (Mori, 2010).

Some pheromonal substances are strictly species-specific, whereas others are found in different insect species (Dewhirst et al., 2010; Roelofs, 2016). For example, various sex pheromones in different families of Lepidoptera and recognition pheromones in eusocial insects partially overlap in their composition (Cervo et al., 2002; Ruther et al., 2002; de Bruyne and Baker, 2008; Roelofs, 2016). In these cases, an appropriate combination and specific ratio of components make the pheromone species-specific (Fleischer and Krieger, 2018).

Sympatric species with closely related pheromone compounds provide illustrate examples of the important specificity of pheromones in interspecific interactions. Thus, for example, the pine engraver *Ips pini* (Say) and *I. latidens* (LeConte) (Coleoptera: Curculionidae) are two broadly sympatric species in western North America that breed in large single-species aggregations, primarily within the phloem of *Pinus contorta* Douglas (Pinaceae). In south-central British Columbia, *I. pini* produces as pheromone ipsdienol with a mean chiral ratio of both enantiomers (S)-(+):(R)-(-) of 66:34 whereas *I. latidens* produces preferentially (S)-(-)-ipsenol (Fig. 1). This species releases this enantiomer to inhibit attraction of *I. pini* to its pheromone ipsdienol in baited funnel traps, which in turn releases (S)-(+)-ipsdienol to inhibit attraction of *I. latidens* to its pheromone (Miller and Borden, 1992).

Pheromones in insects are detected through chemosensory organs present in the antennae. Specialized chemosensory neurons on these organs enable the precise detection of the intraspecific pheromonal signals through the surface of body appendages. Insect olfactory sensilla are morphologically classified as coeloconic, basiconic and trichoid. The olfactory signals received by these sensilla are categorized as chemosensory receptor proteins, which include the odorant receptors (ORs), the gustatory receptors (GRs) and the ionotropic receptors (IRs). Not all olfactory sensilla neurons are responsive to pheromones (Fleischer and Krieger, 2018 and literature cited therein). Pheromonal and odorous molecules enter the sensilla via cuticular pores and move into the aqueous lymph where they bind to water-soluble odorant-binding proteins (OBPs). Pheromone-responsive neurons (PRNs) are extremely selective due to the presence of specific pheromone receptors (PRs) in the ciliary membrane of the insect body surface. Several PRs are known today in major groups of insects such as Lepidoptera. PRNs require a putative co-receptor, called 'sensory neuron membrane proteins' (SNMPs), which enable highly sensitive responses, rapid activation and termination of pheromone-induced activity in insects. As of now, we know the presence of some SNMPs (Fleischer and Krieger, 2018 and literature cited therein).

3 Established uses of pheromones

There are thousands of examples of IPM by using semiochemicals-based management strategies. In Table 1 we have cited the main most-recent examples, with some major reading sources.

3.1 Monitoring

Use of pheromones for monitoring and detection of insect pests is the most successful application of pheromones within IPM programs (Witzgall et al., 2010a; Smart et al., 2014). Monitoring populations allow growers to relate trap captures to the abundance or the level of damage induced by a specific species, thus deciding which type of action should be taken. In this way, farmers can save money and reduce the amount of broad-spectrum pesticides on the environment (Baker, 2008; Witzgall et al., 2010b). It is essential to have a good knowledge of the biology of the pest, purity of the pheromone, type of dispenser, trap design and trap location as well as the geographical distribution of the pest (Witzgall et al., 2010a). A particularly important application of monitoring is the detection and survey of invasive species and beneficial insects (Baker, 2008). For instance, the tomato leafminer *Tuta absoluta* Meyrick (Lepidoptera: Gelechiidae) is an oligophagous herbivore that is considered as one of the most devastating pests of tomato crops ever (Giorgini et al., 2019). Since its introduction in Europe in 2006 from South America, it has rapidly spread throughout many European countries, North, Central and West Africa,

Table 1 Pheromone use for monitoring, mass trapping, mating disruption, attract and kill, and incorporation of semiochemicals to manage insect pests

Use	Major insect-targeted	Location	Major reading source
Monitoring of insect pests of agriculture/ horticulture crops	Lepidoptera *Cydia nigricana* Fabricius *Plutella xylostella* Linnaeus *Pectinophora gossypiella* Saunders *Heliothis virescens* Fabricius *Helicoverpa armigera* Hübner *Helicoverpa zea* Boddie *Earias biplaga* Walker *Earias vittella* *Manduca sexta* Linnaeus *Anthonomus grandis* Boheman *Cylas formicarius* Fabricius *Leucinodes orbonalis* Guenée *Leucinodes orbonalis* Guenée *Chilo infuscatellus* Snellen *Scirpophaga incertulas* Walker *Cydia pomonella* Linnaeus *Adoxophyes orana* Fischer von Röslerstamm *Grapholita funebrana* Treitschke *Grapholita molesta* Busck *Lobesia botrana* Denis & Schiffermüller *Prays oleae* Bernard Hemiptera *Aonidiella aurantii* *Aonidiella citrina* *Quadraspidiotus perniciosus* Comstock Diptera *Ceratitis capitate* Wiedemann *Dacus oleae* Gmelin *Drosophila melanogaster* Meigen *Bactrocera dorsalis* (Hendel) *Bactrocera cucurbitae* Coquillett	Asia, North America, South America, Australia, New Zealand, Europe, Middle East, South Africa	Witzgall et al. (2010a), Mafra-Neto et al. (2014), Hardie and Minks (1999), Islam (2012)
Monitoring of insect pests of forests	*Choristoneura* sp. Lederer *Lymantria dispar* Linnaeus *Panolis flammea* Denis & Schiffermüller *Hypercompe scribonia* Stoll *Dendroctonus frontalis* Zimmermann *Ips typographus* Linnaeus *Dendroctonus ponderosae* Hopkins Several species of Ambrosia beetles	Asia, North America, South America, Australia, New Zealand, Europe, Middle East, South Africa	Skillen et al. (1997), Shea (1995), Evenden and Silk (2016), Nadel et al. (2012)

(*Continued*)

Table 1 (*Continued*)

Use	Major insect-targeted	Location	Major reading source
Monitoring of insect pests of stored-products	*Ephestia elutella* Hübner *Plodia interpunctella* Hübner *Sitophilus zeamais* Motschulsky *Sitophilus oryzae* Linnaeus *Rhyzopertha dominica* Fabricius Several species of Dermestid beetle *Tribolium castaneum* Herbst *Lasioderma serricorne* Fabricius *Oryzaephilus surinamensis* Linnaeus *Musca domestica* Linnaeus Cockroaches	Asia, North America, South America, Australia, New Zealand, Europe, Middle East, South Africa	Burkholder and Ma (1985), Phillips (1997), Trematerra (1997, 2012)
Mass trapping	*Ips typographus* Linnaeus Fir beetles *Anthonomus grandis* Boheman *Rhynchophorus ferrugineus* Olivier *Popillia japonica* Newman *Dacus oleae* Gmelin *Musca domestica* Linnaeus Cockroaches *Lymantria dispar* Linnaeus *Conopomorpha cramerella* Snellen *Spodoptera exigua* Hübner *Scirpophaga incertulas* Walker	Asia, North America, South America, Australia, New Zealand, Europe, Middle East	El-Sayed et al. (2006), Baker (2008, 2011), Witzgall et al. (2010a)
Mating Disruption	*Pectinophora gossypiella* Saunders *Lobesia botrana* Denis and Schiffermüller *Grapholita molesta* Busck *Cydia pomonella* Linnaeus *Scirpophaga incertulas* Walker *Keiferia lycopersicella* Walsingham *Platyptilia carduidactyla* Riley *Homona coffearia* Nietner *Epiphyas postvittana* Walker *Synanthedon tipuliformis* Clerck	California, Arizona and Egypt; Europe and North America; the United States, Europe, Australia and Brazil; Europe and North America; The United States and Mexico; Australia and New Zealand; New Zealand; South America	Mahr and Baker (2001), Welter et al. (2005), Cardé and Minks (1995), Biddinger and Krawczyk (2009), Baker (2011), Mafra-Neto et al. (2014), Islam (2012)
Attract and Kill	*Pectinophora gossypiella* *Musca domestica* *Dacus oleae* Gmelin *Ciratitis capitate*	Asia, North America, South America, Australia, New Zealand, Europe, Middle East, Spain, Greece, Hawaii, Japan	El-Sayed et al. (2009)

Use	Major insect-targeted	Location	Major reading source
Integration of kairomones, allomones and other semiochemicals	Species of bark beetles and pine beetles in forest *Chilo partellus* Swinhoe and other species of stemborers in agriculture *Carpophilus lugubris* Murray *Rhynchophorus phoenicis* L. and other species of palm weevil *Sitophilus oryzae* L. *Anthonomus grandis* Boheman	Asia, North America, South America, Australia, New Zealand, Europe, Middle East, South Africa	Smart et al. (2014), Weatherston and Minks (1995), El-Shafie and Faleiro (2017), Reddy and Guerrero (2004, 2010), Mafra-Neto et al. (2014), Cook et al. (2007), Murali-Baskaran et al. (2018), Nadel et al. (2012), Shea (1995), Khan et al. (2010)

This list includes only major insect pests. Please refer to the cited references for greater details. Pherobase (http://www.pherobase.com) includes all the arthropod taxa for which pheromones are known.

the Middle East, Central America and is a major threat to Mexico, the United States, Canada, China and India (http://tutaabsoluta.com/). The synthetic pheromone is highly attractive to conspecific males in the field and has been successfully used in monitoring populations and detection programs (Caparros Megido et al., 2013; NAPPO, 2013; Roda et al., 2015). New technologies have emerged in the last years for pest detection and monitoring systems, based on acoustic transducers that are able to detect the sound of the insects and then send electronic signals to a central station where the trap locations are identified (Beroza, 2002). Other systems based on an automatic counting of the caught insects fitted with crossing transducers integrated with transmission data systems have been designed (Tabuchi et al., 2006). In an interesting different approach, Guarnieri et al. (2011) developed an automatic electronic trap for monitoring the codling moth by modifying a commercial trap-specific for the insect with the use of a mobile phone camera and transmit the data to a remote server. This type of technology is likely to be most economically feasible for detection of high-impact exotic pests where the first incursion of a new moth is the most valuable task.

3.2 Mass trapping

Mass trapping is one of the methods to reduce or suppress populations of target pest species by capturing as many individuals as possible (Baker, 2008). The attractant can be a synthetic pheromone, a food or host attractant or a combination of both, and traps should be efficient enough to catch a large proportion of the population and retain or kill the captured individuals (Smart et al., 2014). Ideally, the best target species are univoltine, small or isolated populations with low immigration rates, or those with a limited life cycle, host range or flight period (El-Sayed et al., 2006). There are many examples of mass trapping experiments designed to control a range of Coleoptera, Lepidoptera and Diptera species (Baker, 2008; Witzgall et al., 2010b; Smart et al., 2014). The strategy has been particularly successful for coleopteran species, which produce aggregation pheromones attracting both sexes and thereby providing a good opportunity to reduce the population of both sexes at the same time. Examples of successful cases are bark beetles (Schlyter et al., 2001), large weevils in tropical crops such as oil palms, plantains and bananas (Oehlschlager et al., 2002; Reddy et al., 2009; Alpizar et al., 2012) or stored-product pests (Trematerra, 2012). In Lepidoptera, which predominantly use female-produced sex pheromones to attract only males, a large proportion of males should be removed for efficiency due to their capacity for multiple matings. Recent successful cases are the control of the leopard moth *Zeuzera pyrina* (Lepidoptera: Cossidae) in olive orchards (Hegazi et al., 2009), the cerambycid beetle *Prionus californicus* Motschulsky (Coleoptera: Cerambycidae) in hop yards (Maki et al., 2011) and the tomato leafminer *T. absoluta* in solanaceous crops (Chermiti and Abbes, 2012). In Diptera, food odors and host plant volatiles have been used alone or in combination with pheromones to trap both sexes in search of food or oviposition sites (El-Sayed et al., 2009). Floral colors and odors have been used in traps to enhance the efficiency of traps, which is also known as 'biomimicry'. Floral scents can be used as generic insect attractants, especially for invasive pest species (El-Sayed et al., 2008), but in some cases they may also attract non-target beneficial insects that lower the efficiency of this strategy to control the target pest.

3.3 Mating disruption

Mating disruption has become the most widely used application of semiochemicals in insect control (Witzgall et al., 2010a). The technique aims to disrupt the chemical communication channel between organisms to interrupt matings, thus affecting their chance of reproduction. Microencapsulation, hand application, monolithic flowable (SPLAT) formulations and aerial dispensers have been used for dispersal of pheromones (Cardé and Minks, 1995; Welter et al., 2005; Lapointe et al., 2009). Unlike mass trapping, the use of synthetic pheromone mixture is not strictly required since single pheromone components or non-attractive pheromone 'off-blends' can be used (see below) resulting in considerable economic savings (Rodriguez-Saona and Stelinski, 2009). For instance, a repellent blend of sex pheromone and attraction inhibitors was used for mating disruption of the pea moth *Cydia nigricana* F. (Lepidoptera:

Published by Burleigh Dodds Science Publishing Limited, 2020.

Tortricidae) (Bengtsson et al., 1994), and use of (Z,Z)-7,11-16:Ald as a unique pheromone component was shown to effectively disrupt communication of the citrus leafminer *Phyllocnistis citrella* S. (Lepidoptera: Gracillariidae) (Mafi et al., 2005). For practical use, mating disruption generally requires large amounts of semiochemicals, notably higher than in mass trapping, so that insects flying within the area of treatment are diverted from naturally occurring plumes (Trematerra et al., 2011). From the 1990s, the area treated with mating disruption has increased almost exponentially up to ca. 770 000 ha in 2010 and the three species with the highest index of treatment were the gypsy moth, *L. dispar* (L.) (Lepidoptera: Erebidae) in North American forests, the codling moth, *C. pomonella* (Lepidoptera: Tortricidae) in apple and pear trees worldwide and the grapevine moth, *Lobesia botrana* Denis and Schiffermüller (Lepidoptera: Tortricidae) in grape in the EU and Chile (Witzgall et al., 2010a; Ioriatti et al., 2011). The appearance of resistance in mating disruption is a remote risk but the first case of such possibility appeared after the continuous use of (Z)-11-tetradecenyl acetate, the major pheromone component of the smaller tea tortrix *Adoxophyes honmai* Yasuda (Lepidoptera: Tortricidae) (Mochizuki et al., 2002). The efficacy of the treatment decreased after 10 years experimentation but returned to its former level when the complete 4-component pheromone blend was again used (Tabata et al., 2007). For stored-product moth pests a novel auto-confusion system for mating disruption has been developed (Armsworth et al., 2006). The strategy is based on the use of electrostatically charged powders, able to adhere to insect cuticles, as carriers for active ingredients such as pheromones, entomopathogens and insecticides (Trematerra, 2012). Mating disruption has also been efficient to control plant-feeding midge pests that cause important crop losses in forestry, horticultural and fruit crops (Smart et al., 2014). Mating disruption trials using the pheromone of the swede midge *Contarinia nasturtii* Kieffer (Diptera: Cecidomyiidae) on fields of Brussels sprouts, broccoli and cauliflower have resulted in a reduction of damage of 59% in a small scale experiment and 91% in a large scale experiment. In scale of deployment traps, important factors for efficacy are labor, time-consuming and population density of the pest (Samietz et al., 2012; Smart et al., 2014).

3.4 Attract and kill

During attract and kill application, a combination of mass trapping with insecticides, the insect pest is attracted to the semiochemical lure but instead of being trapped it is killed by a toxicant, generally an insecticide (Trematerra, 2012; Tewari et al., 2014). In economic terms, the technique compares favorably with mating disruption since smaller amounts of pheromone are needed and the insecticide is limited to smaller areas. However, the inclusion of the insecticide may pose some environmental problems and public acceptance; therefore, the chemical is subjected to pesticide regulatory processes, especially within the EU (Regulation EU 1107/2009) (Smart et al., 2014). In spite of being confined to small areas, the use of insecticides poses risk health issues to growers, consumers and other beneficial species (Smart et al., 2014). Fungal or viral insect pathogens might be used instead, but they face two additional drawbacks: a slow mode of action and a short life span under field conditions (Witzgall et al., 2010a). The technology has received considerable attention for commercial development since formulations are patentable and many of them are now available. A variety of pests has been controlled worldwide using this technique, among them, the cotton boll weevil *Anthonomus grandis* Boheman (Coleoptera: Curculionidae), fruit flies like *Bactrocera oleae* (Rossi), *B. dorsalis* (Hendel) or *B. cucurbitae* (Coquillett) (Diptera: Tephritidae), *Ceratitis capitata*, (Wiedemann) (Diptera: Tephritidae), and moths like *C. pomonella, G. molesta* or *H. armigera* (Mafra-Neto et al., 2014; El-Shafie and Faleiro, 2017).

3.5 Push-pull strategies

Push-pull experiments using a combination of repellents and attractants could be also a sustainable alternative to conventional insecticides. Repellents could 'push' away insects from vulnerable crops whereas at the same time attractants could 'pull' them into annihilating traps (Cook et al., 2007; Khan et al., 2010; Reddy and Guerrero, 2010; Sharma et al., 2019; Werle et al., 2019). Push components include visual cues, synthetic repellents, non-host volatiles, host-derived semiochemicals, anti-aggregation pheromones, alarm pheromones, antifeedants, oviposition deterrents and oviposition deterring pheromones. In turn, pull components comprise visual stimulants, host volatiles, sex and aggregation pheromones, and gustatory and oviposition stimulants (Cook et al., 2007). Advantages of push-pull strategies over conventional pest control methods include: (a) increased efficiency due to the additive and often synergistic effects of their components; (b) higher potential for the use of antifeedants and oviposition deterrents in IPM programs; (c) reduction of chemical and biological control material as the pest population is concentrated in specific areas and (d) reduction of the possibility of pests to develop insecticide resistance (Cook et al., 2007). Although the strategy has considerable potential in agriculture, horticulture, forestry, veterinary and medical pests, and urban insect pests, among others (Cook et al., 2007), it has not yet been widely adopted. A recent experiment has been aimed at minimizing attacks by ambrosia beetles on vulnerable trees (Werle et al., 2019). These beetles are repelled by verbenone, a known anti-aggregation pheromone for various bark beetles, and strongly attracted by ethanol. The experiment used verbenone emitters to 'push' beetles away from vulnerable trees and ethanol lures to 'pull' them into annihilative traps. The trial, however, was not fully successful and both chemicals should be further optimized (Werle et al., 2019).

3.6 Other strategies

Semiochemicals can also be used in other ways for insect management. They can activate mechanisms for plant defense making the plants more resistant to insect attack and, therefore, some of them have been commercialized for crop protection (Vallad and Goodman, 2004). Flowering plants emit bouquets which are strong attractants for Lepidoptera, Coleoptera, Diptera and Hymenoptera (El-Sayed, 2008). Noctuids also respond to synthetic floral attractants composed of phenylacetaldehyde, β-myrcene and other volatiles (Landolt et al., 2006), and to compounds resulting from fermentation of food sources, such as acetic acid and 3-methylbutanol (Landolt et al., 2007). (Z)-Jasmone is a component of flower volatiles but also released from cotton leaves and flowers upon feeding by lepidopterous larvae. Wheat plants sprayed with (Z)-jasmone became less attractive to aphids but more attractive to their parasitoids in the laboratory and in the field (Bruce et al., 2003). It was felt that the reduction of aphid infestations was in part due to increased parasitism of aphids by parasitoids. Plant volatiles can also stimulate insects to produce pheromones, for example, some female moths deposit pheromones on the host plant, which get re-released and make leaf surfaces more attractive for male moths (Landolt and Heath, 1990). Genetically modified crop plants have been used to produce (E)-β-farnesene, an alarm pheromone released by many aphids that causes other aphids in the vicinity to stop feeding and to move away (Beale et al., 2006). A high level of expression in *Arabidopsis thaliana* (L.) Heynh. (Brassicaceae) plants of an (E)-β-farnesene synthase gene caused emission of pure pheromone to repel aphids and attract the aphid parasitoid *Diaeretiella rapae* (M'Intosh) (Hymenoptera: Ichneumonidae) (Beale et al., 2006).

4 Pheromone inhibitors

Interference of the chemical communication channels of insect pests by semiochemicals is an interesting approach with potential use in IPM programs (Reddy and Guerrero, 2010; Witzgall et al., 2010b). In this context, a large variety of pheromone analogues have been developed by modification of the alkyl chain and/or the polar group in the search for pheromone analogues (inhibitors/antagonists or synergists) that may be useful in mating disruption experiments (Renou and Guerrero, 2000; Plettner, 2002). For instance, use of natural pheromone compounds of other species, synthetic geometric isomers of the pheromone, formate esters instead of parent pheromone aldehydes and so on have been successfully used in mating disruption of *Synanthedon pictipes* Grote and Robinson (Lepidoptera: Sesiidae), *Eupoecilia ambiguella* Hübner (Lepidoptera: Tortricidae), *Planotortrix octo* Dugdale (Lepidoptera: Tortricidae) and *Ostrinia nubilalis* Hübner (Lepidoptera: Crambidae) (Renou and Guerrero, 2000). Pheromone inhibitors should be active at low concentrations similar to the pheromone and should inhibit specific components of the peripheral olfactory system of the insect so that the response threshold to the pheromone is increased (Plettner, 2002). Moreover, pheromone inhibitors should be bound to and transported by PBPs to reach the sensory cells (see Table 2 for a list of representative examples of pheromone communication inhibitors).

Due to the special features of fluorine, particular attention has been paid to the fluorinated pheromone derivatives and analogues in which one or more hydrogen atoms have been replaced by fluorine. Comments of some illustrative cases follow. The pheromone of *Chilo suppressalis* (Walker) (Lepidoptera: Crambidae) is a mixture of aldehydes, (Z)-11-hexadecenal, (Z)-13-octadecenal and (Z)-9-hexadecenal (Cork, 2004). The high instability of aldehydes in the field has induced the development of more stable structurally related compounds to replace them either as attractants or pheromone inhibitors in communication disruption experiments (Tatsuki and Kanno, 1981). The compounds resulted by replacement of the aldehyde function by the corresponding alcohol, acetate and methyl groups were tested. All of them resulted in good inhibitors of the pheromone action suggesting that the pheromone communication of the pest could be disrupted by permeating the atmosphere with pheromone-related chemicals (Tatsuki and Kanno, 1981). Other analogues, such as formate analogues, have been used as substitutes for the corresponding aldehyde pheromones, as in *Heliothis virescens* Fabricius (Lepidoptera: Noctuidae) (Renou and Guerrero, 2000) or in the honeydew moth, *Cryptoblabes gnidiella* Millière (Lepidoptera: Pyralidae) (Sellanes et al., 2010), an economically important exotic pest of vineyards in southern Brazil and Uruguay. Pheromone antagonists may be also used by females to optimize fecundity. Thus, females of *Helicoverpa armigera* Hübner (Lepidoptera: Noctuidae) avoid non-optimal matings by inhibiting males with the pheromone antagonist (Z)-11-hexadecenol, thus ensuring maximum fecundity. Females advertise sexual maturity by regulating Z11-16:OH levels and males have evolved the ability to use this signal to get information on female receptivity to mating. The authors used the CRISPR/Cas9 system to knock out the odorant receptor 16 (OR16), the only pheromone receptor tuned to Z11-16:OH, and found that in mutant males, electrophysiological and behavioral responses to Z11-16:OH were abolished. Inability to detect Z11-16:OH prompted males to mate with immature females, resulting in significantly reduced viability of eggs in comparison to wild-type males which did detect the antagonist (Chang et al., 2017).

Table 2 Representative examples of pheromone communication inhibitors

Compound inhibitor	Insect	References
(Z)-11-hexadecenyl trifluoromethyl ketone	*Sesamia nonagrioides* (Lepidoptera: Noctuidae)	Riba et al. (2001), Quero et al. (2004)
	Ostrinia nubilalis (Lepidoptera: Pyralidae)	Solé et al. (2008)
3-octylthio-1,1,1-trifluoro-2-propanone	*Sesamia nonagrioides* Lepidoptera: Noctuidae	Reddy et al. (2002)
	Spodoptera littoralis Lepidoptera: Noctuidae	Riba et al. (2001)
n-decylthiotrifluoropropanone	*Bombyx mori* (Lepidoptera: Bombycidae)	Pophof et al. (2000)
(E,Z,Z)-3,8,11-tetradecatrienyl trifluoromethyl ketone	*Tuta absoluta* (Lepidoptera: Gelechiidae)	Dominguez et al. (2016)
(E,Z,Z)-3,8,11-tetradecatrienyl methyl ketone	*Tuta absoluta* (Lepidoptera: Gelechiidae)	Dominguez et al. (2016)
(Z)-9-14-formate	*Cryptoblabes gnidiella* (Lepidoptera: Pyralidae)	Sellanes et al. (2010)
(Z)-5-10:Ac	*Autographa gamma* (Lepidoptera: Noctuidae)	Subchev et al. (2009)
(Z)-9-14:Ac	*Autographa gamma* (Lepidoptera: Noctuidae)	Subchev et al. (2009)
(E,E)-8,10-dodecadienyl trifluoromethyl ketone	*Cydia pomonella* (Lepidoptera: Tortricidae)	Giner et al. (2009)
(E,E)-8,10-dodecadienyl methyl ketone	*Cydia pomonella* (Lepidoptera: Tortricidae)	Sans et al. (2013)
(E,E)-11,13-pentadecadien-2,3-dione	*Cydia pomonella* (Lepidoptera: Tortricidae)	Sans et al. (2013)
methyl (E,E)-10,12-2-oxo-tetradeca-dienoate	*Cydia pomonella* (Lepidoptera: Tortricidae)	Sans et al. (2013)
(Z)-11-hexadecenal	*Ostrinia nubilalis* (Lepidoptera: Pyralidae)	Gemeno et al. (2006), Linn et al. (2007)
(Z)-9-tetradecenyl trifluoromethyl ketone	*Spodoptera frugiperda* (Lepidoptera: Noctuidae)	Malo et al. (2013)
(E,Z)-2,13-octadecadienyl trifluoromethyl ketone	*Zeuzera pyrina* (Lepidoptera: Cossidae)	Muñoz et al. (2011)
(E,Z)-3,13-octadecadienyl trifluoromethyl ketone	*Zeuzera pyrina* (Lepidoptera: Cossidae)	Muñoz et al. (2011)
(Z)-11-hexadecenyl trifluoroacetate	*Plutella xylostella*	Wang et al. (2019)
(Z)-11-hexadecenyl pentafluoropropanoate	*Plutella xylostella*	Wang et al. (2019)

It is well known that after interaction with the antennal receptors, volatile organic chemicals, particularly pheromones, are efficiently degraded by odorant/pheromone degrading enzymes (ODEs/PDEs) (Vogt et al., 1985; Prestwich, 1987; Kaissling, 1996). Inhibition of these enzymes can disrupt the chemical communication of insects, and, consequently, antennal enzyme inhibition has been proposed as the basis of a potential approach for pest control (Plettner, 2002; Rosell et al., 2008; Reddy and Guerrero, 2010). Trifluoromethyl ketone (TFMK) analogues of the pheromone are remarkable disruptants of the pheromone communication in a variety of Lepidoptera, that is, *Sesamia nonagrioides* Lefèbvre (Noctuidae), *Spodoptera littoralis* Boisduval (Noctuidae), *T. pityocampa* Denis and Schiffermüller (Thaumetopoeidae), *Mamestra brassicae* L. (Noctuidae), *O. nubilalis* Hübner (Crambidae), *B. mori* (Bombycidae), *C. pomonella* (Tortricidae), *S. frugiperda* Smith (Noctuidae), *Z. pyrina* L. (Cossidae) and *T. absoluta* Meyrick (Gelechiidae), among others (see Dominguez et al., 2016 and references cited therein). These chemicals inhibit a variety of esterases and proteases, particularly ODEs and PDEs (see above), through the formation of a stable hemiacetal of tetrahedral geometry between the serine residue of the esterase with the highly electrophilic carbonyl (Rayó et al., 2010). In the same context, methyl ketone (MK) analogues structurally close to the pheromone have been reported to exert a reversible inhibition of male behavioral responses to the pheromone of *O. nubilalis* Hübner (Lepidoptera: Crambidae) (Solé et al., 2008), *C. pomonella* (Lepidoptera: Tortricidae) (Sans et al., 2013) and *T. absoluta* (Lepidoptera: Gelechiidae) (Dominguez et al., 2016).

5 Advancement and future perspectives

Utilization of pheromones could reduce the use of chemical treatments and insecticides and could play an important role in achieving sustainable production. Pheromones along with other semiochemicals will play a greater role in IPM. Lepidopteran pheromones are the most studied but in the last two decades, researchers have begun focusing on other non-Lepidopteran groups as well, such as Hymenoptera, Isoptera and Hemiptera (Hardie and Minks, 1999). Coleoptera is another major group where the role of pheromones has been well explored in agriculture, forestry and stored-product pests (Skillen et al., 1997; Trematerra, 1997; Trematerra, 2012; Tewari et al., 2014).

For stored-product insect pests, attractant pheromones have been identified for over 40 species and are commercially available for about 20 species, mainly for Coleoptera to be used in monitoring traps. However, more work-incorporating deterrent and repellent semiochemicals is required for achieving better results in managing these pests. Also, biorational alternative strategies are required including temperature management and use of natural enemies.

Pheromone trap captures with new tools are required. For example, computer-assisted monitoring to estimate bionomics, population growth and the spatial distribution of insects in a multitrophic ecosystem with varying environmental factors for different food industries and storage facilities is needed. Also, trap capture data collected over time can be visualized for a high or low probability of encountering a pest insect or infestation. On the basis of comprehensive data, an improved decision can be made to reduce the use of an insecticide. Advanced sampling devices such as acoustic sampling method can be explored further. Also the implication of appropriate sampling time to standardize the 'Economic Injury Level' can be used as an important tool (Burkholder and Ma, 1985; Phillips and Throne, 2010; Trematerra, 2012). Similarly, for forest pests, development of reliable, consistent and efficient delivery systems for more successful release devices and application equipment is required. In forestry landscapes where field conditions are extremely variable, pheromone delivery with consistent release system by air over uneven terrain in selected micro-sites (such as canopy and ground) is required. Improved efficacy of semiochemical-based methods for forest pest management is also needed. More advances in chemical ecology research and applications of semiochemicals, such as 'reverse chemical ecology' in which important chemical signals may be discovered after characterization of OBPs by molecular techniques, are needed to be explored more rigorously (Shea, 1995; Evenden and Silk, 2016). In agriculture systems as well, achieving reductions in pest populations with pheromones has been challenging, due to their volatility, difficulty in application over large areas of the crop and release over extended periods of the crop season. Sometimes, even the best formulations are not strong enough to prevent insect pests from entering the large areas of agricultural systems. In the future, greater incorporation of plant semiochemicals to create push-pull systems will be needed. Also, more research on the use of plant activators to provide limited and variable effects in reducing pest populations should be undertaken. Certain semiochemicals (acting as activators) can enhance both direct and indirect defenses in plants and as a result emission of herbivore-induced volatiles can stimulate the defense pathways of neighboring plants protecting them from insect attack. However, caution should be taken as some activators can be phytotoxic and also a sustained activation of defense mechanism in plants can reduce the yields due to the investment of plant's resources in defense (Smart et al., 2014).

Identification of pheromones is one important task and producing them in a cost-effective manner for commercial use is another aspect which requires intense work to make the practical use of pheromones successful (Baker, 2008). Extraction and chemical characterization of semiochemicals is a labor intensive work, which includes electrophysiological and behavioral bioassays and requires expensive methodologies, such as solid-phase microextraction (SPME), gas chromatography-electroantennography (GC-EAG), gas chromatography-mass spectrometry (GC-MS) and high-field NMR (Baker, 2011; Soroker et al., 2015; El-Shafie and Faleiro, 2017). Semiochemical formulations also incur an elevated cost for deployment of the traps, which can be significantly high in comparison to conventional synthetic pesticides. Physical limitations such as instability, volatility, and sensitivity of the active ingredient to temperature and light, along with inconsistency of the formulation release rate, and short-field longevity should be also considered. SPLAT provides us a pointer for development of better deployment mechanism (El-Shafie and Faleiro, 2017).

In recent years an extensive research on molecular processes related to pheromone reception has been conducted providing us greater insight on the role of PRs for recognition and coding of pheromone signals (Fleischer and Krieger, 2018). However, there are still several points which need to be addressed and explored. In the future, genome sequencing and bioinformatics will help us to explore insect species which are not yet been studied. Also, further exploration of PRs will enlighten the mode of signaling to transduce intraspecific signals. The future of semiochemicals-based insect pest management lies in exploring new formulations for field deployment, optimization of controlled-release technologies and improving trapping efficiency. Crop breeding could allow us to improve crop cultivars. Defense traits from wild relatives with appropriate defense metabolism including the production of antibiotic secondary metabolites as well as volatile semiochemicals to manage insect pest could be explored. Similarly, 'push-pull' mechanism should be explored further by including more research on spatial scale effects on pests (Eigenbrode et al., 2016) and natural enemy population dynamics and increased capability of spatially explicit computer models. These new prospects will allow increased use of semiochemicals-based research to be incorporated in future IPM strategies.

Another aspect which has received attention is the use of novel meso- and mega-pheromone dispensers. Standard hand-applied dispensers release low quantity of pheromone and are applied at 700 per ha (100 μg pheromone/dispenser and are applied at 1000 per ha), whereas 'mega dispensers' release huge amounts of pheromone in few dispensers per ha (336 mg pheromone/dispenser and are applied at a density of only a few per ha), which lowers the costs of the application and is very useful in high canopy crops. Another type of dispenser, called 'meso-dispensers', lies between hand and mega dispensers and releases substantially more pheromone than the hand dispensers, allowing the application of pheromones at much lower rates (25–100 μg pheromone/dispenser and 50–100 devices per ha) and reducing the labor cost which is usually required in hand application. By using these dispensers, about 50% of pheromone can be saved by reducing the overall concentration, rate of emission and period of release (McGhee, 2014). Recently, Michigan State

University researchers have developed a new pheromone delivery system for upper tree canopy, which is being tested in apple orchards for mating disruption of the codling moth, *C. pomonella*. This system, called 'Tangler®' dispenser, is potentially less expensive than the hand-applied disruption formulations (Gut et al., 2015).

The use of biodegradable materials for pheromone traps and lures is a new and promising area of research. Recently, there have been some efforts to use biodegradable materials for pheromone lures and insect traps, such as sawdust (Samet, 2018). Other biodegradable materials such as lacto-polymers and poly-coated recycled papers can remain effective for the cropping period and will eventually biodegrade in the soil. This approach has the potential to reduce the labor required for installation and removal of trapping materials. As cited above, many researchers are involved in the 'attract and kill technology' that growers can effectively use to manage insect pests on a variety of agricultural crops. For example, Reddy et al. (2018) found that pitfall traps baited with lures emitting an aggregation pheromone (4-methyl-3,5-heptanedione) attracted both male and female pea leaf weevil, *Sitona lineatus* L. (Coleoptera: Curculionidae). Further studies are underway using biodegradable plastic capsules impregnated with a pheromone that would last for over a year and could be combined with an insecticide, such as cypermethrin, or an entomopathogenic fungus, such as *Metarhizium brunneum* Petch or *Beauveria bassiana* Bals.-Criv. Such a product could attract and kill adults of *S. lineatus* and remain active over long periods (Fig. 2). These capsules of pheromone and pesticide can be deployed in holes stamped into the soil at a regular spacing, making their deployment cost

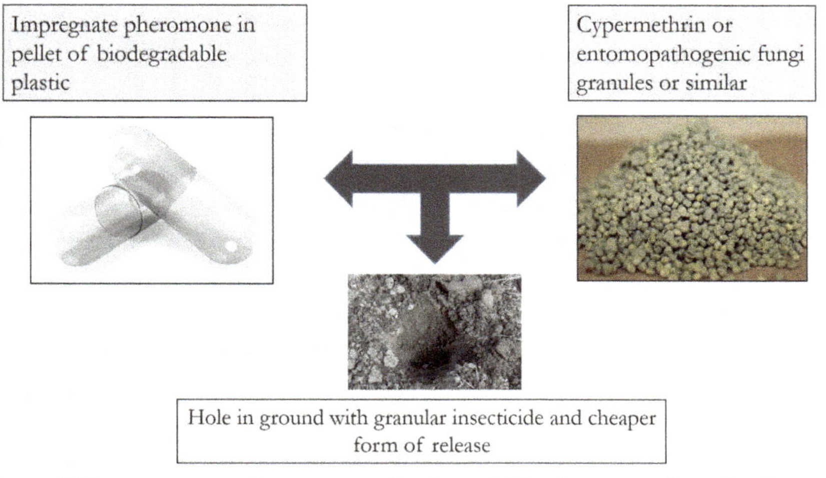

Figure 2 New trapping technique using biodegradable pheromone lures for the pea leaf weevil, *Sitona lineatus*, developed at Montana State University Western Triangle Agricultural Research Center.

effective for large cropping areas. A similar trapping method for click beetles, *Agriotes* sp. (Coleoptera: Elateridae), is being developed at Agriculture and Agri-Food Canada (AAFC) (Agassiz, British Columbia, Canada). Prototype granules with *M. brunneum* and pheromone compounds have shown promising results for attracting *Agriotes* sp. to bait sources (Kabaluk et al., 2015), so the authors hypothesized that the pheromones granules could also be used for *Agriotes* sp. mating disruption. On the other hand, the prototype granules with *M. brunneum* may also contribute to larval mortality of *Agriotes* sp.

By combining pheromones with entomopathogenic fungi, auto-dissemination of the fungi can be achieved. This process enables the use of targeted pests to aid in their dissemination. The process of auto-dissemination can be used for both adults and larvae since adults can get contaminated with fungus and still enable the contamination of larval habitats, thus providing control for larval stages in less accessible habitats such as soil (Klein and Lacey, 1999; Lacey et al., 2001).

Current needs include: (1) Developing a method that can enhance the durability of pheromones. In rural areas, particularly in developing countries lacking cold storage or refrigerators, poor storage is a major limitation in using pheromones efficiently. The role of stabilizers, such as BHT (butylated hydroxytoluene) or other compounds should be explored further to determine how they can be used without compromising the attractiveness of pheromone lures (Divya et al., 2017). (2) The effective usage of long- and short-term pheromones should also be explored further to improve pheromone efficacy. Proper characterization of pheromone range will allow us to create a practical receptive range for various types of pheromones (Renou, 2014). (3) The risk of developing resistance in insects to pheromone compounds is pressing, especially in insect species for which mating disruption is being used for a long period of time (Evenden and Haynes, 2001). (4) To improve the effectiveness of pheromones, other factors should always be explored, such as trap type, trap color, trap size and placement of traps. These factors should be considered at every step for the development of a pheromone lure (Reddy et al., 2005, 2012, 2018; Reddy and Guerrero, 2010; Reddy, 2011). (5) Specific insects where pheromone traps did not work in the past should continue to be evaluated. For example, studies on the mating disruption of *Plutella xylostella* (L.) (Lepidoptera: Plutellidae) showed that sex pheromone use yielded ambiguous results (Schroeder et al., 2000). In these instances, more rigorous field studies are required. (6) Although trifluoromethyl ketone compounds have shown promise in inhibiting insect communication (Reddy et al., 2002), their practical use in managing insect pests has yet to be explored. The review by Holdcraft et al. (2016) was abridged on the evidence and implication of pheromone autodetection, but this is an important area where there is presently a lack of knowledge regarding its use in 'attract and kill' strategies.

6 Conclusion and future trends

The use of pheromones to manage population of insect pests is an integral part of current pest management. In this chapter, we have specifically mentioned the recent advancements in this field. Pheromone traps are presently used to monitor the movement of exotic insect pests into new areas. A second important use of pheromones is to mass trap insects to reduce population levels. A third major application of pheromones is the mating disruption of populations of insect pests. This has been most effectively used with insect pests of agricultural crops. Overall, pheromones play a vital role in IPM for structural, landscape, horticultural, agricultural or forest insect pest problems. Nevertheless, more research is required on the use of pheromones for monitoring and managing insect pests of veterinary and medical importance. A collaboration between chemists and applied biologists to develop new and highly effective pheromone lures would be a necessary point of consideration in insect pest management.

7 Acknowledgement

This work has been supported by The Montana Specialty Crop Block Grant Award # 15SCBGPMT0005, and the USDA National Institute of Food and Agriculture, [Hatch: Developing Integrated Pest Management Programs for Insects in the Western Triangle Agricultural Region of Montana], project accession number 1009746.

8 Where to look for further information section

The complied updated information is available on THE PHEROBASE and The PHEROLIST (https://www.pherobase.com/ and http://www-pherolist.slu.se/index.html).

Centers of expertise include:

- Center for Chemical Ecology, Department of Entomology, College of Agriculture, Penn State, 105, University Park, PA 16802, USA.
- Chemical Ecology and Tropical Entomology Laboratory, Biology Department, University of Nevada, Reno, 89557, NV, USA.
- Chemical Ecology Laboratory, Department of Entomology, University of Wisconsin-Madison, 1630 Linden Drive, Madison, WI 53706 USA.

• Chemical Ecology, Biological Sciences, Simon Fraser University, 8888 University Drive, Burnaby BC, V5A 1S6, Canada.

9 References

Ali, M. F. and Morgan, E. D. 1990. Chemical communication in insect societies: a guide to insect pheromones with special emphasis on social insects. *Biol. Rev.* 65(3), 227–47. doi:10.1111/j.1469-185X.1990.tb01425.x.

Alpizar, D., Fallas, M., Oehlschlager, A. C. and Gonzales, L. M. 2012. Management of *Cosmopolites sordidus* and *Metamasius hemipterus* in banana by pheromone-based mass trapping. *J. Chem. Ecol.* 38(3), 245–52. doi:10.1007/s10886-012-0091-0.

Ando, T. and Yamakawa, R. 2015. Chiral methyl-branched pheromones. *Nat. Prod. Rep.* 32(7), 1007–41. doi:10.1039/C4NP00138A.

Armsworth, C. G., Baxter, I. H., Barton, L. E. E., Poppy, G. M. and Nansen, C. 2006. Effects of adhesive powders on the mating and flight behavior of Mediterranean fruit fly (Diptera: Tephritidae). *J. Econ. Entomol.* 99(4), 1194–202. doi:10.1603/0022-0493-99.4.1194.

Baker, T. C. 2008. Use of pheromones in IPM. In: Radcliffe, E. B., Hutchison, W. D. and Cancelado, R. E. (Eds), *Integrated Pest Management*. Cambridge University Press, Cambridge, UK, pp. 273–85.

Baker, T. C. 2011. Insect pheromones: useful lessons for Crustacean pheromone programs? In: Breithaupt, T. and Thiel, M. (Eds), *Chemical Communication in Crustaceans*. Springer, New York, pp. 531-50.

Bakke, A. 1991. Using pheromones in the management of bark beetle outbreaks. In: Baranchikov, Y. N., Mattson, W. J., Hain, F. P. and Payne, T. L. (Eds), *Forest Insects Guild: Patterns of Interaction with Host Trees*. USDA Forest Service General Technical Report NE-153, pp. 371–7.

Beale, M. H., Birkett, M. A., Bruce, T. J. A., Chamberlain, K., Field, L. M., Huttly, A. K., Martin, J. L., Parker, R., Phillips, A. L., Pickett, J. A., Prosser, I. M., Shewry, P. R., Smart, L. E., Wadhams, L. J., Woodcock, C. M. and Zhang, Y. 2006. Aphid alarm pheromone produced by transgenic plants affects aphid and parasitoid behavior. *Proc. Natl. Acad. Sci. U.S.A.* 103(27), 10509–13. doi:10.1073/pnas.0603998103.

Bengtsson, M., Karg, G., Kirsch, P. A., Löfqvist, J., Sauer, A. and Witzgall, P. 1994. Mating disruption of the pea moth *Cydia nigricana* F. (Lepidoptera: Tortricidae) by a repellent blend of sex pheromone and attraction inhibitors. *J. Chem. Ecol.* 20(4), 871–87. doi:10.1007/BF02059584.

Beroza, M. 2002. More efficient and environmentally desirable means of detecting insects entering lure-baited traps or attractive areas. U.S. Patent No. 0144452.

Biddinger, D. and Krawczyk, G. 2009. Pheromone mating disruption for conventional and organic orchards. Available at: http://extension.psu.edu/ipm/resources/nrcs/programs/treefruit/pheromonemating (accessed on 21 June 2011).

Bruce, T. J. A., Martin, J. L., Pickett, J. A., Pye, B. J., Smart, L. E. and Wadhams, L. J. 2003. cis-Jasmone treatment induces resistance in wheat plants against the grain aphid, *Sitobion avenae* (Fabricius) (Homoptera: Aphididae). *Pest Manag. Sci.* 59(9), 1031–6. doi:10.1002/ps.730.

Burkholder, W. E. 1990. Practical use of pheromones and other attractants for stored-product insects. In: Ridgway, R. L., Silverstein, R. M. and Inscoe, M. N. (Eds),

Behavior-Modifying Chemicals for Insect Management. Marcel Dekker, New York, pp. 497–516.

Burkholder, W. E. and Ma, M. 1985. Pheromones for monitoring and control of stored-product insects. *Annu. Rev. Entomol.* 30(1), 257–72. doi:10.1146/annurev.en.30.010185.001353.

Butenandt, A. 1959. Wirkstoffe des Insektenveiches. *Naturwissenschaften* 46(15), 461–71. doi:10.1007/BF00634039.

Butler, C. G. 1967. Insect pheromones. *Biol. Rev.* 42(1), 42–84. doi:10.1111/j.1469-185X.1967.tb01339.x.

Caparros Megido, R., Haubruge, E. and Verheggen, F. J. 2013. Pheromone-based management strategies to control the tomato leafminer, *Tuta absoluta* (Lepidoptera: Gelechiidae): a review. *Biotechnol. Agron. Soc. Environ.* 17(3), 475–82.

Cardé, R. T. and Minks, A. K. 1995. Control of moth pests by mating disruption: successes and constraints. *Annu. Rev. Entomol.* 40(1), 559–85. doi:10.1146/annurev.en.40.010195.003015.

Cervo, R., Dani, F. R., Zanetti, P., Massolo, A. and Turillazzi, S. 2002. Chemical nestmate recognition in a stenogastrine wasp, *Liostenogaster flavolineata* (Hymenoptera Vespidae). *Ethol. Ecol. Evol.* 14(4), 351–63. doi:10.1080/08927014.2002.9522736.

Chang, H., Liu, Y., Ai, D., Jiang, X., Dong, S. and Wang, G. 2017. A pheromone antagonist regulates optimal mating time in the moth *Helicoverpa armigera*. *Curr. Biol.* 27(11), 1610–1615.e3. doi:10.1016/j.cub.2017.04.035.

Chen, R. Z., Klein, M. G., Sheng, C. F., Li, Q. Y., Li, Y., Li, L. B. and Hung, X. 2014. Mating disruption or mass trapping, compared with chemical insecticides, for suppression of *Chilo suppressalis* (Lepidoptera: Crambidae) in Northeastern China. *J. Econ. Entomol.* 107(5), 1828–38. doi:10.1603/EC14148.

Chermiti, B. and Abbes, K. 2012. Comparison of pheromone lures used in mass trapping to control the tomato leafminer *Tuta absoluta* (Meyrick, 1917) in industrial tomato crops in Kairouan (Tunisia). *Bull. OEPP/EPPO Bull.* 42(2), 241–8. doi:10.1111/epp.2561.

Cook, S. M., Khan, Z. R. and Pickett, J. A. 2007. The use of push-pull strategies in integrated pest management. *Annu. Rev. Entomol.* 52, 375–400. doi:10.1146/annurev.ento.52.110405.091407.

Cork, A. 2004. *Pheromone Manual*. Natural Resources Institute, Chatham Maritime, UK.

de Bruyne, M. and Baker, T. C. 2008. Odor detection in insects: volatile codes. *J. Chem. Ecol.* 34(7), 882–97. doi:10.1007/s10886-008-9485-4.

Dewhirst, S. Y., Pickett, J. A. and Hardie, J. 2010. Aphid pheromones. *Vitam. Horm.* 83, 551–74.

Divya, C., Kanara, H. G. and Jethva, D. M. 2017. Role of pheromones and its antioxidants in IPM. *Adv. Plants Agric. Res.* 6, 30–1.

Dominguez, A., Puigmartí, M., Bosch, M. P., Rosell, G., Crehuet, R., Ortiz, A., Quero, C. and Guerrero, A. 2016. Synthesis, functional assays, electrophysiological activity and field tests of pheromone antagonists of the tomato leafminer *Tuta absoluta*. *J. Agric. Food Chem.* 64(18), 3523–32. doi:10.1021/acs.jafc.6b00674.

Durisko, Z., Kemp, R., Mubasher, R. and Dukas, R. 2014. Dynamics of social behavior in fruit fly larvae. *PLoS ONE* 9(4), e95495. doi:10.1371/journal.pone.0095495.

Eigenbrode, S. D., Birch, A. N. E., Lindzey, S., Meadow, R. and Snyder, W. E. 2016. Review: a mechanistic framework to improve understanding and applications of push-pull systems in pest management. *J. Appl. Ecol.* 53(1), 202–12. doi:10.1111/1365-2664.12556.

Einhorn, J., Guerrero, A., Ducrot, P. H., Boyer, F. D., Gieselmann, M. and Roelofs, W. 1998. Sex pheromone of the oleander scale, *Aspidiotus nerii*: structural characterization and absolute configuration of an unusual functionalized cyclobutane. *Proc. Natl. Acad. Sci. U.S.A.* 95(17), 9867–72. doi:10.1073/pnas.95.17.9867.

El-Sayed, A. M. 2008. The Pherobase: database of insect pheromones and semiochemicals. Available at: www.pherobase.com.

El-Sayed, A. M., Suckling, D. M., Wearing, C. H. and Byers, J. A. 2006. Potential of mass trapping for long-term pest management and eradication of invasive species. *J. Econ. Entomol.* 99(5), 1550–64. doi:10.1603/0022-0493-99.5.1550.

El-Sayed, A. M., Byers, J. A., Manning, L. M., Jürgens, A., Mitchell, V. J. and Suckling, D. M. 2008. Floral scent of Canadian thistle, *Cirsium arvense* (L.) Scop and its potential as a generic insect attractant. *J. Econ. Entomol.* 101(3), 720–7. doi:10.1603/0022-0493(2008)101[720:FSOCTA]2.0.CO;2.

El-Sayed, A. M., Suckling, D. M., Byers, J. A., Jang, E. B. and Wearing, C. H. 2009. Potential of 'lure and kill' in long-term pest management and eradication of invasive species. *J. Econ. Entomol.* 102(3), 815–35. doi:10.1603/029.102.0301.

El-Shafie, H. A. F. and Faleiro, J. R. 2017. Semiochemicals and their potential use in pest management. In: Shields, V. (Ed.), *Biological Control of Pest and Vector Insects*. Available at: https://www.intechopen.com/books/biological-control-ofpest-and-vector-insects/semiochemicals-and-their-potential-use-in-pest-management (accessed on 07 June 2017).

Evenden, M. L. and Haynes, K. F. 2001. Potential for the evolution of resistance to pheromone-based mating disruption tested using two pheromone strains of the cabbage looper, *Trichoplusia ni*. *Entomol. Exp. Appl.* 100(1), 131–4. doi:10.1046/j.1570-7458.2001.00856.x.

Evenden, M. L. and Silk, P. J. 2016. The influence of Canadian research on semiochemical-based management of forest insect pests in Canada. *Can. Entomol.* 148(S1), S170–209. doi:10.4039/tce.2015.17.

Fleischer, J. and Krieger, J. 2018. Insect pheromone receptors – key elements in sensing intraspecific chemical signals. *Front. Cell. Neurosci.* 12, 425. doi:10.3389/fncel.2018.00425.

Gamberale, G. and Tullberg, B. S. 1998. Aposematism and gregariousness: the combined effect of group size and coloration on signal repellence. *Proc. R. Soc. Lond. B* 265(1399), 889–94. doi:10.1098/rspb.1998.0374.

Gemeno, C., Sans, A., López, C., Albajes, R. and Eizaguirre, M. 2006. Pheromone antagonism in the European corn borer moth *Ostrinia nubilalis*. *J. Chem. Ecol.* 32(5), 1071–84. doi:10.1007/s10886-006-9046-7.

Giner, M., Sans, A., Riba, M., Bosch, D., Gago, R., Rayo, J., Rosell, G. and Guerrero, A. 2009. Development and biological activity of a new antagonist of the pheromone of the codling moth *Cydia pomonella*. *J. Agric. Food Chem.* 57(18), 8514–9. doi:10.1021/jf901979k.

Ginzel, M. D. and Blomquist, G. J. 2016. Insect hydrocarbons: biochemistry and chemical ecology. In: Cohen, E. and Moussian, B. (Eds), *Extracellular Composite Matrices in Arthropods*. Springer Verlag, Cham, Switzerland, pp. 221–52.

Giorgini, M., Guerrieri, E., Cascone, P. and Gontijo, L. 2019. Current strategies and future outlook for managing the Neotropical tomato pest *Tuta absoluta* (Meyrick) in the Mediterranean basin. *Neotrop. Entomol.* 48(1), 1–17. doi:10.1007/s13744-018-0636-1.

Guarnieri, A., Maini, S., Molari, G. and Rondelli, V. 2011. Automatic trap for moth detection in integrated pest management. *Bull. Insect.* 64, 247–51.

Gut, L., Huang, J. and McGhee, P. 2015. A novel mating disruption system designed for rapid feployment of reservoir pheromone dispensers. *N. Y. Fruit Q.* 23, 3.

Hardee, D. D. and Harris, F. A. 2003. Eradicating the boll weevil (Coleoptera: Curculionidae): a clash between a highly successful insect, good scientific achievement, and differing agricultural philosophies. *Am. Entomol.* 49(2), 82–97. doi:10.1093/ae/49.2.82.

Hardie, J. and Minks, A. K. 1999. *Pheromones of Non-Lepidopteran Insects Associated with Agricultural Plants.* CABI publishing, Wallingford, UK, 466pp.

Hegazi, E., Khafagi, W. E., Konstantopoulou, M., Ratptopoulos, D., Tawfik, H., Abd El-Aziz, G. M., Abd El-Rahman, S. M., Atwa, A., Aggamy, E. and Showeil, S. 2009. Efficient mass-trapping as an alternative tactic for suppressing populations of leopard moth (Lepidoptera: Cossidae). *Ann. Entomol. Soc. Am.* 102(5), 809–18. doi:10.1603/008.102.0507.

Holdcraft, R., Rodriguez-Saona, C. and Stelinski, L. L. 2016. Pheromone autodetection: evidence and implications. *Insects* 7(2), 17; doi:10.3390/insects7020017.

Ioriatti, C., Anfora, G., Tasin, M., De Cristofaro, A., Witzgall, P. and Lucchi, A. 2011. Chemical ecology and management of *Lobesia botrana* (Lepidoptera: Tortricidae). *J. Econ. Entomol.* 104(4), 1125–37. doi:10.1603/ec10443.

ISCA Technologies. 2010. *SPLAT*. Available at: http://www.iscatech.com/exec/SPLAT.html (accessed on 19 March 2012).

Islam, M. A. 2012. Pheromone use for insect control: present status and prospect in Bangladesh. *Int. J. Agril. Res. Innov. Tech. (IJARIT)* 2(1), 47–55. doi:10.3329/ijarit.v2i1.14005.

Jain, S. C., Dussourd, D. E., Conner, W. E., Eisner, T., Guerrero, A. and Meinwald, J. 1983. Polyene pheromone components from an arctiid moth (*Utetheisa ornatrix*): characterization and synthesis. *J. Org. Chem.* 48(13), 2266–70. doi:10.1021/jo00161a024.

Kabaluk, J. T., Lafaontaine, J. P. and Bordon, J. H. 2015. An attract and kills tactic for click beetles based on *Metarhizium brunneum* and a new formulation of sex pheromone. *J. Pest Sci.* 88(4), 707–16. doi:10.1007/s10340-015-0661-3.

Kaissling, K. E. 1996. Pheripheral mechanisms of pheromone reception in moths. *Chem. Senses* 21(2), 257–68. doi:10.1093/chemse/21.2.257.

Karlson, P. and Lüscher, M. 1959. Pheromones: a new term for a class of biologically active substances. *Nature* 183(4653), 55–6. doi:10.1038/183055a0.

Khan, Z. R., Midega, C. A., Bruce, T. J., Hooper, A. M. and Pickett, J. A. 2010. Exploiting phytochemicals for developing a 'push-pull' crop protection strategy for cereal farmers in Africa. *J. Exp. Bot.* 61(15), 4185–96. doi:10.1093/jxb/erq229.

Klein, M. G. and Lacey, L. A. 1999. An attractant trap for autodissemination of entomopathogenic fungi into populations of Japanese beetle, *Popillia japonica* (Coleoptera: Scarabaeidae). *Biocontr. Sci. Technol.* 9, 151–8.

Lacey, L. A., Frutos, R., Kaya, H. K. and Vail, P. 2001. Insect pathogens as biological control agents: do they have a future? *Biol. Control* 21(3), 230–48. doi:10.1006/bcon.2001.0938.

Landolt, P. J. and Heath, R. R. 1990. Sexual role reversal in mate finding strategies of the cabbage looper moth. *Science* 249(4972), 1026–8. doi:10.1126/science.249.4972.1026.

Landolt, P. J. and Phillips, T. W. 1997. Host plant influences on sex pheromone behavior of phytophagous insects. *Annu. Rev. Entomol.* 42, 371–91. doi:10.1146/annurev.ento.42.1.371.

Landolt, P. J., Adams, T. and Zack, R. S. 2006. Field response of alfalfa looper and cabbage looper moths (Lepidoptera: Noctuidae, Plusiinae) to single and binary blends of floral odorants. *Environ. Entomol.* 35(2), 276–81, doi:10.1603/0046-225X-35.2.276.

Landolt, P. J., Suckling, D. M. and Judd, G. J. R. 2007. Positive interaction of a feeding attractant and a host kairomone for trapping the codling moth, *Cydia pomonella* (L.). *J. Chem. Ecol.* 33(12), 2236–44. doi:10.1007/s10886-007-9391-1.

Lapointe, S. L., Stelinksi, L. L., Evens, T. J., Niedz, R. P., Hall, D. G. and Mafra-Neto, A. 2009. Sensory imbalance as mechanism of mating disruption in the leafminer *Phyllocnistis citrella*: elucidation by multivariate geometric designs and response surface models. *J. Chem. Ecol.* 35(8), 896–903. doi:10.1007/s10886-009-9674-9.

Linn, C. E., Domingue, M. J., Musto, C. J., Baker, T. C. and Roelofs, W. L. 2007. Support for (Z)-11-hexadecenal as a pheromone antagonist in *Ostrinia nubilalis*: flight tunnel and single sensillum studies with a New York population. *J. Chem. Ecol.* 33(5), 909–21. doi:10.1007/s10886-007-9268-3.

Mafi, S. A., Vang, L. V., Nakata, Y., Ohbayashi, N., Yamamoto, M. and Ando, T. 2005. Identification of the sex pheromone of the citrus leafminer (*Phyllocnistis citrella* Stainton, Lepidoptera: Gracillariidae) with a trial of control by the communication disruption method. *J. Pestic. Sci.* 30(4), 361–7. doi:10.1584/jpestics.30.361.

Mafra-Neto, A., Fettig, C. J., Unson, A. S., Rodriguez-Saona, C., Holdcraft, R., Faleiro, J. R., El-Shafie, H., Reinke, M., Bernardi, C. and Villagran, K. M. 2014. Development of specialized pheromone and lure application technologies (SPLAT®) for management of Coleopteran pests in agricultural and forest systems. In: Gross, A. D., Coats, J. R., Duke, S. O. and Seiber, J. N. (Eds), *Biopesticides: State of the Art and Future Opportunities*. ACS Symposium Series: American Chemical Society, Washington DC. doi:10.1021/bk-2014-1172.ch015.

Mahr, D. and Baker, T. C. 2001. Mating disruption for insect control: Where are we? *Proceedings of the 2001 Wisconsin Cranberry School*. University of Wisconsin, Madison, pp. 1–4.

Maki, E. C., Millar, J. G., Rodstein, J., Hanks, L. M. and Barbour, J. D. 2011. Evaluation of mass trapping and mating disruption for managing *Prionus californicus* (Coleoptera: Cerambycidae) in hop production yards. *J. Econ. Entomol.* 104(3), 933–8. doi:10.1603/ec10454.

Malo, E. A., Rojas, J. C., Gago, R. and Guerrero, Á 2013. Inhibition of the responses to sex pheromone of the fall armyworm, *Spodoptera frugiperda*. *J. Insect Sci.* 13, 134. doi:10.1673/031.013.13401.

McGhee, P. S. 2014. Impact of high releasing mating disruption formulations on (male) codling moth, *Cydia pomonella* L., behavior. Dissertation. Michigan State University, East Lansing, MI.

Miller, D. R. and Borden, J. H. 1992. (*S*)-(+)-ipsdienol: interspecific inhibition of *Ips latidens* (Leconte) by *Ips pini* (Say) (Coleoptera: Scolytidae). *J. Chem. Ecol.* 18(9), 1577–82. doi:10.1007/BF00993230.

Mochizuki, F., Fukumoto, T., Noguchi, H., Sugie, H., Morimoto, T. and Ohtani, K. 2002. Resistance to a mating disruptant composed of (Z)-11-tetradecenyl acetate in the smaller tea tortrix, *Adoxophyes honmai* (Yasuda) (Lepidoptera: Tortricidae). *Appl. Entomol. Zool.* 37(2), 299–304. doi:10.1303/aez.2002.299.

Mori, K. 2010. Pheromones in chemical communication. In: Hermann, A. (Ed.), *The Chemistry and Biology of Volatiles*. John Wiley & Sons Ltd., Chichester, pp. 123–49.

Muñoz, L., Bosch, M. P., Batllori, Ll., Rosell, G., Bosch, D., Guerrero, A. and Avilla, J. 2011. Synthesis of allylic trifluoromethyl ketones and their activity as inhibitors of the sex pheromone of the leopard moth, *Zeuzera pyrina* L. (Lepidoptera: Cossidae). *Pest Manag. Sci.* 67(8), 956–64. doi:10.1002/ps.2139.

Murali-Baskaran, R. K., Sharma, K. C., Kaushal, P., Kumar, J., Parthiban, P., Senthil-Nathan, S. and Mankin, R. W. 2018. Role of kairomone in biological control of crop pests: a review. *Physiol. Mol. Plant Pathol.* 101, 3–15. doi:10.1016/j.pmpp.2017.07.004.

Nadel, R. L., Wingfield, M. J., Scholes, M. C., Lawson, S. A. and Slippers, B. 2012. The potential for monitoring and control of insect pests in Southern Hemisphere forestry plantations using semiochemicals. *Ann. For. Sci.* 69(7), 757–67. doi:10.1007/s13595-012-0200-9.

National Academy of Science. 2003. Insect pheromones: mastering communication to control pests. Available at: http://www.nasonline.org/publications/beyond-discovery/insect-pheromones.pdf (accessed on March 2019).

North American Plant Protection Organization-NAPPO. 2013. *Surveillance Protocol for the Tomato Leaf Miner, Tuta absoluta, for NAPPO Member Countries*. The Secretariat of NAPPO, Ottawa, Ontario, Canada.

Oehlschlager, A. C., Chinchilla, C., Castillo, G. and Gonzalez, L. 2002. Control of red ring disease by mass trapping of *Rhynchophorus palmarum* (Coleoptera: Curculionidae). *Fla. Entomol.* 85(3), 507–13. doi:10.1653/0015-4040(2002)085[0507:CORRDB]2.0.CO;2.

Phillips, T. W. 1997. Semiochemicals of stored-product insects: research and applications. *J. Stored Prod. Res.* 33(1), 17–30. doi:10.1016/S0022-474X(96)00039-2.

Phillips, T. W. and Throne, J. E. 2010. Biorational approaches to managing stored-product insects. *Annu. Rev. Entomol.* 55, 375–97. doi:10.1146/annurev.ento.54.110807.090451.

Plettner, E. 2002. Insect pheromone olfaction: new targets for the design of species-selective pest control agents. *Curr. Med. Chem.* 9(10), 1075–85. doi:10.2174/0929867024606669.

Pophof, B., Gebauer, T. and Ziegelberger, G. 2000. Decyl-thio-trifluoropropanone, a competitive inhibitor of moth pheromone receptors. *J. Comp. Physiol. A* 186(3), 315–23. doi:10.1007/s003590050432.

Prado, E. and Tjallingii, W. F. 1997. Effects of previous plant infestation on sieve element acceptance by two aphids. *Entomol. Exp. Appl.* 82(2), 189–200. doi:10.1046/j.1570-7458.1997.00130.x.

Prestwich, G. D. 1987. Chemical studies of pheromone reception and catabolism. In: Prestwich, G. D. and Blomquist, G. J. (Eds), *Pheromone Biochemistry*. Academic Press, Inc., New York, pp. 473–527.

Quero, C., Malo, E. A., Fabriàs, G., Camps, F., Lucas, P., Renou, M. and Guerrero, A. 1997. Reinvestigation of female sex pheromone of processionary moth (*Thaumetopoea pityocampa*): no evidence for minor components. *J. Chem. Ecol.* 23(3), 713–26. doi:10.1023/B:JOEC.0000006406.30444.69.

Quero, C., Bau, J., Guerrero, A. and Renou, M. 2004. Responses of the olfactory receptor neurons of the corn stalk borer *Sesamia nonagrioides* to components of the pheromone blend and their inhibition by a trifluoromethyl ketone analogue of the main component. *Pest Manag. Sci.* 60(7), 719–26. doi:10.1002/ps.867.

Rayó, J., Muñoz, L., Rosell, G., Hammock, B. D., Guerrero, A., Luque, F. J. and Pouplana, R. 2010. Reactivity versus steric effects in fluorinated ketones as esterase inhibitors: a quantum mechanical and molecular dynamics study. *J. Mol. Model.* 16(11), 1753–64. doi:10.1007/s00894-010-0807-4.

Reddy, G. V. P. 2011. Development of semiochemical-based strategies for old-house borer, *Hylotrupes bajulus* (Coleoptera: Cerambycidae). *J. Entomol. Acarol. Res.* 43(2), 111–15. doi:10.4081/jear.2011.111.

Reddy, G. V. P. and Guerrero, A. 2004. Interactions of insect pheromones and plant semiochemicals. *Trends Plant Sci.* 9(5), 253–61. doi:10.1016/j.tplants.2004.03.009.

Reddy, G. V. P. and Guerrero, A. 2010. New pheromones and insect control strategies. *Vitam. Horm.* 83, 493–519.

Reddy, G. V. P., Quero, C. and Guerrero, A. 2002. Activity of octylthiotrifluoropropan-2-one, a potent esterase inhibitor, on growth, development and intraspecific communication in *Spodoptera littoralis* and *Sesamia nonagrioides*. *J. Agric. Food Chem.* 50(24), 7062–8. doi:10.1021/jf020255r.

Reddy, G. V. P., Fettköther, R., Noldt, U. and Dettner, K. 2005. Capture of female *Hylotrupes bajulus* as influenced by trap type and pheromone blend. *J. Chem. Ecol.* 31(9), 2169–77. doi:10.1007/s10886-005-6083-6.

Reddy, G. V. P., Cruz, Z. T. and Guerrero, A. 2009. Development of an efficient pheromone-based trapping method for the banana root borer *Cosmopolites sordidus*. *J. Chem. Ecol.* 35(1), 111–7. doi:10.1007/s10886-008-9580-6.

Reddy, G. V. P., Gadi, N. and Taianao, A. J. 2012. Efficient sex pheromone trapping: catching the sweetpotato weevil *Cylas formicarius*. *J. Chem. Ecol.* 38(7), 846–53. doi:10.1007/s10886-012-0160-4.

Reddy, G. V. P., Shrestha, G., Miller, D. A. and Oehlschlager, A. C. 2018. Pheromone-trap monitoring system for pea leaf weevil, *Sitona lineatus*: effects of trap type, lure type and trap placement within fields. *Insects* 9(3), 75; doi:10.3390/insects9030075.

Renou, M. 2014. Pheromones and general odor perception in insects. In: Mucignat-Caretta, C. (Ed.), *Neurobiology of Chemical Communication*. CRC Press/Taylor & Francis, Boca Raton, FL. Available at: https://www.ncbi.nlm.nih.gov/books/NBK200986/.

Renou, M. and Guerrero, A. 2000. Insect parapheromones and other compounds as behavior-modifying agents in pest control strategy. *Annu. Rev. Entomol.* 48, 605–30.

Riba, M., Sans, A., Bau, P., Grolleau, G., Renou, M. and Guerrero, A. 2001. Pheromone response inhibitors of the corn stalk borer *Sesamia nonagrioides*. Biological evaluation and toxicology. *J. Chem. Ecol.* 27(9), 1879–97. doi:10.1023/a:1010468911352.

Riipi, M., Alatalo, R. V., Lindstrom, L. and Mappes, J. 2001. Multiple benefits of gregariousness cover detectability costs in aposematic aggregations. *Nature* 413(6855), 512–4. doi:10.1038/35097061.

Roda, A. L., Brambila, J., Barria, J., Euceda, X. and Korytkowski, C. 2015. Efficiency of trapping systems for detecting *Tuta absoluta* (Lepidoptera: Gelechiidae). *J. Econ. Entomol.* 108(6), 2648–54. doi:10.1093/jee/tov248.

Rodriguez-Saona, C. R. and Stelinski, L. L. 2009. Behavior modifying strategies in IPM: theory and practice. In: Peshin, R. D. and Dhawan, A. K. (Eds), *Integrated Pest Management: Innovation-Development Process*. Springer, The Netherlands, pp. 263–315.

Roelofs, W. L. 2016. Reminiscence of the early days. In: Alisson, J. D. and Cardé, R. (Eds), *Pheromone Communication in Moths: Evolution, Behavior, and Application*. University of California Press, Oakland, CA, pp. 3-10.
Rosell, G., Quero, C., Coll, J. and Guerrero, A. 2008. Biorational insecticides in pest management. *J. Pestic. Sci.* 33(2), 103-21. doi:10.1584/jpestics.R08-01.
Ruther, J., Sieben, S. and Schricker, B. 2002. Nestmate recognition in social wasps: manipulation of hydrocarbon profiles induces aggression in the European hornet. *Naturwissenschaften* 89(3), 111-4. doi:10.1007/s00114-001-0292-9.
Samet, C. 2018. Biodegradable insect traps. US Patent Application #20180042213.
Samietz, J., Baur, R. and Hillbur, Y. 2012. Potential of synthetic sex pheromone blend for mating disruption of the Swede midge, *Contarinia nasturtii*. *J. Chem. Ecol.* 38(9), 1171-7. doi:10.1007/s10886-012-0180-0.
Sans, A., Gago, R., Mingot, A., García, W., Bosch, D., Coll, J., Rosell, G., Bosch, M. P., Riba, M. and Guerrero, A. 2013. Electrophilic derivatives antagonize pheromone attraction in *Cydia pomonella*. *Pest Manag. Sci.* 69, 1280-90.
Schlyter, F., Zhang, Q.-H., Liu, G.-T. and Ji, L.-Z. 2001. A successful case of pheromone mass trapping of the bark beetle *Ips duplicatus* in a forest island, analyzed by 20-year time series data. *Integr. Pest Manag. Rev.* 6(3/4), 185-96. doi:10.1023/A:1025767217376.
Schroeder, P. C., Shelton, A. M., Ferguson, C. S., Hoffmann, M. P. and Petzoldt, C. H. 2000. Application of synthetic sex pheromone for management of diamondback moth, *Pluteall xylostella*, in cabbage. *Entomol. Exp. Appl.* 94(3), 243-8. doi:10.1046/j.1570-7458.2000.00626.x.
Sellanes, C., Rossini, C. and González, A. 2010. Formate analogs as antagonists of the sex pheromone of the honeydew moth, *Cryptoblabes gnidiella*: electrophysiological, behavioral and field evidence. *J. Chem. Ecol.* 36(11), 1234-40. doi:10.1007/s10886-010-9861-8.
Sharma, A., Shrestha, G. and Reddy, G. V. P. 2019. Trap crops: how far we are from using them in cereal crops? *Ann. Entomol. Soc. Am.* 112(4), 330-9. doi:10.1093/aesa/say047.
Shea, P. J. 1995. Use of insect pheromones to manage forest insects: present and future. In: Hall, F. R. and Barry, J. W. (Eds), Biorational Pest Control Agents' Formulation and Delivery. American Chemical Society Symposium Series 0097-6156, pp. 272-83.
Skillen, E. L., Berisford, C. W., Camann, M. A. and Reardon, R. C. 1997. Semiochemicals of forest and shade tree insects in North America and management applications. Forest Health Technology Enterprise Team, FHTET 96 - 15. USDA Forest Service, Morgantown, WV.
Smart, L. E., Aradottir, G. I. and Bruce, T. J. A. 2014. Role of semiochemicals in Integrated Pest Management. In: Abrol, D. P. (Ed.), *Integrated Pest Management, Current Concepts and Ecological Perspective*. Academic Press, Cambridge, MA, pp. 93-109.
Solé, J., Sans, A., Riba, M., Rosa, E., Bosch, M. P., Barrot, M., Palència, J., Castellà, J. and Guerrero, A. 2008. Reduction of damage by the Mediterranean corn borer *Sesamia nonagrioides* and the European corn borer *Ostrinia nubilalis* in maize fields by a trifluoromethyl ketone pheromone analogue. *Entomol. Exp. Appl.* 126, 28-39.
Soroker, V., Harari, A. and Faleiro, J. R. 2015. The role of semiochemicals in date pest management. In: Wakil, W., Faleiro, J. R. and Miller, T. (Eds), *Sustainable Pest Management in Date Palm: Current Status and Emerging Challenges*. Springer International Publishing, Switzerland, pp. 315-46.

Su, M. W., Fang, Y. L., Tao, W. Q., Yan, G. Z., Ma, W. E. and Zhang, Z. 2008. Identification and field evaluation of the sex pheromone of an invasive pest, the fall webworm *Hyphantria cunea* in China. *Chin. Sci. Bull.* 53(4), 555–60. doi:10.1007/s11434-008-0124-9.

Subchev, M. A., Voerman, S. and Milkova, T. S. 2009. Inhibitors for *Autographa gamma* L. (Lep., Noctuidae) sex pheromone. *J. Appl. Entomol.* 104, 425-9.

Tabata, J., Noguchi, H., Kainoh, Y., Mochizuki, F. and Sugie, H. 2007. Behavioral response to sex pheromone-component blends in the mating disruption-resistant strain of the smaller tea tortrix, *Adoxophyes honmai* Yasuda (Lepidoptera: Tortricidae), and its mode of inheritance. *Appl. Entomol. Zool.* 42(4), 675–83. doi:10.1303/aez.2007.675.

Tabuchi, K., Moriya, S., Mizutani, N. and Ito, K. 2006. Recording the occurrence of the bean bug *Riptortus clavatus* (Thunberg) (Heteroptera: Alydidae) using an automatic counting trap. *Jpn. J. Appl. Entomol. Zool.* 50(2), 123–9. doi:10.1303/jjaez.2006.123.

Tatsuki, S. and Kanno, H. 1981. Disruption of sex pheromone communication in *Chilo suppressalis* with pheromone and analogs. In: Mitchell, E. R. (Ed.), *Management of Insect Pests with Semiochemicals*. Springer, Boston, MA, pp. 313–25.

Tewari, S., Leskey, T. C., Nielsen, A. L., Piñero, J. C. and Rodriguez-Saona, C. R. 2014. Use of pheromones in insect pest management, with special attention to weevil pheromones. In: Abrol, D. P. (Ed.), *Integrated Pest Management: Current Concepts and Ecological Perspective*. Elsevier Inc., San Diego, CA, pp. 141–68.

Trematerra, P. 1997. Integrated pest management of stored-product insects: practical utilization of pheromones. *Anz. Schadlingskde., Pflanzenschutz, Umweltschutz* 70(3), 41–4. doi:10.1007/BF01996919.

Trematerra, P. 2012. Advances in the use of pheromones for stored-product protection. *J. Pest Sci.* 85(3), 285–99. doi:10.1007/s10340-011-0407-9.

Trematerra, P., Athanassiou, C., Stejskal, V., Sciarretta, A., Kavallieratos, N. and Palyvos, N. 2011. Large scale mating disruption of *Ephestia* spp. and *Plodia interpunctella* in Czech Republic, Greece and Italy. *J. Appl. Entomol.* 135(10), 749–62. doi:10.1111/j.1439-0418.2011.01632.x.

USDA. 2012. Pheromone (crops). Evaluation Report. Available at: https://www.ams.usda.gov/sites/default/files/media/Pheromones%20TR.pdf.

Vallad, G. E. and Goodman, R. M. 2004. Systemic acquired resistance and induced systemic resistance in conventional agriculture. *Crop Sci.* 44(6), 1920–34. doi:10.2135/cropsci2004.1920.

Vogt, R. G., Riddiford, L. M. and Prestwich, G. D. 1985. Kinetic properties of a pheromone-degrading enzyme: the sensillar esterase of *Antheraea polyphemus*. *Proc. Natl. Acad. Sci. U.S.A.* 82(24), 8827–31. doi:10.1073/pnas.82.24.8827.

Wang, A. J., Zhang, K. X., Gao, Y. L., Weng, A. Z., Wang, L. Y., Zhang, Y. H., Zhang, Z., She, D. M., Ning, J. and Mei, X. D. 2019. Synthesis and bioactivity studies of sex pheromone analogs of the diamond back moth, *Plutella xylostella*. *Pest. Manag. Sci.* 75(4), 1045–55. doi:10.1002/ps.5214.

Weatherston, I. and Minks, A. K. 1995. Regulation of semiochemicals- global aspects. *Integr. Pest Manag. Rev.* 1(1), 1–13. doi:10.1007/BF00140330.

Welter, S. C., Pickel, C., Millar, J., Cave, F., Van Steenwyk, R. A. and Dunley, J. 2005. Pheromone mating disruption offers selective management options for key pests. *Calif. Agric.* 59(1), 16–22. doi:10.3733/ca.v059n01p16.

Werle, C. T., Ranger, C. M., Schultz, P. B., Reding, M. E., Addesso, K. M., Oliver, J. B. and Sampson, B. J. 2019. Integrating repellent and attractant semiochemicals into a push-pull strategy for ambrosia beetles (Coleoptera: Curculionidae). *J. Appl. Entomol.* 143(4), 333–43. doi:10.1111/jen.12594.

Wertheim, B., van Baalen, E. J. A., Dicke, M. and Vet, L. E. 2005. Pheromone-mediated aggregation in nonsocial arthropods: an evolutionary ecological perspective. *Annu. Rev. Entomol.* 50, 321–46. doi:10.1146/annurev.ento.49.061802.123329.

Wilson, E. O. 1963. Pheromones. *Sci. Am.* 208(5), 100–14. doi:10.1038/scientificamerican0563-100.

Wilson, E. O. and Regnier, F. E. 1971. Evolution of alarm-defense system in formicine ants. *Am. Nat.* 105(943), 279–89. doi:10.1086/282724.

Witzgall, P., Kirsch, P. and Cork, A. 2010a. Sex pheromones and their impact on pest management. *J. Chem. Ecol.* 36(1), 80–100. doi:10.1007/s10886-009-9737-y.

Witzgall, P., Trematerra, P., Liblikas, I., Bengtsson, M. and Unelius, C. R. 2010b. Pheromone communication channels in tortricid moths: lower specificity of alcohol vs. acetate geometric isomer blends. *Bull. Entomol. Res.* 100(2), 225–30. doi:10.1017/S0007485309990186.

Yew, J. Y. and Chung, H. 2015. Insect pheromones: an overview of function, form, and discovery. *Prog. Lipid Res.* 59, 88–105. doi:10.1016/j.plipres.2015.06.001.

Chapter 12

Possible use of allelochemicals in integrated pest management (IPM)

Toby Bruce, Keele University, UK

1 Introduction

Allelochemicals are specialised chemicals produced and released into the environment by one species that influence the fitness of another. Unlike pheromones, which are intraspecific signals, allelochemicals influence interactions between species. Earlier use of the term was quite narrow; 'allelochemical' is still defined in the *Merriam-Webster* dictionary as a 'toxic chemical produced by a plant in order to defend itself against herbivores or competing plants'. However, more recently the term has been used in a wider sense as any secreted chemical that has an effect on other species. Yoneyama and Natsume (2010) state that allelochemicals are 'chemicals released from donor organisms into the environment and affect the growth and development of receiver organisms'. This wider definition includes chemicals that can have positive as well as negative effects on the receiver. It also includes semiochemicals or infochemicals that carry information from one species to another.

Allelochemicals can be classified according to which organism benefits. A kairomone is an allelochemical that benefits the receiver but not the emitter

http://dx.doi.org/10.19103/AS.2020.0073.13

(Metcalf and Metcalf, 1992), for example a plant volatile used by a herbivore for host recognition. Allomones benefit the producer but have no effect on the receiver whereas synomones benefit both the producer and the receiver, for example a floral odour that attracts pollinators (Nordlund and Lewis, 1976). Synomones are involved in tritrophic interactions where herbivore-infested plants emit volatile signals to attract natural enemies of the herbivore - release of the cue or signal benefits both the receiver and the producer. Puzzlingly, there is no specific term for an allelochemical which benefits the emitter but is harmful to the receiver. The nomenclature based on the benefit to the producer or receiver is also challenged when the same phytochemical can have multiple functions and simply calling signalling chemicals as infochemicals is an alternative approach (Dicke and Sabelis, 1988). However, infochemical is similar to semiochemical and, as the name suggests, is for information carrying chemicals and not chemicals that have a toxic effect on the recipient.

Insects use volatile compounds for orientation, communication and decision making, for example as cues for host identification and the assessment of host quality. A 'signal' is a stimulus that is emitted to elicit certain responses in a specific intended receiver which feed back to the emitter (Greenfield, 2002; Searcy and Nowicki, 2005). By contrast, a 'cue' represents a by-product of a primarily endogenous process that carries reliable information on the identity, localisation or physiological condition of an individual and, thus, can be used for decision making by another individual (Diggle et al., 2007).

2 Allelochemicals for pest management

There is a huge diversity of phytochemicals in nature and many of them have evolved to defend plants from attacking organisms. Conversely, adapted herbivores have evolved to withstand allelochemicals from their host plants (Ali and Agrawal, 2012). In this section we are considering allelochemicals for use in pest management. Perhaps the most famous toxic allelochemical is azadirachtin which has been widely used for crop protection against herbivore pests. Another good example is the root exudates from desmodium plants that cause suicidal germination of parasitic striga weed seeds before they contact host maize roots in the push-pull system (Khan et al., 2010).

Semiochemical allelochemicals influence the behaviour of recipient organisms. Attractants can be used in pest management as insect lures in traps or as trap crops. Repellents can divert pests from crops. There are few practical examples of their use in integrated pest management (IPM), perhaps because the reality of the field is different from short-duration olfactometer bioassays in the lab. In a field setting there are natural background odours that can interfere with insect responses. Also, with prolonged exposure insects can change their responses through associative learning. The use of allelochemicals in

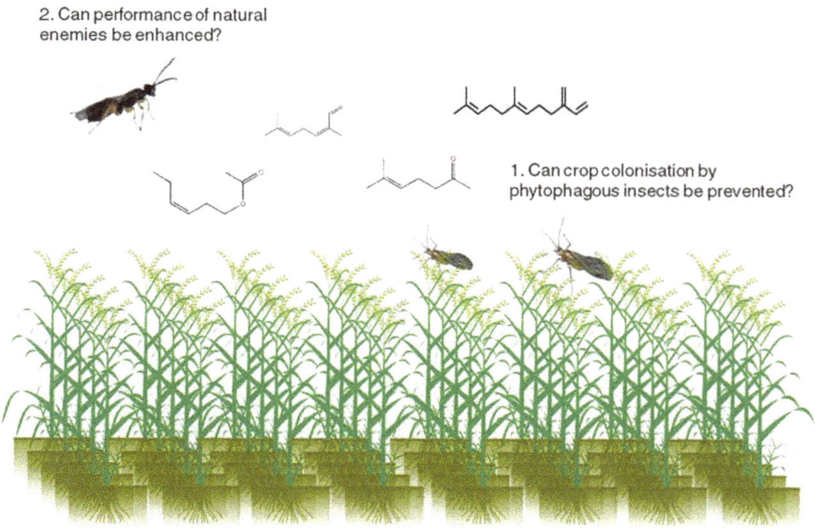

Figure 1 Strategies for use of volatile plant allelochemicals in IPM can 1. directly target the pest as repellents, antifeedants or antibiotics, or 2. can indirectly target the pest by attracting its natural enemies.

agriculture has some way to go before it becomes as effective as the use of chemicals by wild plants in nature. Strategies intended to reduce herbivore pressure on crops by deployment of volatiles are ambitious because they seek to intervene with insects that are highly adapted for locating hosts. Any artificial mimics need to be better than natural odour sources if insects are to be tricked. Strategies either involve disruption of host plant colonisation by the herbivore or enhancing the performance of natural enemies in attacking their herbivore hosts (Fig. 1). These interactions have been fine-tuned by millions of years of co-evolution (Bruce et al., 2005) and insects have a highly sophisticated sensory system and neuroethological processes that are fine-tuned to respond to the external environment to maximise the insect's chances of finding suitable hosts and mates to optimise their fitness. This means that artificial interventions may be more challenging than anticipated.

3 Botanical pesticides

Traditional pest management, before synthetic pesticides became available, used botanical pesticides based on plants such as *Derris*, tobacco (*Nicotiana* spp.) and *Ryania* for centauries (Dubey et al., 2010; Duke et al., 2010). Botanical pesticides are based on obtaining allelochemical extracts from one plant used to protect a crop plant. Although sometimes assumed to be safer because of their natural origin, this is not necessarily the case and should be verified (Biondi et al., 2012). Neem (*Azadirachta indica*), with its active ingredient azadirachtin, is the most famous botanical pesticide but there are others, including natural pyrethrum from Chrysanthemums (*Chrysanthemum cinerariifolium*) from which the synthetic pyrethoids were developed. Azadirachtin has antifeedant, growth inhibition, moulting inhibition and antibiotic properties (Mordue and Blackwell, 1993) and pyrethrum has neurotoxic activity against sodium channels in the insect nervous system. Other plants with insecticidal properties include sabadilla (*Schoenocaulon officinale*), which produces cevadine; *Derris* (Leguminosae), which produces rotenone; and tobacco which, of course, produces nicotine (Duke et al., 2010).

Botanical insecticides were reviewed by Isman (2006) who highlighted neem and pyrethrum but wrote that commercialisation of other plant extracts is limited. The main bottleneck to commercialisation is registration (Regnault-Roger et al., 2012). Plant defence chemicals with activity against insects are reviewed by Mithöfer and Boland (2012). Many feeding inhibitors from plant sources have given excellent results in laboratory conditions but only a few of them, such as neem, are effective in field situations (Koul, 2008). There are natural products, evolved for plant defence, with a toxic mode of action that show promise, for example extracts of pepper plants (*Piper nigrum* and *P. tuberculatum*) that can be used against Colorado potato beetle (Scott et al., 2003). Akhtar and Isman (2004) found that extracts of *Melia volkensin* had antifeedant and growth inhibitor activity against cabbage looper (*Trichoplusia ni*), and the armyworm (*Pseudaletia unipuncta*). However, difficulties with formulation, stability, application method and regulatory constraints impede commercialisation.

4 Repellents

There are few successful examples of field deployment of insect repellents (Wallingford et al., 2017). Only a limited number of studies have provided field plot experiments with plant volatile treatments. Although some studies (Mallinger et al., 2011; Simpson et al., 2011; Braasch et al., 2012; Gadino et al., 2012) reported significant reductions in herbivore insect populations, or increases in populations of natural enemies, the reductions tended not to be substantial and in the order of less than 20% change when compared to control treatments. This approach is not used commercially. An exception was Kessler and Baldwin (2001) who found that MeJA treatment of *Nicotiana attenuata* resulted in a substantial (>90%) reduction in *Manduca sexta* survival. However, this could have been due to MeJA acting as an inducing agent and activating the direct as well as indirect plant defences. Possible reasons why the use of repellents has been limited are that point source artificial dispensers may not mimic natural release from plant foliage well and that continuous emission of volatiles over extended time periods is unnatural. Furthermore, associative learning could lead to a change in behaviour if an insect learns to associate the odour of a compound with its food source.

5 Attractants

There are many more examples of the use of attractants in crop pest management than repellents (Wallingford et al., 2017). However, the most successful examples of attractants to attract and kill traps often use pheromones or a combination of pheromones and host plant volatiles rather than host plant volatiles alone (El-Sayed et al., 2006). Simplified plant volatile mimics are likely to be swamped by similar background plant volatiles. Competing and more realistic natural sources of plant volatiles may be preferred by the insects. Also, plant volatiles are less species specific and are usually weaker attractants than pheromones. Attractant-baited traps have been used for monitoring and suppression of fruit fly populations but these use protein hydrolysate and yeast extract baits rather than plant volatiles (e.g. Vargas et al., 2002).

Many studies (e.g. James, 2003; James and Grasswitz, 2005; Zhu and Park, 2005; Yu et al., 2008; Rodrigues Vieira et al., 2014) report trapping beneficial insects in traps baited with plant volatiles. While these demonstrate that predators/parasitoids are attracted to particular volatile baits in the field, it does not actually show if this translates into reduced herbivore populations or into crop yield or plant fitness being increased. A study with both traps and herbivore population assessments in field plots (Lee, 2010) shows that predators' numbers sampled on traps can increase without any reduction in herbivore population. James (2003) found that MeSA-baited cards in hop yards caught lacewings (*Chrysopa nigricornis*) which could enhance predation of aphids and mites but herbivore populations were not measured. Similarly, James and Grasswitz (2005) found MeSA, MeJA or Z3HA caught more natural enemies (*Metaphycus* and *Anagrus*) but herbivore populations were not measured. Zhu and Park (2005) found that MeSA-baited traps attracted *Coccinella septempunctata* and 2-phenylethanol but herbivore populations were not measured. Again, Yu et al. (2008) found significantly greater numbers of *Coccinella septempunctata* on sticky cards baited with Z3HA but herbivore populations were not measured. Rodrigues Vieira et al. (2014) found increased trap catches of egg parasitoid *Trissolcus* with (*E*)-2-hexenal in short-term 1-week experiments but not in longer 6-week experiments and that field application of (*E*)-2-hexenal had no effect on parasitism rates. Some studies use plant volatiles to trap herbivores (e.g. Sun et al., 2014) but it is hard to envisage how a sufficient herbivore population reduction could be obtained. A huge number would have to be caught in traps to have an impact on herbivore population levels and the synthetic plant volatiles would need to be more attractive than background plant odours and perhaps coupled with a killing agent in a lure-and-kill approach.

6 Defence activators

The ability of plants to respond to allelochemicals that are associated with insect or pathogen attack allows them to fine-tune their metabolism according to the likelihood of exposure to biotic stress factors. For example, emission of herbivore-induced volatiles from neighbouring plants can lead to activation of defence pathways that make a plant more resistant to insect attack (Farmer and Ryan, 1990; Karban et al., 2000; Baldwin et al., 2006). The concentration of volatiles that is required for this may mean that under natural conditions plant-plant interactions mediated by volatile signals occur over relatively short distances. However, in an agricultural situation artificial application of plant semiochemicals can overcome these distance limitations and there are opportunities to exploit natural plant signalling processes in alternative crop protection strategies that trigger priming or induction of defence pathways in crop plants. In this context it should also be possible to use semiochemicals that are more active.

cis-Jasmone was tested as a plant defence activator and found to cause significant but not substantial reductions in cereal aphid infestation in wheat (Bruce et al., 2003). Lee (2010) tested MeSA lures in strawberry plots and found that lacewings (Chrysopidae) and *Orius tristicolor* were attracted to baited traps in treated plots, however, there was no significant effect on levels of herbivores in the adjacent field plots. Braasch et al. (2012) found significant responses to HIPVs but these were context dependent with crop type having a strong influence on how arthropods responded. In some cases, (mirid bug in corn, some hymenoptera in soybean) the HIPVs even had repellent effects. They suggested that crop background altered attraction. Kelly et al. (2014) found that aggregation pheromone helped retain predatory stink bugs released in tomato field plots but MeSA did not.

Intercrops, grown between rows of a main crop, can be used as a source of allelochemicals for use in IPM. For example, Desmodium secretes root exudates that cause suicidal germination of striga weed (Khan et al., 2010) and plant parasitic nematodes can be suppressed by a cover crop of *Chromolaena odorata* (Odeyemi et al., 2013). There is potential for further use of this approach provided that plants with suitable bioactivity can be found. There are important agronomic considerations such as ensuring that the companion crop does not interfere with growth of the main crop. Ideally the companion crop should be of economic use. For example, desmodium can be used as an animal forage crop and is also nitrogen fixing. Trap crops, planted at the border of main crop areas, are used in certain agroecosystems and are reviewed by Hokkanen (1991) and

Shelton and Badenes-Perez (2006). For example, alfalfa can be used as a trap crop for Lygus bugs, in cotton, in California.

The 'push-pull' system of IPM uses locally available companion plants instead of synthetic chemicals to deliver semiochemicals in the field (Cook et al., 2007). This approach was developed for smallholder agriculture and has been used with much success in maize and sorghum in eastern Africa (Hassanali et al., 2008; Khan et al., 2008b) against two main pest problems, stem or stalk borers and the African witchweed, *Striga*. The use of constitutive emission of relevant semiochemicals from plant sources provides an appropriate solution for smallholder African agriculture where synthetic chemicals would be too costly and logistically difficult to deliver.

The system used a combination of an intercrop which releases semiochemicals that are repellent to pests (push) and a trap crop grown around the edges of the plot which releases semiochemicals that are attractive to pests of the main crop (pull). Semiochemicals released by the intercrop also attract natural enemies of pests (Khan et al., 1997). Intercrops constitutively release volatiles such as (*E*)-ocimene and (*E*)-4,8-dimethyl-1,3,7-nonatriene that are typically released from maize when attacked by chewing herbivores. These act to repel pests but attract their natural enemies. The main semiochemcials released from trap crops that make them attractive are green leaf volatiles which are produced in much larger amounts than in maize or sorghum, at nightfall when stem borer moths are active (Chamberlain et al., 2006). For stem borer control alone, the intercrop is molasses grass, but for controlling the *Striga* weed, the intercrop is silverleaf or greenleaf, two cattle forage legumes of *Desmodium* species, which also repel stem borers. The mechanism of *Striga* control consists of release of root exudate allelochmicals from *Desmodium* that induce suicidal germination of *Striga* seeds. The trap crop is preferably Napier grass, but Sudan grass can also be used. Besides controlling stem borers and the *Striga* weed, all of these companion crops are valuable as cattle forage, improving thereby livestock holdings in addition to producing a sustainable cereal harvest protected against pests and weeds.

7 Overcoming barriers to improve use of allelochemicals in integrated pest management (IPM)

To date, applications of allelochemicals in IPM are quite limited. There are more examples for the use of pheromones (reviewed by Witzgall et al., 2010). Why is this? For volatile repellents and attractants, the insect response is often plastic and strongly context-dependent. Associative learning behaviour can allow insects to change how they respond. The responses of insects to pheromones are innate whereas responses to host cues are usually learned or fine-tuned through learning (Turlings and Wäckers, 2004). Insects can learn to respond to blends of ubiquitously occurring plant volatiles and recognize particular combinations of volatiles and particular concentrations rather than individual compounds (Bruce et al., 2005; Webster et al., 2010). This allows them to concentrate their foraging on the most profitable environments, but it also means they are well able to change their responses if a volatile is artificially presented in the wrong context and without a 'reward' in terms of prey or hosts. Insects can learn to ignore allelochemicals that carry a deceptive message unlike pheromones which are more hardwired and species-specific signals. Another major obstacle to the use of volatile repellents and attractants are background odours: a promising semiochemical when tested against clean air in the lab may not work so well in a noisier field environment with natural sources of background odours which may have an overriding effect.

For toxic allelochemicals used as botanical insecticides, there are challenges with formulation of plant compounds that are often unstable, volatile or have low water solubility. This means repeated applications are often required which makes them expensive and time consuming. There are opportunities to overcome these constraints with improved formulations. Microencapsulation can provide control over release rate and protect the active ingredients from breakdown. Nanoformulation has even greater potential (Athanassiou et al., 2018) and we currently have a project in the lab evaluating nanoformulations of essential oils and phytochemicals as treatments against aphids. Use of nanotechnology in the agricultural sector is relatively recent when compared to drug delivery and pharmaceuticals but it is an expanding new research field (Athanassiou et al., 2018).

For formulation of botanical insecticides, nanoparticles have a number of distinct advantages including improved efficacy due to higher surface area, higher solubility and elimination of organic solvents (de Oliveira et al., 2014). Indeed, nanotechnology could overcome many of the current constraints which are limiting the development of botanical insecticides. Due to their higher ratio of droplet surface area per mass unit, nanoemulsions typically have increased kinetic stability and higher encapsulation capacity (Golden et al., 2018). There is also scope to improve target specificity and minimise run-off of plant protection products into the environment beyond the crop being treated. Problems with low stability that were obstacles to commercialisation of natural products and plant essential oils as biopesticides can be solved. There are examples of improved longevity of nanoformulations: For example, Ebadollahi et al. (2017) found that loading of *Thymus eriocalyx* essential oil in a MCM-41 nanoformulation significantly increased durability and fumigation persistence from 6 days up to 20 days, when tested against two-spotted spider mite. Similarly, Golden et al. (2018) observed high efficacy, against two stored products pest species, of nanoemulsions of pulegone for 5 weeks compared to only 1 week with a coarse emulsion. Even after 200 days, the pulegone concentration in nano emulsions was still high. Oliveira et al. (2017) found that Procetyl® AWS nanoformulations of the essential oil of *Lippia sidoides* and its major compound, thymol, could be stored for up to 7 months without losing activity. Likewise, Yang et al. (2009) achieved 5-month control efficacy against adult *T. castaneum* with a nanoformulation of garlic essential oil which was associated with the slow and persistent release of the active components from the nanoparticles. These results are encouraging and show that the stability of bioactive phytochemicals can be extended by using appropriate nanoformulations.

Perhaps one of the most famous classes of phytochemicals with insecticidal activity is the pyrethrins, naturally produced by chrysanthemums, from which the synthetic pyrethroid insecticides were developed (Elliott and Janes, 1978). The natural pyrethrins are notoriously unstable but it has recently been shown that a nanotech water-in-oil micro-emulsion formulation of pyrethrin has enhanced bioactivity against *Aphis gossypii* aphids compared to conventional suspension concentrates (Papanikolaou et al., 2018). The authors attributed this to increased longevity as well as improved dispersion, deposition and leaf adhesion of droplets of the spray solution on leaf surfaces, resulting in an increased insecticide bioavailability and easier penetration due to the increased interfacial area of the system.

There are regulatory requirements that need to be fulfilled to use these substances as botanical pesticides. This requires investment and a regulatory system that finds the right balance between assessing risks and allowing benefits to be realised. Natural products are sometimes assumed to be safer than synthetic chemical but this is not necessarily the case and needs to be verified, for example, plants can produce cyanide for defence (Mithöfer and Boland, 2012). Some natural products have a history of safe use as food ingredients, medical treatments or in aromatherapy. If allelochemicals that have such history of same use could be fast tracked it might improve the prospects of making new treatments commercially available to farmers.

8 Conclusion

Some progress has been made with the use of allelochemicals but there is still a long way to go before they can become widely available, reliable tools that farmers can use. There is, however, considerable potential for further future development. With reduced availability of conventional toxic insecticides, there is increasing demand for new IPM tools. The arsenal of phytochemicals available in nature potentially provides a source of new interventions but we need to learn how to use them better. There are few examples of practical use of volatile repellents and attractants. Consideration of insect learning behaviour may allow development of strategies that either attract and kill herbivores or provide honest signals to lure and reward natural enemies. This would make it less likely that behavioural change would weaken the strategy. To allow better use of allelochemicals it might be worth reconsidering allelochemicals that have a role in plant defence as toxins or antifeedants. It is likely that improvements in means of deployment will improve prospects for use of allelochemicals in pest management. There are opportunities to use advanced formulations that will allow better deployment.

9 Where to look for further information

The development of allelochemicals for use in IPM is still largely at an experimental stage, except for notable exceptions such as the Push-Pull companion cropping system (www.push-pull.net). However, there is increasing demand for innovation and implementation of alternative approaches to pest management, due to the lack of availability of new conventional pesticides. Therefore, it is possible that there will be increased commercialisation and launch of new products, provided that regulatory hurdles can be overcome.

- For further information about any new developments, we recommend that readers check the farming press (such as 'Farmers Weekly' or 'Farmers Guardian' in the United Kingdom).
- Another source of practical information is levy boards that provide knowledge to farmers (such as the Agriculture and Horticulture Development Board in the United Kingdom: https://ahdb.org.uk).
- Certain interventions, such as companion plants, may be developed as an agronomic advice on which plants grow together, which means that agronomy organisations could provide vital information. A UK based example is the Association of Independent Crop Consultants (AICC).
- The scientific journal, 'Pest Management Science', provides scientific reports on precommercial as well as commercial crop treatments.
- The underpinning science of chemical ecology is covered by the 'Journal of Chemical Ecology', the specialist scientific journal dedicated to this field. The International Society of Chemical Ecology (ISCE), which publishes the journal, also host regular international scientific conferences (https://www.chemecol.org).

10 References

Akhtar, Y. and Isman, M. B. 2004. Comparative growth inhibitory and antifeedant effects of plant extracts and pure allelochemicals on four phytophagous insect species. *Journal of Applied Entomology* 128(1), 32–8. doi:10.1046/j.1439-0418.2003.00806.x.

Ali, J. G. and Agrawal, A. A. 2012. Specialist versus generalist insect herbivores and plant defense. *Trends in Plant Science* 17(5), 293–302. doi:10.1016/j.tplants.2012.02.006.

Athanassiou, C. G., Kavallieratos, N. G., Benelli, G., Losic, D., Rani, P. U. and Desneux, N. 2018. Nanoparticles for pest control: current status and future perspectives. *Journal of Pest Science* 91(1), 1-15. doi:10.1007/s10340-017-0898-0.

Baldwin, I. T., Halitschke, R., Paschold, A., Von Dahl, C. C. and Preston, C. A. 2006. Volatile signaling in plant-plant interactions: "talking trees" in the genomics era. *Science* 311(5762), 812-5. doi:10.1126/science.1118446.

Biondi, A., Desneux, N., Siscaro, G. and Zappalà, L. 2012. Using organic-certified rather than synthetic pesticides may not be safer for biological control agents: selectivity and side effects of 14 pesticides on the predator *Orius laevigatus*. *Chemosphere* 87(7), 803-12. doi:10.1016/j.chemosphere.2011.12.082.

Braasch, J., Wimp, G. M. and Kaplan, I. 2012. Testing for phytochemical synergism: arthropod community responses to induced plant volatile blends across crops. *Journal of Chemical Ecology* 38(10), 1264-75. doi:10.1007/s10886-012-0202-y.

Bruce, T. J. A., Martin, J. L., Pickett, J. A., Pye, B. J., Smart, L. E. and Wadhams, L. J. 2003. cis-Jasmone treatment induces resistance in wheat plants against the grain aphid, *Sitobion avenae* (Fabricius) (Homoptera: Aphididae). *Pest Management Science* 59(9), 1031-6. doi:10.1002/ps.730.

Bruce, T. J., Wadhams, L. J. and Woodcock, C. M. 2005. Insect host location: a volatile situation. *Trends in Plant Science* 10(6), 269-74. doi:10.1016/j.tplants.2005.04.003.

Chamberlain, K., Khan, Z. R., Pickett, J. A., Toshova, T. and Wadhams, L. J. 2006. Diel periodicity in the production of green leaf volatiles by wild and cultivated host plants of stemborer moths, *Chilo partellus* and *Busseola fusca*. *Journal of Chemical Ecology* 32(3), 565-77.

Cook, S. M., Khan, Z. R. and Pickett, J. A. 2007. The use of push-pull strategies in integrated pest management. *Annual Review of Entomology* 52, 375-400. doi:10.1146/annurev.ento.52.110405.091407.

de Oliveira, J. L., Campos, E. V. R., Bakshi, M., Abhilash, P. C. and Fraceto, L. F. 2014. Application of nanotechnology for the encapsulation of botanical insecticides for sustainable agriculture: prospects and promises. *Biotechnology Advances* 32(8), 1550-61. doi:10.1016/j.biotechadv.2014.10.010.

Dicke, M. and Sabelis, M. W. 1988. Infochemical terminology: based on cost-benefit analysis rather than origin of compounds? *Functional Ecology* 2(2), 131-9. doi:10.2307/2389687.

Diggle, S. P., Gardner, A., West, S. A. and Griffin, A. S. 2007. Evolutionary theory of bacterial quorum sensing: when is a signal not a signal? *Philosophical Transactions of the Royal Society B: Biological Sciences* 362(1483), 1241-9. doi:10.1098/rstb.2007.2049.

Dubey, N. K., Shukla, R., Kumar, A., Singh, P. and Prakash, B. 2010. Prospects of botanical pesticides in sustainable agriculture. *Current Science* 98(4), 479-80.

Duke, S. O., Cantrell, C. L., Meepagala, K. M., Wedge, D. E., Tabanca, N. and Schrader, K. K. 2010. Natural toxins for use in pest management. *Toxins* 2(8), 1943-62. doi:10.3390/toxins2081943.

Ebadollahi, A., Sendi, J. J. and Aliakbar, A. 2017. Efficacy of nanoencapsulated *Thymus eriocalyx* and *Thymus kotschyanus* essential oils by a mesoporous material MCM-41 against *Tetranychus urticae* (Acari: Tetranychidae). *Journal of Economic Entomology* 110(6), 2413-20. doi:10.1093/jee/tox234.

Elliott, M. and Janes, N. F. 1978. Synthetic pyrethroids-a new class of insecticide. *Chemical Society Reviews* 7(4), 473-505. doi:10.1039/cs9780700473.

El-Sayed, A. M., Suckling, D. M., Wearing, C. H. and Byers, J. A. 2006. Potential of mass trapping for long-term pest management and eradication of invasive species. *Journal of Economic Entomology* 99(5), 1550–64. doi:10.1603/0022-0493-99.5.1550.

Farmer, E. E. and Ryan, C. A. 1990. Interplant communication: airborne methyl jasmonate induces synthesis of proteinase inhibitors in plant leaves. *Proceedings of the National Academy of Sciences of the United States of America* 87(19), 7713–6. doi:10.1073/pnas.87.19.7713.

Gadino, A. N., Walton, V. M. and Lee, J. C. 2012. Evaluation of methyl salicylate lures on populations of *Typhlodromus pyri* (Acari: Phytoseiidae) and other natural enemies in western Oregon vineyards. *Biological Control* 63(1), 48–55. doi:10.1016/j.biocontrol.2012.06.006.

Golden, G., Quinn, E., Shaaya, E., Kostyukovsky, M. and Poverenov, E. 2018. Coarse and nano emulsions for effective delivery of the natural pest control agent pulegone for stored grain protection. *Pest Management Science* 74(4), 820–7. doi:10.1002/ps.4787.

Greenfield, M. D. 2002. *Signalers and Receivers: Mechanisms and Evolution of Arthropod Communication*. Oxford University Press, New York, NY.

Hassanali, A., Herren, H., Khan, Z. R., Pickett, J. A. and Woodcock, C. M. 2008. Integrated pest management: the push-pull approach for controlling insect pests and weeds of cereals, and its potential for other agricultural systems including animal husbandry. *Philosophical Transactions of the Royal Society B: Biological Sciences* 363(1491), 611–21.

Hokkanen, H. M. T. 1991. Trap cropping in pest management. *Annual Review of Entomology* 36(1), 119–38. doi:10.1146/annurev.en.36.010191.001003.

Isman, M. B. 2006. Botanical insecticides, deterrents, and repellents in modern agriculture and an increasingly regulated world. *Annual Review of Entomology* 51, 45–66. doi:10.1146/annurev.ento.51.110104.151146.

James, D. G. 2003. Field evaluation of herbivore-induced plant volatiles as attractants for beneficial insects: methyl salicylate and the green lacewing, *Chrysopa nigricornis*. *Journal of Chemical Ecology* 29(7), 1601–9. doi:10.1023/a:1024270713493.

James, D. G. and Grasswitz, T. R. 2005. Synthetic herbivore-induced plant volatiles increase field captures of parasitic wasps. *BioControl* 50(6), 871–80. doi:10.1007/s10526-005-3313-3.

Karban, R., Baldwin, I. T., Baxter, K. J., Laue, G. and Felton, G. W. 2000. Communication between plants: induced resistance in wild tobacco plants following clipping of neighboring sagebrush. *Oecologia* 125(1), 66–71. doi:10.1007/PL00008892.

Kelly, J. L., Hagler, J. R. and Kaplan, I. 2014. Semiochemical lures reduce emigration and enhance pest control services in open-field predator augmentation. *Biological Control* 71, 70–7.

Kessler, A. and Baldwin, I. T. 2001. Defensive function of herbivore-induced plant volatile emissions in nature. *Science* 291(5511), 2141–4. doi:10.1126/science.291.5511.2141.

Khan, Z. R., Ampong-Nyarko, K., Chiliswa, P., Hassanali, A., Kimani, S., Lwande, W., Overholt, W. A., Overholt, W. A., Picketta, J. A., Smart, L. E. and Woodcock, C. M. 1997. Intercropping increases parasitism of pests. *Nature* 388(6643), 631–2. doi:10.1038/41681.

Khan, Z. R., Midega, C. A. O., Amudavi, D. M., Hassanali, A. and Pickett, J. A. 2008. On-farm evaluation of the 'push-pull' technology for the control of stemborers and striga weed on maize in western Kenya. *Field Crops Research* 106(3), 224–33. doi:10.1016/j.fcr.2007.12.002.

Khan, Z. R., Midega, C. A., Bruce, T. J., Hooper, A. M. and Pickett, J. A. 2010. Exploiting phytochemicals for developing a 'push-pull' crop protection strategy for cereal farmers in Africa. *Journal of Experimental Botany* 61(15), 4185–96. doi:10.1093/jxb/erq229.

Koul, O. 2008. Phytochemicals and insect control: an antifeedant approach. *Critical Reviews in Plant Sciences* 27(1), 1–24. doi:10.1080/07352680802053908.

Lee, J. C. 2010. Effect of methyl salicylate-based lures on beneficial and pest arthropods in strawberry. *Environmental Entomology* 39(2), 653–60. doi:10.1603/EN09279.

Mallinger, R. E., Hogg, D. B. and Gratton, C. 2011. Methyl salicylate attracts natural enemies and reduces populations of soybean aphids (Hemiptera: Aphididae) in soybean agroecosystems. *Journal of Economic Entomology* 104(1), 115–24. doi:10.1603/ec10253.

Metcalf, R. L. and Metcalf, E. R. 1992. *Plant Kairomones in Insect Ecology and Control*. Chapman and Hall Ltd, New York, NY and London, UK.

Mithöfer, A. and Boland, W. 2012. Plant defense against herbivores: chemical aspects. *Annual Review of Plant Biology* 63, 431–50. doi:10.1146/annurev-arplant-042110-103854.

Mordue, A. J. and Blackwell, A. 1993. Azadirachtin: an update. *Journal of Insect Physiology* 39(11), 903–24. doi:10.1016/0022-1910(93)90001-8.

Nordlund, D. A. and Lewis, W. J. 1976. Terminology of chemical releasing stimuli in intraspecific and interspecific interactions. *Journal of Chemical Ecology* 2(2), 211–20. doi:10.1007/BF00987744.

Odeyemi, I. S., Afolami, S. O. and Adigun, J. A. 2013. Plant parasitic nematode relative abundance and population suppression under *Chromolaena odorata* (Asteraceae) fallow. *International Journal of Pest Management* 59(1), 79–88. doi:10.1080/09670874.2013.766776.

Oliveira, A. P., Santana, A. S., Santana, E. D. R., Lima, A. P. S., Faro, R. R. N., Nunes, R. S., Lima, A. D., Blank, A. F., Araújo, A. P. A., Cristaldo, P. F. and Bacci, L. 2017. Nanoformulation prototype of the essential oil of *Lippia sidoides* and thymol to population management of *Sitophilus zeamais* (Coleoptera: Curculionidae). *Industrial Crops and Products* 107, 198–205. doi:10.1016/j.indcrop.2017.05.046.

Papanikolaou, N. E., Kalaitzaki, A., Karamaouna, F., Michaelakis, A., Papadimitriou, V., Dourtoglou, V. and Papachristos, D. P. 2018. Nano-formulation enhances insecticidal activity of natural pyrethrins against *Aphis gossypii* (Hemiptera: Aphididae) and retains their harmless effect to non-target predators. *Environmental Science and Pollution Research International* 25(11), 10243–9. doi:10.1007/s11356-017-8596-2.

Regnault-Roger, C., Vincent, C. and Arnason, J. T. 2012. Essential oils in insect control: low-risk products in a high-stakes world. *Annual Review of Entomology* 57, 405–24. doi:10.1146/annurev-ento-120710-100554.

Scott, I. M., Jensen, H., Scott, J. G., Isman, M. B., Arnason, J. T. and Philogène, B. J. R. 2003. Botanical insecticides for controlling agricultural pests: piperamides and the Colorado potato beetle *Leptinotarsa decemlineata* Say (Coleoptera: Chrysomelidae). *Archives of Insect Biochemistry and Physiology* 54(4), 212–25. doi:10.1002/arch.10118.

Searcy, W. A. and Nowicki, S. 2005. *The Evolution of Animal Communication: Reliability and Deception in Signaling Systems*. Princeton University Press, Princeton, NJ.

Shelton, A. M. and Badenes-Perez, F. R. 2006. Concepts and applications of trap cropping in pest management. *Annual Review of Entomology* 51, 285–308. doi:10.1146/annurev.ento.51.110104.150959.

Simpson, M., Gurr, G. M., Simmons, A. T., Wratten, S. D., James, D. G., Leeson, G. and Nicol, H. I. 2011. Insect attraction to synthetic herbivore-induced plant volatile-treated field crops. *Agricultural and Forest Entomology* 13(1), 45–57. doi:10.1111/j.1461-9563.2010.00496.x.

Sun, Y. F., Yu, H., Zhou, J. J., Pickett, J. A. and Wu, K. M. 2014. Plant volatile analogues strengthen attractiveness to insect. *PLoS ONE* 9(6), e99142. doi:10.1371/journal.pone.0099142.

Turlings, T. C. and Wäckers, F. 2004. Recruitment of predators and parasitoids by herbivore-injured plants. In: Cardé, R. and Millar, J. (Eds), *Advances in Insect Chemical Ecology*. Cambridge University Press, Cambridge, UK, pp. 21–75.

Vargas, R. I., Miller, N. W. and Prokopy, R. J. 2002. Attraction and feeding responses of Mediterranean fruit fly and a natural enemy to protein baits laced with two novel toxins, phloxine B and spinosad. *Entomologia Experimentalis et Applicata* 102(3), 273–82. doi:10.1046/j.1570-7458.2002.00948.x.

Vieira, C. R., Blassioli-Moraes, M. C., Borges, M., Pires, C. S. S., Sujii, E. R. and Laumann, R. A. 2014. Field evaluation of (E)-2-hexenal efficacy for behavioral manipulation of egg parasitoids in soybean. *BioControl* 59(5), 525–37. doi:10.1007/s10526-014-9592-9.

Wallingford, A. K., Cha, D. H., Linn Jr., C. E., Wolfin, M. S. and Loeb, G. M. 2017. Robust manipulations of pest insect behavior using repellents and practical application for integrated pest management. *Environmental Entomology* 46(5), 1041–50. doi:10.1093/ee/nvx125.

Webster, B., Bruce, T., Pickett, J. and Hardie, J. 2010. Volatiles functioning as host cues in a blend become nonhost cues when presented alone to the black bean aphid. *Animal Behaviour* 79(2), 451–7. doi:10.1016/j.anbehav.2009.11.028.

Witzgall, P., Kirsch, P. and Cork, A. 2010. Sex pheromones and their impact on pest management. *Journal of Chemical Ecology* 36(1), 80–100. doi:10.1007/s10886-009-9737-y.

Yang, F. L., Li, X. G., Zhu, F. and Lei, C. L. 2009. Structural characterization of nanoparticles loaded with garlic essential oil and their insecticidal activity against *Tribolium castaneum* (Herbst)(Coleoptera: Tenebrionidae). *Journal of Agricultural and Food Chemistry* 57(21), 10156–62. doi:10.1021/jf9023118.

Yoneyama, K. and Natsume, M. 2010. Allelochemicals for plant-plant and plant-microbe interactions. In: *Comprehensive Natural Products II: Chemistry and Biology*. Elsevier, pp. 539–61.

Yu, H., Zhang, Y., Wu, K., Gao, X. W. and Guo, Y. Y. 2008. Field-testing of synthetic herbivore-induced plant volatiles as attractants for beneficial insects. *Environmental Entomology* 37(6), 1410–5. doi:10.1603/0046-225X-37.6.1410.

Zhu, J. W. and Park, K. C. 2005. *Methyl salicylate*, a soybean aphid-induced plant volatile attractive to the predator *Coccinella septempunctata*. *Journal of Chemical Ecology* 31(8), 1733–46. doi:10.1007/s10886-005-5923-8.

Chapter 13

Peptides as novel biopesticides

Lin Bao, Robert M. Kennedy, Kyle Schneider, Alvar Carlson and Andy Renz, Vestaron Corporation, USA

1 Introduction

The global pesticide market was estimated to be nearly $56 billion in 2012, with a $9 billion market in the United States alone (Atwood and Paisely-Jones, 2013). Currently, synthetic agrochemical insecticides occupy the majority of the agro-insecticide market. Synthetic agrochemical insecticides are cheap, fast acting and have high efficacy. However, they are usually stable and persistent in the environment and act broadly across classes of insects. This results in environment pollution, a negative impact on beneficial insects and human health concerns. Additionally, agrochemical insecticides usually target only one of the five major neuronal targets of insects. The limited targets and overuse have inevitably resulted in widespread target-site and detoxification-based resistances in insects. As a result, a lot of agrochemical insecticides have been de-registered and withdrawn from the market. For example, between January 2005 and December 2009, 169 agrochemical insecticides were de-registered or cancelled, with only 9 new insecticides registered (Windley et al., 2012). Therefore, the need for new effective, biodegradable and safer biopesticides has increased dramatically. The global biopesticide market has grown from $1.6 billion in 2009 to $3.3 billion in 2014, at a 15.6% compound

http://dx.doi.org/10.19103/AS.2020.0073.14

annual growth rate (CAGR) (Aneja et al., 2016; Damalas and Koutroubas, 2018).

According to the definition by the US Environmental Protection Agency (EPA), biopesticides include:

- Naturally occurring substances that control pests (biochemical pesticides)
- microorganisms that control pests (microbial pesticides)
- pesticidal substances produced by plants containing added genetic material

Naturally occurring insecticidal proteins or peptides are gaining more and more attention in pest control. Currently, the leading biopesticides are Bt toxin-based formulations, which are based on the toxic proteins produced from the bacterium *Bacillus thuringiensis*, which bind to and disrupt insect gut linings. However, decades of worldwide application of Bt agents as spray-on product and as a GM trait in Bt crops has led to the evolution of Bt resistance in previously susceptible insects (Schnepf et al., 1998). There is now a need for new classes of biopesticide to replace Bt agents.

2 Venom peptides as biopesticides

Venom peptides are small toxic proteins originated from animal venoms. They have attracted great interest as a route to develop the next generation of novel, effective and safe biopesticides. The major prey of some venomous predators, such as spiders, scorpions, centipedes, wasps, ants etc. are insects. Their venom contains hundreds of components, including toxic peptides, which work together to quickly knock-down their prey (paralyse or kill) by a single strike. Some of the toxic peptides in their venom are insect-specific, and are promising as safe, effective biopesticides.

These animal venoms are a gold mine for novel biopesticides. The venom from a single spider contains more than 1000 distinct peptides (Escoubas et al., 2006). As over 42 000 species of spiders have been described so far, and as many as 4 times the number of species are predicted to exist (Coddington and Levi, 1991), spider venoms may contain more than 10 million bioactive peptides (Windley et al., 2012). It is not surprising that agricultural companies have put significant R&D effort into venom peptides as next generation biopesticides.

Overall, compared to traditional chemical insecticides, insect-specific insecticidal venom peptides are safe without toxic effects to humans and vertebrates (Windley et al., 2012). They are small peptides and hence biodegradable. Therefore, they will not accumulate and cause adverse environment impacts. Even though venom peptides may target the same insecticidal receptors as chemical insecticides, they likely bind at different sites, to which insects have not developed any resistances.

However, there are some challenges to commercialize venom peptides as biopesticides. Key challenges are manufacturing costs, product stability (in the field and shelf life), and bioavailability. Fermentation-based manufacturing of peptide-based products is in general more costly than chemical synthesis of synthetic pesticides. Peptides may degrade after application, before effectively protecting the crops from insects. As a consequence more applications per season may be necessary, which increases the growers' pesticides costs per season.

The major challenge is how to deliver venom peptides to their target sites efficiently. In nature, venom peptides are delivered by direct 'injection' into the insect hemolymph from a predator's strike. They are not designed to be active as insecticide when taken up orally by the target insect. As a result they may not survive protease digestion in the insect gut and may lose their efficacy before they reach the target sites. Peptides cannot efficiently penetrate the insect gut wall or cuticle to get into the hemolymph.

Agrochemical companies proposed to deliver venom peptides by genetic engineering into the genome of a baculovirus. Baculoviruses are insect-species-specific, which dramatically affects the peptide toxic spectrum.

Once costs and barriers to efficient delivery are overcome, venom peptides are likely to play a key role in the pesticide market of the future.

3 The discovery of novel venom peptides

FMC Corporation (Food Machinery Corporation) is an American chemical manufacturing company headquartered in Philadelphia, Pennsylvania (USA). In the 1990s FMC put significant effort into discovering novel insecticidal venom peptides from spider venoms and filed patents for a number of venom peptides as potential biopesticides. NPS Pharmaceuticals Inc. was established as a rare disease-focussed biopharmaceuticals company located in Bedminster, NJ (USA), and acquired by Shire Plc. in 2015. NPS collaborated with FMC working on the discovery of novel venom peptides from spiders for biopesticides in the 1990s. In 1995, FMC collaborated with NPS Pharmaceuticals in discovering two new venom peptides from the hob spider venom (*Tengenaria agrestis*), NPS-326 and NPS-331, which were later re-designated as U1-agatoxin-Ta1a and U1-agatoxin-Ta1b, respectively, by ArachnoServer, a spider toxin database (Herzig et al., 2011). Both of them showed broad insecticidal activity against multiple species in the order of Diptera and Lepidoptera (Krapcho et al., 1995a).

In 1995, FMC also patented 4 venom proteins discovered from the Southern house spider (*Filistata hibernalis*), designated as FIL-376, FIL-377, FIL-501and FIL-502, and reported to be active against tobacco budworm, beet armyworm, cabbage looper, fall armyworm, European corn borer, soybean looper, diamondback moth and corn earworm (Jackson et al., 1995). In the same year FMC also patented three other spider venom peptides discovered from desert bush spider (*Diguetia canities*), DK9.2, DK11 and DK12, which had insecticidal activity against multiple Lepidoptera species: tobacco budworm, beet armyworm, cabbage looper, fall armyworm, European corn borer, soybean looper, and diamondback moth (Krapcho et al., 1995b). In 1998, two more spider venom peptides were filed for patent protection by FMC, discovered from the daring jumping spider (*Phidippus audax*), and designated as BW2 and BW5, which later were re-designated as U1-slatitoxin by ArachnoServer. Both showed insecticidal activity against cabbage looper and tobacco budworm (Jackson et al., 1998). FMC proposed to formulate these venom peptides in a baculovirus system for potential application.

In 1997, NPS filed two patent applications for peptide-A from the venom of the spider, *Segestria* sp., which was re-designated as μ-segestritoxin-Ssp1a in ArachnoServer, and peptide-A, B & C from the venom of the spider, *Calisoga* sp., which was re-designated as U1-nemetoxin-Csp1a, Csp1b & Csp1c. All 4 toxin peptides had excitatory toxic effects on tobacco budworm (Johnson and Kral, 1997; Johnson et al., 1997). In 1998, NPS reported five venom proteins discovered from *Bracon hebetor*, which could paralyse tobacco budworm and beet armyworm by injection. NPS also developed a transgenic baculovirus expressing these engineered toxins with oral insecticidal activity against tobacco budworm and beet armyworm (Johnson et al., 1999).

Syngenta AG is an agrochemical and seed company based in Basel, Switzerland. It was formed in 2000 as a merger of Novartis Agribusiness and Zeneca Agrochemicals, and was acquired in 2017 by the Chinese state-owned chemical company ChemChina. In 2001 Syngenta discovered two insecticidal proteins, Brh-I and Brh-V, from a parasitic wasp, *Bracon hebetor*. Both were active against tobacco hornworm, tobacco budworm, corn earworm, beet armyworm, wax moth, cabbage looper, southern corn rootworm (Leisy et al., 2001). In 2001 Novartis AG, a Swiss company, filed a patent to protect nine insecticidal peptides discovered from the venom of the primitive hunting spider (*Plectreurys tristis*), Plt-V, Plt-VI, Plt-VIII, Plt-XI, Plt-XII, Plt-XIII, Plt-XIV, Plt-IX and Plt-X. They were found to be active against tobacco hornworm, tobacco budworm and beet armyworm (Quistad and Leisy, 2001). It was proposed to use a baculovirus system for the delivery of those peptides in field applications.

A few other companies have also filed patents for insecticidal venom peptides. For example, Zeneca Limited, a British pharmaceutical company, filed a patent in 1999 on a series of peptides discovered from the venom of cellar spider, *Segestria florentina*, which is toxic to beet armyworm (Windass et al., 1999). Soricimed Biopharma Inc., a Canadian pharmaceutical company filed a patent in 2011 relating to insecticidal peptides from the submaxilary glands of shrew which could paralyse mealworms (Stewart et al., 2011).

4 The first commercialized venom peptide-based biopesticide: SPEAR®

SPEAR®, with its active ingredient GS-omega/kappa-Hxtx-Hv1a, is the first example of a commercialized insecticidal cystine knot peptide for crop protection. This product was developed by Vestaron Corporation. Vestaron was established in 2007 by Prof. Glenn King, University of Connecticut, based on a patent for a family of peptides isolated from the Blue Mountain funnel web spider of Australia (King et al., 2008). This aggressive spider was a significant health issue. As a result the Australian government funded the milking of the venom of a colony of spiders to support the development of an anti-venom to treat human victims of spider bites.

While the venom comprised of a major component of potent sodium channel activity against mammals, Prof. King's insight was to isolate several minor components which showed no mammalian activity but did evidence broad insecticidal activity (Fletcher et al., 1997). The lack of mammalian activity was later on confirmed by Vestaron Corporation as part of the EPA registration of the SPEAR® bioinsecticide (Federal Food, Drug, and Cosmetics Act (FFDCA) Considerations for GS-omega/kappa-Hxtx-Hv1a, 2013). Initially, the activity of GS-omega/kappa-Hxtx-Hv1a was thought to derive from calcium and potassium channel blockade. However, it has now been shown that these three families of peptides, the so-called omega, kappa and hybrid families, derive their insecticidal effects from positive allosteric modulation of nicotinic acetylcholine receptors (Chambers et al., 2019). Since no competitive binding or cross resistance has been observed in either the neonicotinic or Spinosad classes, IRAC has assigned SPEAR® a novel code, 32.

To develop the biopesticide, Vestaron had to surmount three hurdles. First, Vestaron developed a yeast-based expression system that secrets the SPEAR® peptide in multi-gram per litre levels. The filtered and concentrated fermentation broth is preserved and bottled for use as a foliar spray. Second, by combining SPEAR® with Btk products, an oral formulation was obtained for protection of crops from lepidoptera. The SPEAR® product alone is useful as a contact spray against thrips, spider mites, white flies, aphids and spotted-wing Drosophila (Kund et al., 2019; Kund and Trumble, 2019; Wise et al., 2018; Fanning et al., 2018). Finally, a regulatory path was obtained initially as a microbial-based insecticide subject to biochemical requirements. EPA's Biopesticides and Pollution Prevention Division (BPPD) granted SPEAR®-T and SPEAR®-Lep full section 3 registrations on 19 September 2018 after an extensive review of the compounds data package (EPA Registration Number: 88847-6).

EPA-approved labels were posted to their Pesticide Product and Label System (PPLS) which can be found at https://iaspub.epa.gov/apex/pesticides/f?p=PPLS:1. A new department was later established within the BPPD of the EPA, the Emerging Technologies Branch, specifically to handle products such as SPEAR®.

5 Other commercialized peptides: harpin and trypsin modulating oostatic factor (TMOF)

Microbial infection causes plants to activate an immune defense system termed the systemic acquired resistance (SAR). Upon recognition of a pathogen, the plant uses chemical signals to induce expression of resistance genes which results in heightened resistance to a wide variety of pests and can also improve plant vigor (Malamy and Klessig, 1992; Ward et al., 1991). Application of these chemical signals to plants was shown to be sufficient to upregulate acquired resistance and protect plants from bacterial, viral or fungal pathogens. Researchers have more recently focussed on the upregulation of plant SAR-inducing genes through transgenic approaches in a wide range of fruit and vegetable crops. A particularly well-studied gene used for modifying plants is the *Arabidopsis thaliana* NPR1 (nonexpressor of pathogenesis-related genes) gene (reviewed in Silva et al., 2018), which can induce the SAR in a wide range of economic crops and protect the plant from a wide range of pathogens (fungal, bacterial and viral) and even pests such as soybean cyst nematode (Matthews et al., 2014) and common cutworm (Meur et al., 2008). Notably, over-induction of SAR has been shown to impart a fitness cost on the plant, suggesting careful tuning may be necessary to maximize the benefit of SAR induction (Fitzgerald et al., 2004; Silva et al., 2018).

Plants can specifically recognize pathogen components, such as chitin, glucans or lipopolysaccharides, to elicit the SAR response (Dangl and Jones, 2001). One intriguing pathogen inducer is a family of bacterial infection genes termed harpins. Harpins are essential components of the type 3 secretion system (T3SS) of gram-negative plant pathogens. Upon plant infection, the T3SS injects virulence factors into the host through a needle-like appendage, which ultimately elicits the hypersensitive response (HR) that is characterized by a region of dead plant tissue forming a barrier against the further spread of infection (Horsfall and Cowling, 1982). Importantly, the HR gives rise to the SAR response in a diverse range of plants, suggesting that exogenous application of T3SS components could be used as a pesticide (Wei et al., 1992).

In the early 1990s, a component of the T3SS from the plant pathogen, *Erwinia amylovora*, termed HrpN, was isolated and determined to be sufficient to elicit the HR and SAR responses in plants (Wei et al., 1992). Fragments of this family of proteins were shown to be able to induce the SAR response in a wide variety of plants outside the pathogen host range when a small amount was topically applied, indicating a potential use as a broad-spectrum biopesticide and plant stimulant (He et al., 1993; Dong et al., 1999). From this work, Eden Bioscience was formed and in 2000, it received EPA approval for a commercial formulation of $HrpN_{EA}$ produced heterologously in *E. coli*. A few years later, Eden developed and received approval for an improved harpin called $HrpN_{\alpha\beta}$ which is the result of replacing sections of $HrpN_{EA}$ with orthologous protein fragments from related pathogens, thereby creating a 42 kDa chimeric protein with improved spectrums of activity (Wei, 2009; Environmental Protection Agency, 2005).

Products containing $HrpN_{\alpha\beta}$, now marketed by Plant Health Care (Raleigh, NC), are applied as a foliar spray or seed treatment to improve plant stamina and vigor for a wide range of crops (Environmental Protection Agency, 2013). In 2018, Plant Health Care reported sales of products containing $HrpN_{\alpha\beta}$ in excess of $5 million. Partially due to the instability of HrpN in aqueous solutions, Plant Health Care has further developed short, synthetic peptide mimics of harpins which are more stable and can elicit the HR and SAR similar to HrpN (Wei and Zornetzer, 2017). Regulatory filing for the first of these peptide products was made in 2018 with expected product roll-out as early as 2020 (Plant Health Care, 2018).

Upon taking a bloodmeal, mosquitos and related blood-sucking insects produce a burst of digestive proteases responsible for breaking down the protein-rich meal. To counteract the overproduction of these enzymes, the insects then release a decapeptide hormone called trypsin-modulating oostatic factor (TMOF) that downregulates the production of trypsin-like proteases (Borovsky, 1988; Borovsky et al., 1990). Upon discovery of TMOF in the 1980s, researchers at the University of Florida showed it could be useful as an insecticide when supplied exogenously due to its stability in the insect gut and ability to traverse to its target in the hemolymph (Borovsky and Mahmood, 1995). Mosquito larvae exposed to TMOF were subsequently unable to digest protein due to the lack of protease activity and die of starvation before reaching maturity (Borovsky and Meolo, 2004).

The prospect of a novel biopesticide against mosquito larvae led to the formation of the company Insect Biotechnology, Inc. (Durham, NC - USA). They explored commercial formulations of TMOF manufactured by *Pichia pastoris* and received EPA approval in 2004 for the treatment of mosquito larvae in water using sterilized TMOF-*Pichia* solutions (Environmental Protection Agency, 2004). Due to a lack of investor funds, however, Insect Biotechnology, Inc. ceased operations prior to making formulations of TMOF commercially available.

6 Plant expression of venom peptides

In the search for insecticidal proteins to be expressed in genetically modified (GM)-plants there were a number of potential protein sources to start with, but many scientists began to look at the genes encoding insecticidal peptides among venom proteins. The large scorpions and spiders became the first likely organisms to search due to the ease of milking venom to fractionate and isolate the peptides. The resulting venom peptides were toxic when delivered via a stinger or fang in the wild, and by a needle in the lab, but they were not as insecticidal when fed orally as they had to pass the gut wall barrier to reach their site of action. This led to a research focus of fusing those initial target insecticidal peptide candidates to proteins that aided in transcytosis across the gut epithelium barrier to increase their concentration near their insect nervous system targets.

The first publication of a spider venom peptide being expressed in plants and exhibiting insecticidal activity to *Heliothis armigera* (cotton bollworm) was over twenty years ago (Jiang et al., 1996). Later, rice was transformed with a gene identified as spider insecticide (Sp1) and it conferred enhanced resistance to *Cnaphalocrasis medinalis* and *Chilo suppressalis* (Huang et al., 2001). When the pharmacophore of ω-hexatoxin-Hv1a (Hv1a) insecticidal peptide from *Hadronyche versuta* was defined it gained popularity among a number of plant transformation groups (Tedford et al., 2004). Hv1a was expressed in GM-tobacco under constitutive promoters (Khan et al., 2006) and phloem-specific promoters (Shah et al., 2011) where it provided insecticidal activity against *H. armigera* in three-day long bioassays. It was also expressed in GM-birch (*Betula platyphylla*) and provided insecticidal protection from *Lymantria dispar* (gypsy moth; Zeng et al., 2009). The Hv1a sequence was also fused into one reading frame with a Bt and gave higher *L. dispar* mortality than the controls and less leaf damage in GM-poplar (Cao et al., 2010).

When GM-tobacco and GM-cotton expressing Hv1a were compared, tobacco controlled 100% of *Spodoptera littoralis* (African cotton leafworm) and *Heliothis virescens* (tobacco budworm), but the GM-cotton had no insecticidal effect (Ullah et al., 2015). When expressing Hv1a and another insecticidal protein, the onion leaf lectin, in two expression cassettes in one GM-tobacco plant, there was a more rapid insecticidal activity to three hemipteran pests, namely *Phenacoccus solenopsis* (cotton mealybug), *Myzus persicae* (peach potato aphid) and *Bemisia tabac* (silverleaf whitefly; Javaid et al., 2016) and two lepidopteran pests, namely *H. armigera* and *Spodoptera litura* (tobacco cutworm; Rauf et al., 2019). The Magi6 gene, discovered in another funnel web spider, *Macrothele gigas* (Japanese giant funnelweb spider), exhibited insecticidal activity when expressed in GM-tobacco in bioassays against *Spodoptera frugiperda* (fall armyworm; Hernández-Campuzano et al., 2009). A final spider venom peptide, lycotoxin-I from *Lycosa carolinensis*, was expressed in GM-tobacco and significantly enhanced resistance to the lepidopteran larvae *Helicoverpa zea* (corn armyworm) and Coleopteran *Lasioderma serricorne* (cigarette beetle; Johnson et al., 2014).

The first scorpion venom peptide transgene (BamkIT, from *Buthus martensii*) was expressed in GM-oilseed rape (*Brassica napus*) and it provided reduced leaf damage by lepidopteran *Plutella xylostella* (diamondback moth; Wang et al., 2005). The *Androctonus australis* scorpion insecticidal toxin (AaIT) has been expressed in GM-cotton and gave insecticidal activity against *H. armigera* (Wu et al., 2008). Finally, the LqhIT2 from *Leiurus quinquestriatus* was modified (maintaining 85% amino acid homology) to make a synthetic peptide renamed LMX that exhibits improved stability and insecticidal activity in GM-rice bioassays against *Cnaphalacrosis medinalis* (rice leafroller; Tianpei et al., 2014).

A number of researchers have attempted to increase the delivery of the orally supplied venom peptide Hv1a via its fusion to proteins that aid in transcytosis across the gut epithelium. The first group to achieve success used a luteovirid coat protein (CP) to fuse Hv1a in a CP-Hv1a orientation (Bonning et al., 2014). When the fusion was expressed in GM-*Arabidopsis*, it reduced aphid populations, compared to the controls, by an order of magnitude after 17 days of infestation. When the CP-Hv1a expressing plants were tested with a lepidopteran (*Heliothis virescens*) it did not have any detectable insecticidal activity, likely due to the aphid-specific nature of Luteovirus transmission.

Two more groups attempted *Galanthus nivalis* agglutinin (GNA) and Hv1a fusions in transgenic plants (Nakasu et al., 2014; Alam, 2014). The first group, began with research on a yeast-expressed Hv1a and GNA fusion (Fitches et al., 2012) and established that the Hv1a-GNA had equal insecticidal activity to Hv1a alone when injected it into the hemolymph of the lepidopteran *Mamestra brassicae* (cabbage moth). When they provided the Hv1a-GNA fusion orally (leaf disks coated) it was more insecticidal than the control treatments. Two years later that research group took the Hv1a-GNA fusion protein, expressed it in GM-*Arabidopsis* (Nakasu et al., 2014) and established that it reduced aphid populations, compared to the controls, by 30% after 7 days of infestation. Despite previous knowledge of fusion proteins having lepidopteran activity, no lepidopteran bioassays using the GM-*Arabidopsis* plants were published.

The second group to fuse GNA and Hv1a was published in a PhD thesis (Alam, 2014) but the lepidopteran bioassays on GM-*Arabidopsis* were limited by a toxic level of endogenous glucosinolates present in wild-type plants. When using a cyp79B2 cyp79B myb28 myb29 quadruple Arabidopsis mutant background that lacks endogenous glucosinolates, the GNA-Hv1a and Hv1a-GNA orientation gave increased insecticidal activity, compared to the controls, in a lepidopteran bioassay (*Helicoverpa armigera*).

A final group attempted an *Allium cepa* lectin (ACL) and a Hv1a fusion in transgenic plants (Javaid et al., 2018). While ACL has not been established to accumulate in the hemolyph of a lepidopteran, a related (76% identical) *Allium sativum* agglutinin does achieve transcytosis across the gut (Upadhyay et al.,

2010). A 20–40% increase of insecticidal activity of both ACL-Hv1a and Hv1a-ACL fusions, compared to the controls, was achieved after 6 days of infestation of GM-tobacco by a hemipteran (*Phenacoccus solenopsis*). The tobacco was not a stable GM-tobacco, rather it used a transient potato virus X (PVX) expression system to express the transgenes. Finally, the AaIT fused to the GNA was expressed in three plant species and showed increased insecticidal activity compared to the controls in lepidopteran *H. armigera* bioassays using GM-tobacco and GM-*Arabidopsis* and in hemipteran *Bemisia tabaci* (silverleaf whitefly) bioassays using GM-tobacco and *Nilaparvata lugens* (brown planthopper) bioassays using GM-rice (Liu et al., 2016).

7 Plant defensin and other antifungal peptides

In the previous sections we described the structure and function of animal-derived ICK-type peptides, exemplified by the GS-omega/kappa-Hxtx-Hv1a (SPEAR®) peptide and two other commercialized peptides (Harpin and TMOF). Venom-derived ICK-type peptides are structurally and functionally closely related to antimicrobial peptides (AMPs).

A large variety of AMPs are produced by plants. These are cysteine-rich proteins (CRPs) that can be classified into thionins, defensins, hevein-like peptides, knottin-type peptides (linear and cyclic), lipid transfer proteins, alpha-hairpins and snakins (Tam et al., 2015). Plant defensins have in vitro and in vivo efficacy against relevant fungal pathogens, are non-toxic to humans and other mammals, and resistances develop at very low frequencies (Del Poeta et al., 2014). For these reasons they have been studied extensively and are considered promising candidates for disease resistance traits in crops (Kaur et al., 2011), and have been discussed for development of biological fungicides (Parisi et al., 2019).

Plant defensins are an essential part of the plant innate immune system (Lay and Anderson, 2005), and are characterized as small cationic peptides with a highly conserved tertiary structure, consisting of one alpha-helix and three beta-sheets stabilized by four disulfide bridges (Cools et al., 2017). A common feature and first step in the antifungal mode of action of plant defensins is the binding to fungal membrane targets, namely sphingolipids and phospholipids, followed by fungal cell death through the production of reactive oxygen species, apoptosis and necrosis. Plant defensin-specific mechanisms of action include oligomerization of defensin-lipid complexes, internalization into the fungal pathogen, mitochondrial functionality loss, disruption of divalent cation homeostasis, cell wall stress, septin mis-localization and cell cycle arrest (Cools et al., 2017).

A well-studied plant defensin with high efficacy in vitro and in vivo against relevant fungal plant pathogens is MtDef4. The research group around Dilip Shah from the Donald Danforth Plant Science Center identified a RGFRRR motif to govern the entry of the peptide into the fungal cell (Sagaram et al., 2013). Interestingly the mode of action of MtDef4 against *Neurospora crassa* (a type of red bread mould, used as a model organism) differs from that against *Fusarium graminearum* (one of the causal agents of head blight disease in wheat and barley), which indicates that one particular plant defensin can have multiple antifungal actions depending on the fungal pathogen species (El-Mounadi et al., 2016). The same research group demonstrated the functionality of MtDef4 in plants: expression of an apoplast-targeted plant defensin MtDef4.2 conferred resistance to a leaf rust pathogen, but did not affect mycorrhizal symbiosis in transgenic wheat plants (Kaur et al., 2017).

A very innovative, interdisciplinary approach to develop biological fungicides based on AMPs was published by the research groups around Uwe Conrath and Ulrich Schwaneberg from RWTH Aachen. They showed that a bifunctional dermaseptin–thanatin dipeptide combines antifungal activity of the antimicrobial peptide dermaseptin 01 with the wax-binding properties of thanatin, resulting in a biological fungicide against Asian Soybean Rust with exceptional rainfastness (Schwinges et al., 2019).

8 Summary and future trends

Peptides combine high selectivity/specificity with excellent toxicological/environmental safety profiles and are thus promising candidates for the development of drugs and pesticides. In the pharmaceutical industry peptides are the fastest growing class of drugs with an estimated doubling of the market share between 2011 and 2018, more than 100 peptide-based therapeutic products on the market, and the same number of product candidates in clinical trials (Fosgerau and Hoffmann, 2015). In fact, the term 'biologicals' in pharma refers mostly to peptide- and protein-based products. In agriculture, on the other hand, most 'biologicals' today are living microbes or plant-based extracts. So far peptides do not play a major role in crop protection. Factors limiting the development of peptide-based pesticides have been limited bioavailability and stability, high manufacturing costs, and regulatory uncertainties.

Inhibitory cysteine knot (ICK)-type peptides have a strong selectivity/specificity, are stable and can be produced at high yields in a fermentation-based manufacturing process. In 2018, for the first time a peptide-based insecticide was approved by the EPA, and the product SPEAR® has been launched by Vestaron Corporation. SPEAR® has a novel mode-of-action, and combines the specificity and efficacy of synthetic, small molecule-based insecticides with the safety profile of biologicals.

Due to its efficacy and safety profile, SPEAR® has the potential to be a successful product for use in greenhouses and on field-grown high-value crops. The lack of mammalian activity makes it an attractive and safe product, particularly for greenhouses (worker safety) and food crops (consumer safety). Since SPEAR® is safe for honey bees and other beneficial insects such as ladybird beetles, it is a safe and versatile partner in integrated pest management programs. Improvements in manufacturing and formulation can potentially expand the market from high-value crops into row crops.

ICK-type peptides have the potential to 're-drug' the insecticide world, i.e. it may be possible to identify peptides that bind to all commercialized, neuromuscular insecticide targets in a novel way, resulting in unique mode-of-action classifications - as it was demonstrated for SPEAR®. Similar to ICK-type insecticidal peptides, antifungal peptides could be developed as novel biofungicides.

9 References

Alam, S. 2014. Engineering insect-resistant plants by transgenic expression of an insecticidal spider-venom peptide. Thesis submitted a Doctor of Philosophy. The University of Queensland.

Aneja, K. R., Khan, S. A. and Aneja, A. 2016. Biopesticides an eco-friendly pest management approach in agriculture: status and prospects. *Kavaka* 47, 145-54.

Atwood, D. and Paisely-Jones, C. 2013. Pesticides industry sales and usage: 2008-2012 market estimates. United States Environmental Protection Agency.

Bonning, B. C., Pal, N., Liu, S., Wang, Z., Sivakumar, S., Dixon, P. M., King, G. F. and Miller, W. A. 2014. Toxin delivery by the coat protein of an aphid-vectored plant virus provides plant resistance to aphids. *Nature Biotechnology* 32(1), 102-5. doi:10.1038/nbt.2753.

Borovsky, D. 1988. Oostatic hormone inhibits biosynthesis of midgut proteolytic enzymes and egg development in mosquitoes. *Archives of Insect Biochemistry and Physiology* 7(3), 187-210. doi:10.1002/arch.940070305.

Borovsky, D. and Mahmood, F. 1995. Feeding the mosquito *Aedes aegypti* with TMOF and its analogs; effect on trypsin biosynthesis and egg development. *Regulatory Peptides* 57(3), 273-81. doi:10.1016/0167-0115(95)00041-9.

Borovsky, D. and Meolo, S. M. 2004. Biochemical and cytoimmunological evidence for the control of *Aedes aegypti* larval trypsin with Aea-TMOF. *Archives of Insect Biochemistry and Physiology* 55(3), 124-39. doi:10.1002/arch.10132.

Borovsky, D., Carlson, D. A., Griffin, P. R., Shabanowitz, J. and Hunt, D. F. 1990. Mosquito oostatic factor: a novel decapeptide modulating trypsin-like enzyme biosynthesis in the midgut. *FASEB Journal: Official Publication of the Federation of American Societies for Experimental Biology* 4(12), 3015-20. doi:10.1096/fasebj.4.12.2394318.

Cao, C. W., Liu, G. F., Wang, Z. Y., Yan, S. C., Ma, L. and Yang, C. P. 2010. Response of the gypsy moth, *Lymantria dispar* to transgenic poplar, *Populus simonii* × *P. nigra*, expressing fusion protein gene of the spider insecticidal peptide and Bt-toxin C-peptide. *Journal of Insect Science* 10(1), 200, doi:10.1673/031.010.20001.

Chambers, C., Cutler, P., Huang, Y. H., Goodchild, J. A., Blythe, J., Wang, C. K., Bigot, A., Kaas, Q., Craik, D. J., Sabbadin, D. and Earley, F. G. 2019. Insecticidal spider toxins are high affinity positive allosteric modulators of the nicotinic acetylcholine receptor. *FEBS Letters* 593(12), 1336-50. doi:10.1002/1873-3468.13435.

Coddington, J. A. and Levi, H. W. 1991. Systematics and evolution of spiders (*Araneae*). *Annual Review of Ecology and Systematics* 22(1), 565–92. doi:10.1146/annurev.es.22.110191.003025.

Cools, T. L., Struyfs, C., Cammue, B. P. A. and Thevissen, K. 2017. Antifungal plant defensins: increased insight in their mode of action as a basis for their use to combat fungal infections. *Future Microbiology* 12(5), 441–54. doi:10.2217/fmb-2016-0181.

Damalas, C. A. and Koutroubas, S. D. 2018. Current status and recent developments in biopesticide use. *Agriculture* 8(3), 1–6.

Dangl, J. L. and Jones, J. D. 2001. Plant pathogens and integrated defence responses to infection. *Nature* 411(6839), 826–33. doi:10.1038/35081161.

Del Poeta, M., Nimrichter, L., Rodrigues, M. L. and Luberto, C. 2014. Synthesis and biological properties of fungal glucosylceramide. *PLoS Pathogens* 10(1), e1003832. doi:10.1371/journal.ppat.1003832.

Dong, H., Delaney, T. P., Bauer, D. W. and Beer, S. V. 1999. Harpin induces disease resistance in *Arabidopsis* through the systemic acquired resistance pathway mediated by salicylic acid and the NIM1 gene. *The Plant Journal: for Cell and Molecular Biology* 20(2), 207–15. doi:10.1046/j.1365-313x.1999.00595.x.

El-Mounadi, K., Islam, K. T., Hernandez-Ortiz, P., Read, N. D. and Shah, D. M. 2016. Antifungal mechanisms of a plant defensin MtDef4 are not conserved between the ascomycete fungi *Neurospora crassa* and *Fusarium graminearum*. *Molecular Microbiology* 100(3), 542–59. doi:10.1111/mmi.13333.

Environmental Protection Agency. 2004. Trypsin Modulating Oostatic Factor (TMOF). Biopesticide Registration Action Document.

Environmental Protection Agency. 2005. HARPINαβ PROTEIN. Biopesticide Regulatory Action Document.

Environmental Protection Agency. 2013. ProAct Master Label US, s.l.: s.n.

Escoubas, P., Sollod, B. and King, G. F. 2006. Venom landscapes: mining the complexity of spider venoms via a combined cDNA and mass spectrometric approach. *Toxicon: Official Journal of the International Society on Toxinology* 47(6), 650–63. doi:10.1016/j.toxicon.2006.01.018.

Fanning, P. D., VanWoerkom, A., Wise, J. C. and Isaacs, R. 2018. Assessment of a commercial spider venom peptide against spotted-wing *Drosophila* and interaction with adjuvants. *Journal of Pest Science* 91(4), 1279–90. doi:10.1007/s10340-018-1016-7.

Federal Food, Drug, and Cosmetics Act (FFDCA) Considerations for GS-omega/kappa-Hxtx-Hv1a. 2013. Docket ID Number: EPA-HQ-OPP-2012-0389, 20 December 2013. Available at: www.regulations.gov/document?D=EPA-HQ-OPP-2012-0389-0005.

Fitches, E. C., Pyati, P., King, G. F. and Gatehouse, J. A. 2012. Fusion to snowdrop lectin magnifies the oral activity of insecticidal ω-Hexatoxin-Hv1a peptide by enabling its delivery to the central nervous system. *PLoS ONE* 7(6), e39389. doi:10.1371/journal.pone.0039389.

Fitzgerald, H. A., Chern, M. S., Navarre, R. and Ronald, P. C. 2004. Overexpression of (At) NPR1 in rice leads to a BTH- and environment-induced lesion-mimic/cell death phenotype. *Molecular Plant-Microbe Interactions: MPMI* 17(2), 140–51. doi:10.1094/MPMI.2004.17.2.140.

Fletcher, J. I., Smith, R., O'Donoghue, S. I., Nilges, M., Connor, M., Howden, M. E. H., Christie, M. J. and King, G. F. 1997. The structure of a novel insecticidal neurotoxin ω-atracotoxin-HV1, from the venom of an Australian funnel web spider. *Nature Structural Biology* 4(7), 559–66. doi:10.1038/nsb0797-559.

Fosgerau, K. and Hoffmann, T. 2015. Peptide therapeutics: current status and future directions. *Drug Discovery Today* 20(1), 122–8. doi:10.1016/j.drudis.2014.10.003.
He, S. Y., Huang, H. C. and Collmer, A. 1993. *Pseudomonas syringae pv. syringae* harpinPss: a protein that is secreted via the Hrp pathway and elicits the hypersensitive response in plants. *Cell* 73(7), 1255–66. doi:10.1016/0092-8674(93)90354-s.
Hernández-Campuzano, B., Suárez, R., Lina, L., Hernández, V., Villegas, E., Corzo, G. and Iturriaga, G. 2009.Expression of a spider venom peptide in transgenic tobacco confers insect resistance. *Toxicon: Official Journal of the International Society on Toxinology* 53(1), 122–8. doi:10.1016/j.toxicon.2008.10.020.
Herzig, V., Wood, D. L., Newell, F., Chaumeil, P. A., Kaas, Q., Binford, G. J., Nicholson, G. M., Gorse, D. and King, G. F. 2011. ArachnoServer 2.0, an updated online resource for spider toxin sequences and structures. *Nucleic Acids Research* 39(Database issue), D653–7. doi:10.1093/nar/gkq1058.
Horsfall, J. G. and Cowling, E. B. 1982. Hypersensitivity. In: Mount, M. S. and Lacy, G. H. (Eds), *Phytopathogenic Prokaryotes*. Academic Press, New York, NY, pp. 149–77.
Huang, J. Q., Wel, Z. M., An, H. L. and Zhu, Y. X. 2001. *Agrobacterium tumefaciens*-mediated transformation of rice with the spider insecticidal gene conferring resistance to leaffolder and striped stem borer. *Cell Research* 11(2), 149–55. doi:10.1038/sj.cr.7290080.
Jackson, J. R. H., Karpcho, K. J., Johnson, J. H. and Kral, R. M. 1995. Insecticidally effective spider toxin. United Stated Patent, US5457178.
Jackson, J. R. H., DelMar, E., Johnson, J. H. and Kral, R. M. 1998. Insecticidally effective peptides isolatable from Phidippus spider venom. United States Patent, US5756459.
Javaid, S., Amin, I., Jander, G., Mukhtar, Z., Saeed, N. A. and Mansoor, S. 2016. A transgenic approach to control hemipteran insects by expressing insecticidal genes under phloem-specific promoters. *Scientific Reports* 6, 34706. doi:10.1038/srep34706.
Javaid, S., Naz, S., Amin, I., Jander, G., Ul-Haq, Z. and Mansoor, S. 2018. Computational and biological characterization of fusion proteins of two insecticidal proteins for control of insect pests. *Scientific Reports* 8(1), 4837. doi:10.1038/s41598-018-23138-3.
Jiang, H., Zhu, Y. and Chen, Z. 1996. Insect resistance of transformed tobacco plants with gene of the spider insecticidal peptide. *Acta Botanica Sinica* 38(2), 95–9.
Johnson, J. H. and Kral, R. M. 1997. Insecticidal peptides from *Segestria sp.* spider venom. United States Patent, US5674846.
Johnson, J. H., Kral, R. M. and Krapcho, K. 1997. Insecticidal peptides from spider venom. United States Patent, US5688764.
Johnson, J. H., Kral, R. M. and Krapcho, K. 1999. Insecticidal toxins from *Bracon heretor* nucleic acid encoding said toxin and methods of use. United States Patent, US5874298.
Johnson, E. T., Dowd, P. F. and Hughes, S. R. 2014. Expression of a wolf spider toxin in tobacco inhibits the growth of microbes and insects. *Biotechnology Letters* 36(8), 1735–42. doi:10.1007/s10529-014-1536-z.
Kaur, J., Sagaram, U. S. and Shah, D. 2011. Can plant defensins be used to engineer durable commercially useful fungal resistance in crop plants? *Fungal Biology Reviews* 25(3), 128–35. doi:10.1016/j.fbr.2011.07.004.
Kaur, J., Fellers, J., Adholeya, A., Velivelli S. L. S., El-Mounadi, K., Nersesian, N., Clemente T., Shah, D. 2017. Expression of apoplast-targeted plant defensin MtDef4.2 confers resistance to leaf rust pathogen Puccinia triticina but does not affect mycorrhizal

symbiosis in transgenic wheat. *Transgenic Research 26*, 37-49. doi: 10.1007/s11248-016-9978-9.

Khan, S. A., Zafar, Y., Briddon, R. W., Malik, K. A. and Mukhtar, Z. 2006. Spider venom toxin protects plants from insect attack. *Transgenic Research* 15(3), 349-57. doi:10.1007/s11248-006-0007-2.

King, G. F., McFarland, B. S., Nicholson, G. M. and Gunning, S. J. 2008. Insecticidal polypeptides and methods of use thereof. United States Patent, US7,354,993 B2.

Krapcho, K. J., Jackson, J. R. H., Johnson, J. H., DelMar, E. G. and Kral, R. M. 1995a. Insecticidally effective peptides. United States Patent, US5441934.

Krapcho, K. J., Jackson, J. R. H., VanWagenen, B. C. and Kral, R. M. 1995b. Insecticidally effective peptides. United States Patent, US5461032.

Kund, G. S. and Trumble, J. T. 2019. Effect of insecticides on celery insects, 2017. *Arthropod Management Tests* 44(1). doi:10.1093/amt/tsz017.

Kund, G. S., Jones, M. B. and Trumble, J. T. 2019. Effect of insecticides on celery insects, 2016. *Arthropod Management Tests* 44(1). doi:10.1093/amt/tsz035.

Lay, F. T. and Anderson, M. A. 2005. Defensins – components of the innate immune system in plants. *Current Protein and Peptide Science* 6(1), 85–101. doi:10.2174/1389203053027575.

Leisy, D. J., Quistad, G. B. and Skinner, W. S. 2001. Insecticidal plectoxins from *Plectreurys tristis*. United States Patent, US6265376.

Liu, S. M., Li, J., Zhu, J. Q., Wang, X. W., Wang, C. S., Liu, S. S., Chen, X. X. and Li, S. 2016. Transgenic plants expressing the AaIT/GNA fusion protein show increased resistance and toxicity to both chewing and sucking pests. *Insect Science* 23(2), 265–76. doi:10.1111/1744-7917.12203.

Malamy, J. and Klessig, D. F. 1992. Salicylic acid and plant disease resistance. *The Plant Journal* 2(5), 643–54. doi:10.1111/j.1365-313X.1992.tb00133.x.

Matthews, B. F., Beard, H., Brewer, E., Kabir, S., MacDonald, M. H. and Youssef, R. M. 2014. *Arabidopsis* genes, AtNPR1, AtTGA2 and AtPR-5, confer partial resistance to soybean cyst nematode (*Heterodera glycines*) when overexpressed in transgenic soybean roots. *BMC Plant Biol.* 14, 96. doi:10.1186/1471-2229-14-96.

Meur, G., Budathe, M., Srinivasan, T., Rajesh Kumar, K. R., Dutta Gupta, A. and Kirti, P. B. 2008. Constitutive expression of Arabidopsis NPR1 confers enhanced resistance to the early instance of *Spodoptera litusa* in transgenic tobacco. *Physiol. Plant.* 133, 765–76. doi:10.1111/j.1399-3054.2008.01098.x.

Nakasu, E. Y., Edwards, M. G., Fitches, E., Gatehouse, J. A. and Gatehouse, A. M. 2014. Transgenic plants expressing omega-ACTX-Hv1a and snowdrop lectin (GNA) fusion protein show enhanced resistance to aphids. *Frontiers in Plant Science* 5(673), 673. doi:10.3389/fpls.2014.00673.

Parisi, K., Shafee, T. M. A., Quimbar, P., van der Weerden, N. L., Bleakley, M. R. and Anderson, M. A. 2019. The evolution, function and mechanisms of action for plant defensins. *Seminars in Cell and Developmental Biology* 88, 107–18. doi:10.1016/j.semcdb.2018.02.004.

Plant Health Care. 2018. Annual Reports and Accounts 2018, s.l.: s.n.

Quistad, G. B. and Leisy, D. J. 2001. Insecticidal toxins from the parasitic wasp bracon hebeter. United States Patent, US6251862.

Rauf, I., Javaid, S., Naqvi, R. Z., Mustafa, T., Amin, I., Mukhtar, Z., Jander, G. and Mansoor, S. 2019. In-planta expression of insecticidal proteins provides protection against lepidopteran insects. *Scientific Reports* 9(1), 6745. doi:10.1038/s41598-019-41833-7.

Sagaram, U. S., El-Mounadi, K., Buchko, G. W., Berg, H. R., Kaur, J., Pandurangi, R. S., Smith, T. J. and Shah, D. M. 2013. Structural and functional studies of a phosphatidic acid-binding antifungal plant defensin MtDef4: identification of an RGFRRR motif governing fungal cell entry. *PLoS ONE* 8(12), e82485. doi:10.1371/journal.pone.0082485.

Schnepf, E., Crickmore, N., Van Rie, J., Lereclus, D., Baum, J., Feitelson, J., Zeigler, D. R. and Dean, D. H. 1998. *Bacillus thuringiensis* and its pesticidal crystal proteins. *Microbiology and Molecular Biology Reviews: MMBR* 62(3), 775–806. doi:10.1128/MMBR.62.3.775-806.1998.

Schwinges, P., Pariyar, S., Jakob, F., Rahimi, M., Apitius, L., Hunsche, M., Schmitt, L., Noga, G., Langenbach, C., Schwaneberg, U. and Conrath, U. 2019. A bifunctional dermaseptin-thanatin dipeptide functionalizes the crop surface for sustainable pest management. *Green Chemistry* 21(9), 2316–25. doi:10.1039/C9GC00457B.

Shah, A. D., Ahmed, M., Mukhtar, Z., Khan, S. A., Habib, I., Malik, Z. A., Mansoor, S. and Saeed, N. A. 2011. Spider toxin (Hvt) gene cloned under phloem specific RSs1 and RolC promoters provides resistance against American bollworm (*Heliothis armigera*). *Biotechnology Letters* 33(7), 1457–63. doi:10.1007/s10529-011-0572-1.

Silva, K. J. P., Mahan, N., Mou, Z. and Folta, K. M. 2018. NPR1 as a transgenic crop protection strategy in horticultural species. *Hort. Res.* 5(15), 15. doi:10.1038/s41438-018-0026-1.

Stewart, J. M., Steeves, B. J. and Vernes, K. 2011. Paralytic peptide for use as a insecticide. United States Patent, US8003754.

Tam, J. P., Wang, S., Wong, K. H. and Tan, W. L. 2015. Antimicrobial peptides from plants. *Pharmaceuticals* 8(4), 711–57. doi:10.3390/ph8040711.

Tedford, H. W., Gilles, N., Menez, A., Doering, C. J., Zamponi, G. W. and King, G. F. 2004. Scanning mutagenesis of x-atracotoxin-Hv1a reveals a spatially restricted epitope that confers selective activity against insect calcium channels. *The Journal of Biological Chemistry* 279(42), 44133–40. doi:10.1074/jbc.M404006200.

Tianpei, X., Zhu, Y. and Li, S. 2014. Optimized scorpion polypeptide LMX: a pest control protein effective against rice leaf folder. *PLoS ONE* 9(6), e100232. doi:10.1371/journal.pone.0100232.

Ullah, I., Hagenbucher, S., Álvarez-Alfageme, F., Ashfaq, M. and Romeis, J. 2015. Target and non-target effects of a spider venom toxin produced in transgenic cotton and tobacco plants. *Journal of Applied Entomology* 139(5), 321–32. doi:10.1111/jen.12156.

Upadhyay, S. K., Mishra, M., Singh, H., Ranjan, A., Chandrashekar, K., Verma, P. C., Singh, P. K. and Tuli, R. 2010. Interaction of *Allium sativum* leaf agglutinin with midgut brush border membrane vesicles proteins and its stability in *Helicoverpa armigera*. *Proteomics* 10(24), 4431–40. doi:10.1002/pmic.201000152.

van der Weerden, N. L., Bleackley, M. R. and Anderson, M. A. 2013. Properties and mechanisms of action of naturally occurring antifungal peptides. *Cellular and Molecular Life Sciences: CMLS* 70(19), 3545–70. doi:10.1007/s00018-013-1260-1.

Wang, J., Chen, Z., Du, J., Sun, Y. and Liang, A. 2005. Novel insect resistance in *Brassica napus* developed by transformation of chitinase and scorpion toxin genes. *Plant Cell Reports* 24(9), 549–55. doi:10.1007/s00299-005-0967-3.

Ward, E. R., Uknes, S. J., Williams, S. C., Dincher, S. S., Wiederhold, D. L., Alexander, D. C., Ahl-Goy, P., Metraux, J. P. and Ryals, J. A. 1991. Coordinate gene activity in response to agents that induce systemic acquired resistance. *The Plant Cell* 3(10), 1085–94. doi:10.1105/tpc.3.10.1085.

Wei, Z. 2009. Production, formulation, and uses of stable liquid harpin protein formulations. USA Patent No. 20100043095A1.

Wei, Z. and Zornetzer, G. A. 2017. Elicitor-derived peptides and use thereof. USA Patent No. 20180362993A1.

Wei, Z. M., Laby, R. J., Zumoff, C. H., Bauer, D. W., He, S. Y., Collmer, A. and Beer, S. V. 1992. Harpin, elicitor of the hypersensitive response produced by the plant pathogen *Erwinia amylovora*. *Science* 257(5066), 85–8. doi:10.1126/science.1621099.

Windass, J. D., Blake, A. N., Grishin, E. V., Nosyreva, E. D., Koslov, S. A. and Lipkin, A. V. 1999. Insecticidal compounds. International Patent under PCT, WO99/49035

Windley, M. J., Herzig, V., Dziemborowicz, S. A., Hardy, M. C., King, G. F. and Nicholson, G. M. 2012. Spider-venom peptides as bioinsecticides. *Toxins* 4(3), 191–227. doi:10.3390/toxins4030191.

Wise, J. C., VanWoerkom, A. H., Wheeler, C. E. and Isaacs, R. 2018. Grape berry moth control in concord grape, 2017. *Arthropod Management Tests* 43(1). doi:10.1093/amt/tsy065.

Wu, J., Luo, X., Wang, Z., Tian, Y., Liang, A. and Sun, Y. 2008. Transgenic cotton expressing synthesized scorpion insect toxin AaHIT gene confers enhanced resistance to cotton bollworm (*Heliothis armigera*) larvae. *Biotechnology Letters* 30(3), 547–54. doi:10.1007/s10529-007-9555-7.

Zeng, F., Zhan, Y., Nan, N., Xin, Y., Qi, F. and Yang, C. 2009. Expression of bgt gene in transgenic birch (*Betula platyphylla Suk.*). *African Journal of Biotechnology* 8(15), 3392–8.

Chapter 14

Development of plant-derived compounds as biopesticides

Barbara Thuerig and Lucius Tamm, Research Institute of Organic Agriculture (FiBL), Switzerland

1 Introduction

Extracts of plants have been reportedly used to control noxious organisms in agricultural crops since 4000 years B.C., for example, extracts of the neem tree (*Azadirachta indica*) (containing azadirachtins). Other traditionally used plant extracts against arthropod pests include false hellebore (*Veratrum* sp.) (containing alkaloids), common yew (*Taxus baccata*) (containing alkaloids), Dalmatian chrysanthemum (*Chrysanthemum cinerariaefolium)* (containing pyrethrins), tobacco (*Nicotiana* sp.) (containing the alkaloid nicotine), derris *(Derris sp.)* and lancepods (*Lonchocarpus* sp.) (containing rotenone), ryania (*Ryania speciosa*) (containing ryanodine) and a range of plant oils (Soloway, 1976; Isman, 2000; Philogène et al., 2008). Fungicides were developed much later than insecticides, around the time when micro-organisms were identified as causal agents of plant diseases in the nineteenth century. Among the first fungicides developed were elemental sulphur and copper hydroxide (Finckh et al., 2015); however, these are not plant-based. Only much later, in the 1970s, natural products' researchers became interested in the use of plant extracts against plant pathogens. Extracts of the species Douglas' falsewillow (*Baccharis glutinosa*) (DiSalvo, 1974), pink morning glory (*Ipomoea carnea*) (Narain, 1972), common bean (*Phaseolus vulgaris*) (Wiens, 1978) and the neem tree (*A. indica*) (Singh et al., 1980) as well as some plant oils (Mukherjee, 1976) were among

http://dx.doi.org/10.19103/AS.2020.0073.15

the first plant-based substances with reported activity against plant pathogens. Undoubtedly, many more plants were used to protect crops according to the locally available knowledge in many societies worldwide, but documentation is not easily accessible. Today, plant extracts (botanicals) are produced at a commercial scale and used in professional agriculture worldwide, especially in horticultural crops. Botanicals are used against insects and other pests (e.g. pyrethrum, rape seed and other plant oils, quassia extract, neem), against plant diseases (e.g. lecithin, coconut soap, laminarine, fennel oil) and against weeds (e.g. pine oil) (Finckh et al., 2015; Isman, 2014). Other uses include sprouting inhibition (e.g. caraway seed and mint oil) and the use as additives in formulations of plant protection products (e.g. as spreaders/stickers to improve wetting of/adhesion to surfaces) (e.g. pine oil). Botanicals are currently mainly used in low-input farming systems such as organic agriculture or traditional subsistence farming, but also there is an increasing interest in IPM and 'low residue' production systems. In today's mainstream agriculture, synthetized molecules (often derivatives of naturally occurring plant compounds) represent the vast majority of all plant protection products used (European Commission, 2019).

The 1980s saw an increasing interest of researchers in plant extracts, and several plant extracts have been developed to commercial products. However, recently, the implementation of the European Union's review of existing pesticide-active substances (Regulation (EC) No 2229/2004) (Anon., 2004) ultimately led to a drastic reduction of authorized botanicals in the European market from previously 60 to around 20 authorized botanicals (European Commission, 2019). The main reasons for withdrawal of formerly allowed botanicals were requirements for registration of plant protection products that were inappropriate to evaluate plant extracts and lack of economic feasibility due to very high costs for registration. Regulation (EU) 1107/2009 sets out the requirements, procedure and timeframes for authorization of plant protection products (PPPs). Dossier requirements have been developed for synthetic, pure chemicals, but not for complex mixtures such as plant extracts containing many potentially interacting compounds. A guidance document was developed by the European commission in 2014 (Anon., 2014). This document defines that a 'technical grade' extract (i.e. an extract produced from a defined source and by a described manufacturing process to ensure a constant quality) is regarded as the 'active substance'. Yet, many uncertainties persist, for example, it remains unclear down to what level extracts have to be chemically characterized and which degree of variation between batches is considered acceptable by authorities. It is generally believed that the legal framework may also have substantially contributed to the fact that only few new botanicals have been registered in the last decade in the EU (Ehlers, 2011; Anon., 2018), even though a large body of literature on the use of botanicals against pests and pathogens published in recent years highlights the large potential of botanicals in IPM and reflects the public interest in sustainable plant protection and use of crop protection products which are safe for the environment and for human or animal consumption.

In addition to the difficult regulatory situation, other technical and economic challenges may contribute to the limited number of commercial plant-derived plant protection products. Besides good efficacy against target organisms under controlled and many field conditions, plant extracts have to fulfil several more criteria to be valid candidates for the development of commercial products, including a sufficiently high stability of the active compound(s) under field and glasshouse conditions, a favourable toxicological profile (including any negative effects on beneficial organisms including biological control agents and pollinators), and availability of affordable and sustainably produced raw materials. Upscalable, cost-efficient and sustainable extraction processes yielding extracts of a constant quality with high concentrations of the active compounds have to be developed. Furthermore, preferably the plant protection properties of the candidate plant extract or purified natural product molecule have not been described in detail before, allowing intellectual property (IP) protection needed for commercial development. IP protection is often the prerequisite for an engagement of industrial partners, justifying substantial investments required not only for the previously described research activities, but also for the preparation and submission of a registration dossier against specific target pests and diseases of cultivated plants (agricultural/horticultural food and fodder crops, flowers).

In this chapter, we will discuss the requirements and hurdles for the development of commercial plant-derived plant protection products and discuss the necessary steps to overcome bottlenecks. We will give an overview of the regulatory framework of authorization of botanicals in plant protection and address the key aspects of a successful development of a botanical from the initial research through the developmental stages needed to finally supply the market with a safe, useful and economically feasible plant protection tool which is usually compatible with IPM for conventional and organic production systems.

2 Legal framework and authorization of plant-derived plant protection products: an overview

In recent years, society has become more and more aware that the agricultural intensification has put pressure on the environment, driving pests to become resistant to current crop protection products, depleting populations of beneficial natural enemies of pests, threatening pollinators and creating threats to human health. In this context, the European Union has developed the 'Sustainable Use of Pesticides Directive' (Directive 128/2009/EC) (Anon., 2009a). This Directive focusses on an approach to minimize harm to human health and the environment caused by pesticide overuse while producing good-quality food for an increasing population. To implement the Directive, bioprotection technologies are believed to be ideal tools. Bioprotection technologies aiming to protect crops from pests and diseases are of natural origin or nature-identical if synthesized, produced from renewable resources and should have a low impact on the environment and human health to be compatible with the IPM for conventional and organic production. They include, for example, biocontrol organisms, semiochemicals and botanicals (Anon., 2018).

Botanicals, as any other synthetic plant protection products, have to go through a registration process to guarantee a safe use for humans, farmed animals and the wider environment (especially for species under threat and essential for agriculture, including pollinators and natural enemies that help to regulate pest or disease outbreaks). Current mechanisms for approval of bioprotection technologies (including botanicals) in the European Union have evolved from regulations for conventional synthetic active substances. Yet, unlike synthetic substances, plant extracts often contain a large range of substances, only few of which are active against the targeted pathogens/pests. The mixture of compounds is determined by the raw material (where and how the source plants are grown) but also by an extraction and purification technology. Thus, the term 'pure active substance,' as used in regulatory affairs, does not typically apply to most botanical active substances as it implies that the 'active substance' consists of more than 90% of the pure active molecule and only small fractions of 'impurities' resulting from the synthesis process.

To meet the special properties of plant extracts, a guidance document (OECD, 2017) has been developed within the framework of the OECD Bio Pesticides Steering Group (BPSG), a sub-group of the OECD Working Group on Pesticides (WGP) (http://www.oecd.org/env/ehs/pesticides-biocides/biological-pesticides.htm). The European Union served as the initial author of the guidance document (Anon., 2014), which has been reviewed and further developed by the BPSG. The BPSG includes representatives from OECD member countries and the regulated industry. This document defines that a 'technical grade' extract is regarded as the 'active substance' in the sense of Regulation (EC) No 1107/2009 (Anon., 2009b). A technical grade extract is an extract produced from a defined source and by a described manufacturing process to ensure its quality specifications are constant in the formulated product within set specifications. This means that neither the types and sources of raw material nor the extraction processes can be easily and significantly changed after submission of the registration dossier. All toxicity studies (human and eco toxicity) required for the registration dossier should usually be performed with this technical grade extract to ensure consistency and repeatability of test results.

The characterization of technical grade active substances is still under debate. Among the active substances and other compounds there might be some compounds of concern. The legislative terminology aims at a distinction between the compounds relevant for risk assessment and the compounds of no concern to environment or human health. To characterize technical grade extracts, the provision of 'fingerprints' (a spectroscopic and/or chromatographic profile matched against a reference sample) has been discussed. However, it remains open as to what level compounds of an extract need to be specified for the registration dossier.

Thus, despite some adaptations, the registration process for botanicals in the European Union remains challenging. This is reflected by the fact that the review of existing pesticide active substances in the EU ultimately led to a drastic reduction of authorized botanicals in the European market, from previously 60 to 20 authorized botanicals. In few cases such as the insecticide rotenone which is very poisonous to fish, withdrawal was due to a problematic toxicological profile; for the majority, no attempt was made to develop a dossier for re-evaluation. Furthermore, only very few botanicals have been newly approved in the EU in the last decade (EU Pesticides database (European Commission, 2019)). Applicants are often deterred by the long and expensive registration process, as well as by uncertainties in the interpretation of data requirements (Anon., 2018). For example, it is often not clear whether published literature and/or a reasoned approach can be accepted, to what level compounds in extracts have to be analysed and quantified (minor compounds), and how similar different batches of technical grade of the same product have to be. Organizations such as the Dutch Board for the Authorisation of Plant Protection Products (ctgb) have taken the initiative to facilitate the registration of novel compounds (https://english.ctgb.nl/topics/green-deal).

In some particular cases, registration as a 'basic substance' can be achieved in the EU. Basic substances such as lecithine, beer or mustard seed powder are substances which have a predominant safe-use history for another purpose (e.g. as food ingredients) but may also be used for plant protection (Speiser et al., 2018). Registration for basic substances is faster, but an exclusivity of commercialization to the applicant is not given. Therefore, registration requests are often made by non-profit organizations.

Unlike the EU, many botanicals have been registered in the United States in recent decades, as a result of an active support of the industry by a programme (IR-4 programme, https://www.ir4project.org/) which aims at facilitating registration of sustainable pest management technologies, as well as appropriate registration requirements (Anon., 2018). The USA EPA process for regulating biopesticides was established in 1994 with the sole aim to regulate bioprotection products (https://www.epa.gov/pesticide-registration/biopesticide-registration). Registration dossiers are rigorously evaluated by dedicated evaluators with relevant expertise within a year after submission. Evaluation follows a tiered approach for mammalian toxicity and non-target organism testing, with adopted data requirements depending on current knowledge of the bioprotection product under evaluation. For example, if no adverse effects are detected in tier I of non-target organisms testing (short-term studies on non-target birds, aquatic organisms, insects, plants), no further testing is required, whereas adverse effects in tier I trigger tier II studies (environmental fate), and adverse effects in tier II trigger tier III studies (non-target wildlife studies) (Leahy et al., 2014). The FAO recommends similar regulatory approaches, which have been adopted by several countries including Australia, Brazil, Canada and China. The International Biocontrol Manufacturers Association (IBMA) has made several suggestions on how to improve the situation in the European Union (Anon., 2018). For example, it has been suggested to apply similar procedures as already implemented, for example, by the European Medicines Agency or by other centralised EU regulatory systems (e.g. feed additives, biocides, cosmetics) for evaluation of herbal products, or procedures as applied by the EU regulation 'REACH' (Registration, Evaluation, and Authorisation of Chemicals) for chemicals (EC 1907/2006) (Anon., 2006). REACH data requirements are proportional to risk and quantity of a chemical used (Anon., 2018). This facilitates niche solutions being made available with reduced regulatory time and expenses. Applied to approval of plant protection products, this approach would be especially desirable for biopesticidal products with smaller markets than synthetic pesticides used on major crops on a national or international scale.

When developing new plant-derived plant protection products, data requirements should be considered during all development steps in order to avoid unnecessary costs and delays. For example, some preliminary toxicity studies (e.g. QSAR studies, studies using cell cultures) with preliminary technical grade extracts can be useful at early stages of product development, to anticipate future complications. In the present paper, legal aspects are discussed for all relevant steps.

3 Development of plant-derived plant protection products

Plant protection products in general need to fulfil a range of criteria that also apply to botanicals: Products need to control target organisms efficiently (proven efficacy); they need to be proven safe for non-target organisms and they have to be available in sufficient quantities at a competitive price. Since investments in R&D, manufacturing and dossier development are substantial, investments need to be protected by appropriate intellectual property protection, trade secrets and product-specific authorization.

3.1 Identification of plant-derived substances with activity against plant pathogens

A pre-requisite for any development consists obviously in the identification of activity of a candidate plant extract against specified target organisms. Various strategies have been applied to identify plant extracts with plant protection properties, including screening of literature for former reports on bioactivity (against plant pests and pathogens or in other related fields such as cosmetics or medicine), searching for additional uses of by-products available in large quantities, or by systematically screening libraries of plant extracts.

Based on previous reports on bioactivity in plant protection or related fields such as medicine or cosmetics, it is often possible to identify relatively quickly several plant species whose extracts show good activity under controlled conditions. However, only a fraction of such candidates have been developed to commercial products due to limitations in one or several of the requirements for a plant-derived plant protection product described in detail below (e.g. limited efficacy in the field, non-target toxicity, availability of the raw material, low content of active compounds, stability under realistic conditions of use, economic feasibility). For example, sage extract from *Salvia officinalis* has an excellent activity against *Plasmopara viticola* (Dagostin et al., 2010), but high production costs limited its potential for further development. Unfortunately, such information is rarely published and due to lack of information exchange, several research teams may invest resources and eventually come to similar conclusions. However, even if a candidate plant extract might be highly promising and economically competitive, publication of the bioactivities often prevents IP protection, and the lack of options for IP protection is effectively prohibitive to any commercial interest in its further development (see also Section 3.3).

Plant protection properties of some botanicals have been detected when searching for additional uses of commodity substances or by-products. Lecithine, pinolene or rapeseed oil are commodities used at a large scale in the food industry or cosmetics. In such a case, a registration as a 'basic substance' might be feasible in the EU (see Section 2). There are also obvious advantages of using by-products as raw material for extraction: for these materials, production processes are fully established at a commercial scale, supply chains and quality control are in place and costs are often relatively low. Examples include spruce knot extract (by-product of paper industry) (Pietarinen et al., 2006; Välimaa et al., 2007), orange oil (a by-product of citrus industry) (Pullen, 2012), Larixyne (candidate plant-derived plant protection product extracted from larch bark)

(Thuerig et al., 2018a) and a candidate plant-derived plant protection product extracted from by-products of the liquorice industry (Schuster et al., 2010; Scherf et al., 2012).

Systematic screening of a library of candidate substances/plant extracts is another strategy to identify candidates with desired bioactivity. This strategy has a successful history in the pharmaceutical industry (Potterat and Hamburger, 2006), with many medicinal drugs derived from natural products. In the plant protection industry, there is a long tradition of systematic screening of libraries of synthesized substances. However, screening of libraries of plant extracts has not been often applied yet or results are not available for open access especially if performed by industry, but some results have been published by academics. Recently, researchers have screened large collections of plant extracts for activity against selected plant pathogens. For example, Thuerig et al. (2016) have screened a botanical library comprising more than 3000 plant extracts which had been produced by successively extracting plant material with solvents of decreasing polarity, using a standardized extraction protocol (Potterat and Hamburger, 2014). Several extracts showed a high activity against *Plasmopara viticola*, *Venturia inaequalis* and *Phytophthora infestans*, including plants with previously reported and new fungicidal activity (Thuerig et al., 2016, 2018b; Ramseyer et al., 2017). The creation of a botanical library is quite laborious. Yet, once established, botanical libraries allow screening for bioactivities in different fields of interest such as agriculture, medicine and cosmetics.

Recently, the ratification of the Nagoya protocol has caused uncertainties when working with plant extracts. The Nagoya protocol on access and benefit sharing was developed to allow 'the fair and equitable sharing of benefits arising out of the utilization of genetic resources, thereby contributing to the conservation and sustainable use of biodiversity' (Anon., 2011). It was adopted in 2010, entered into force on 12 October 2014, and has been ratified by 120 parties (status November 2019; https://www.cbd.int/abs/nagoya-protocol/signatories/). However, since the implementation of the Nagoya Protocol is carried out at the national level, legal situations differ from country to country and many uncertainties persist. For example, the national competent authority has only been assigned in 64 out of 120 countries. Legislative, administrative or policy measures have been taken in 60 countries, and national websites or databases have been created in 30 countries (status November 2019; https://absch.cbd.int/countries). To comply with the Nagoya protocol, dealing with plant material from foreign countries requires detailed legal knowledge (e.g. national legal frameworks in the country of origin of the plant material as well as the countries of research) and substantial administrative efforts (e.g. identification of responsible authorities of specific procedures and data requirements, negotiations with authorities). At this stage, the implementation of the Nagoya protocol slows down the development of novel botanicals in

Europe since researchers limit themselves to plant extract collections completed before the Nagoya protocol entered into force in October 2014. Worse still, many researchers avoid working with plant material from other countries, and even reject collaboration with partner institutes in countries with unclear legal situations (Welch et al., 2013; Cressey, 2014).

However, despite some administrative and legal hurdles related to the adoption of the Nagoya protocol, there are obvious advantages to systematically screen libraries for candidate substances: The main advantage of library screenings is the relatively high probability to identify substances with previously unknown or reported activities, allowing IP protection (see also Section 3.3).

3.2 Efficacy of botanicals

The obvious pre-requisite for a plant-derived plant protection product is its ability to protect plants against specified target organisms. Typically, programmes for development of botanicals first examine efficacy of extracts in vitro, followed by plant-pathogen or plant-pest bioassays under specific controlled conditions (e.g. lab, glasshouse, semi-field or field). There is a wide range of expectations with respect to the minimal required activity. Concentrations of extract necessary for an inhibitory effect vary considerably between studies, depending on plant species, conditions of plant cultivation, extraction technique, and plant-pathogen/pest bioassays. For example, EC_{50} values against *Plasmopara viticola* and *Venturia inaequalis* on grapevine and apple seedlings ranged between 120 and 200 μg mL^{-1} for extracts of *Juncus effusus*, *Magnolia officinalis* and *Verbesina lanata* when tested under identical conditions (Thuerig et al., 2016, 2018b; Ramseyer et al., 2017). An extract of *Glycyrrhiza glabra* showed EC_{50} values between 1000 and 5000 μg mL^{-1} against cucumber downy mildew and bean rust on potted beans (Scherf et al., 2012; Schuster et al., 2010). ED_{90} values between 125 and 10000 μg mL^{-1} were found for an extract of *Inula viscosa* against different diseases under controlled conditions (Cohen et al., 2006; Wang et al., 2004). For extracts of *Yucca schidigera*, a concentration of 5000 μg mL^{-1} was necessary for maximal protection of apple plants (Bengtsson et al., 2009).

Identification of the active compounds in an extract and evaluation of their potency is necessary for most forthcoming steps in product development, for example, for risk assessment (e.g. human toxicity and environmental risk assessment), quality control of raw material and extract, development of efficient extraction techniques, economic feasibility assessments, IP protection and registration. Tracking bioactivities in a complex matrix such as plant extracts is a challenging task (Potterat and Hamburger, 2006). While the classical process involving several consecutive steps of preparative chromatographic separation while tracking bio-activity can be slow, laborious and expensive, other processes like bio-guided fractionation are much faster to discover active compound(s) (Potterat and Hamburger, 2006, 2014; Thuerig et al., 2016).

Despite excellent efficacy under controlled conditions, many extracts fail to be effective under more realistic glasshouse or field conditions. For example, Dagostin et al. (2011) tested 16 commercial or near-commercial plant extracts against *Plasmopara viticola* on grapevine in the field and found no, low or inconsistent efficacy for most products, even though efficacy under controlled conditions was previously demonstrated for all of these extracts in the same study. Limited efficacy in the field may result from several factors, including low UV stability, low persistence on leaves due to microbial degradation, or limited performance due to inappropriate agronomic properties (e.g. low rainfastness or solubility, phytotoxicity) (Dagostin et al., 2011). Many natural compounds are degraded rather quickly under natural conditions, due to UV, temperature extremes or natural degradation processes (Coimbra et al., 2011). On the one hand, this instability is highly desirable to avoid residues on food or feed products. Yet, if degraded too fast, efficacy is no longer guaranteed against target pests and pathogens, requiring more frequent applications, adding to farmers' production costs. To overcome lack of persistence, synthetic derivatives of natural compounds showing lower effective concentrations and longer persistence were developed to replace naturally occurring compounds, for example, pyrethrins or strobilurins (Sauter et al., 1999).

Formulation of purified active substances or technical grade plant extracts with additives (emulsifiers, spreaders, stickers) is key to dissolve products in a sprayable solution, to provide rain fastness and to provide sufficient UV protection. To provide rainfastness is particularly challenging for polar extracts, due to their high solubility in water. For example, low efficacy of sage extract against *Plasmopara viticola* under field conditions was assigned to low rainfastness (Dagostin et al., 2010). In contrast, apolar extracts cannot easily be dissolved in water and typically precipitate or block the spray equipment if not formulated appropriately. For example, an apolar extract of the bark of *Magnolia officinalis* failed to protect grapevine from *Plasmopara viticola* in the field when applied as non-formulated crude extract, but showed efficacy up to 75% when applied in a formulation (Thuerig et al., 2018b). Formulations have to be continuously improved throughout product development and adapted each time the extraction processes are altered. Extraction and formulation both require equipment, financial resources and highly specialized expertise.

3.3 Novelty of invention and IP protection

Bringing a plant protection product – including botanicals – to the market requires substantial investments. Even though investments in the development of botanicals are believed to be lower than for novel chemicals, investments are substantial nevertheless. Kleeberg estimated a total cost of 10–12 million Euros for the development of a novel botanical for dossier submission, even though costs for supply chain development were not included (https://www.oecd.org/env/ehs/pesticides-biocides/env-jm-mono-2012-36-ppt.pdf). While the first research steps (e.g. efficacy evaluation, identification of active compounds) - if not performed by industry - can often be financed by public funds or projects, further developmental steps such as upscaling of extraction, building up of supply chains, chemical profiling of the extract, toxicity and environmental fate studies and dossier preparation are rarely publicly funded. The gap between promising publicly funded research results and the successful commitment of industry is occasionally referred to as the 'death valley of innovation', that is, the well-known difficulties to take research discovery and innovation to commercial exploitation. Since collaboration of academia with one or more industry partners and financial investment are necessary at some stage of product development, and industry depends on future return of investment, exclusivity has a high priority. Exclusivity can either be granted by IP protection by a patent or by retaining information as trade secrets. Only novel findings which include an inventive step can be patented (https://www.epo.org/applying/basics.html). Therefore, literature and patent databases as well as other publicly available sources have to be checked for former reports on the intended use of the selected plant extract and its active compound(s) preferentially at an early stage of product development. Results have to be kept secret until filing of a patent or abandoning of the project. The patent search will also reveal whether the planned usage does not infringe existing patents ('freedom to operate') and is sufficiently novel to be protected. Such limitations on publication can deter some publically funded scientists from this type of research, since career progress is often related to peer-reviewed, public access publication of results.

3.4 Raw material and supply chains

Crops and horticultural flowering plants requiring plant protection measures to avoid considerable yield losses cover very large areas worldwide, and crops have often to be treated repeatedly during the season to avoid and restrict development of diseases. For example, spray programmes for protection of grapevine against downy mildew in Europe include up to 14 treatments in organic and integrated production (Kuflik et al., 2009), and the total area under vine extends to 32000 km^2 (status 2015) (https://ec.europa.eu/eurostat/statistics-explained/index.php/Vineyards_in_the_EU_-_statistic). Furthermore, plant protection costs per ha and season need to be economically feasible, depending on crop value. Acceptable costs therefore vary significantly between large production area arable crops and high-value speciality crops, i.e. horticulture, flowers and viticulture. Therefore, availability of the raw material of a uniform standard in large quantities at low costs is a prerequisite for plant-derived plant protection products. By-products from large-scale industries are promising candidates. For example, pine oil (pulp industry) and orange oil (food industry) are both successfully used in agriculture as adjuvants or insecticides. Alternatively, plants can be specifically cultivated. For example, pyrethrum used as an insecticide is gained from the dried flowers of various *Tanacetum* species. Raw materials may also be obtained by collection from wild plants, whilst not threatening their survival and abundance in natural habitats. An example is the nuts of the tree *Azadirachta indica* which are collected from wild and cultivated trees to produce the insecticide neem (National Research Council, 1992). Short transport distances for voluminous raw materials can contribute to keep costs as low as possible.

Not only the quantity but also the quality of the raw material is an important issue including aspects of external as well as internal quality (Zhang et al., 2012). External quality challenges include contaminations with heavy metals or pesticides, adulteration and misidentification of the plant material. Contaminations with heavy metals and pesticide residues have often been reported for herbal medicines (Harris et al., 2011; Zhang et al., 2012). Some plants are known to be hyper-accumulators of heavy metals (Pollard et al., 2014), or contaminations may occur during processing. Pesticide residues may either come from current spray programmes against diseases or from slowly degrading pesticides applied many years ago (Harris et al., 2011). For example, organochlorine pesticides have been banned since more than 30 years in many countries, but persist in the soil and can still be found in quantities above the threshold in many natural products (Xue et al., 2008). From a safety point of view, the levels of contamination might be less problematic for plant-derived plant protection products than for herbal medicine destined for direct human consumption. Yet, the aim to contribute to a more sustainable plant protection by the use of botanicals can only be reached if the raw material is produced and harvested in a sustainable way.

The internal quality aspects include complexity and non-uniformity of the compounds in plant material and extracts. Plants and their extracts can contain hundreds of known as well as unknown compounds, and their concentration can vary from batch to batch as a result of variable growth conditions in different years (e.g. rainfall, sunshine, temperature, humidity, UV intensity) and at different sites (e.g. soil type, altitude, geographic region) (Kaushik et al., 2007; Zhang et al., 2012; National Research Council, 1992). Furthermore, genetic background of plants, cultural practices (e.g. planting densities, sowing time, applied fertilizers, pesticides) and storage can affect chemical composition. For example, marker compounds in Ginseng powders and extracts varied up to 200fold between commercially available samples (Harkey et al., 2001). Variability between different batches can be reduced, for example, by standardization of production, blending of batches and/or by an extraction process selectively enriching the active compounds.

Complexity and variability of compounds are relevant issues for registration of botanicals (see also Sections 2 and 3.5). For registration, at least five representative batches from industrial scale production have to be analysed for their content of the active compounds as well as for 'impurities, additives and each further component other than additives, as appropriate' ('five batch analysis') (Anon., 2014). In certain cases (where a safe use of a botanical has been demonstrated before, e.g. food grade plant material), it is only required to show that 'each sample of botanical active substance is similar in its composition and comparable to the existing safe use specification, with variation within defined acceptable margins' (similar 'fingerprints' or chemical profiles). Despite these adaptations, uncertainties persist, since it has not yet been defined to what level minor compounds have to be identified and quantified, and which degree of variability in the fingerprint is considered acceptable by the authorities.

3.5 Extraction process

Processing of plants is generally necessary to produce a plant-derived plant protection product, including pressing, milling, crushing, distillation and/or extraction. Often, extraction steps are required to produce extracts containing sufficient amounts of the active compounds. Furthermore, extraction and further purification may be necessary to reduce the volume of the extract, to exclude compounds with unwanted effects (e.g. phytotoxicity), or to produce material of constant quality within agreed technical specifications. Since most bioactive compounds are secondary metabolites, plant material often only contains limited amounts. For example, concentration of active compounds are below 1% in seeds of the Neem tree (*Azadirachta indica*) (azadirachtin active against many insects) (Isman, 2006), below 2% in *Tanacetum* sp. (a.i. pyrethrine active against many insects) (Li et al., 2014), and up to 7% in bark of *Magnolia officinalis* (a.i. magnolol and honokiol active against several pathogens) (Thuerig et al., 2018b; Zai-kang et al., 2002). Concentrations of the compounds larixyl acetate and larixol are between 0.2% and 0.5% in wood and bark of the European larch (*Larix decidua*) and around 40% in the oleoresin collected from living larch trees after injuring the bark (Thuerig et al., 2018a; Zule et al., 2015). Often these bio-active compounds vary in concentration depending on plant tissue, age and health of plants being extracted. These variables also need to be taken into account to maximize production of a uniform and effective end product.

Quantity and quality of technical grade extracts are significantly affected by the solvent (e.g. Hasan et al., 2012; Michielin et al., 2009; Mulholland et al., 2017; Ostroukhova et al., 2012; Thuerig et al., 2018a; Złotek et al., 2016). For example, when larch bark was extracted at laboratory scale, the extract yield varied between 0.5% and 7% and the concentration of the active compounds of larixyl acetate and larixol in the extracts varied between 0.18% and 16.2% depending on the solvents used (Thuerig et al., 2018a). While hydrophilic compounds such as saponins can be obtained by extraction with water, steam or ethanol (Güçlü-Üstündağ and Mazza, 2007), many bioactive compounds are hydrophobic and thus solvents such as acetone, methanol or hexane have to be used for an efficient extraction. Alternatively, extraction with supercritical CO_2, which is currently mainly used in the food (e.g. hops, cacao, coffee) or pharmaceutical (e.g. algae, hemp, mistletoe) industry, might also be highly advantageous to produce botanical active substances, since it produces solvent-free extracts and potentially useful plant residues (i.e. plant material after extraction which can be used e.g. as feed or fertilizer) while being environmentally safe (Gómez et al., 1996; Mendes et al., 2003; Ramsey et al., 2009).

During product development, up-scalable extraction protocols have to be developed to produce large quantities of high-quality extracts in an economic way. The economy of extraction is affected by many factors, including costs of equipment, consumables (e.g. solvents, energy), manpower and safety requirements. For example, extraction with hot water is very energy intensive while being safe, whereas extraction with explosive solvents such as acetone or hexane requires expensive safety measurements. Preferentially, the extraction process is also optimized to remove unwanted compounds of concern (e.g. human toxicity) or with undesirable effects (e.g. phytotoxicity) in as few steps as possible to keep costs low. Downstream uses of plant residues - occurring in large amounts during industrial-scale production - will also affect costs: In the best case, plant residues will have an added-value for other uses. In the worst case, residues have to be disposed of at high costs. For example, sunflower and rapeseed cake (leftovers after oil extraction) have been used as animal feed, fertilizer or, recently, as a microbial substrate in enzyme or antibiotics productions (Lomascolo et al., 2012). Solvent-free residues for example from extraction with water or supercritical CO_2 can be directly used as food or feed if suitable.

The industrial-scale process has to be developed and finalized before performing expensive studies (e.g. chemical characterization with validated analytical methods, toxicity studies and environmental fate studies) necessary for the registration of any plant protection product. This is necessary because in the case of botanical active substances, a 'technical grade extract' is assessed instead of a pure compound, and the 'technical grade' is defined as 'an extract produced from the defined source(s) and by the described manufacturing processes' (see also section 2) (Anon., 2014).

4 Conclusion

There is a growing societal interest in safe and environmentally friendly produced agricultural products including food, feed, and flowers, and sustainable plant protection technologies including macrobials (e.g. natural enemies of pests including nematodes, predatory mites, parasitic insects), biocontrol organisms, semiochemicals and botanicals are cornerstones to reach this goal. Plant-derived plant protection products have a long history of safe and successful use and can fill gaps which cannot be filled by other sustainable plant protection technologies, and are often compatible with other IPM tools (e.g. biocontrol) and with pollinators. To be valid candidates, botanicals must fulfil a range of criteria, including high efficacy, appropriate concentration and stability of the active compounds, high availability of the raw material at low prices, sustainable production of the raw material, efficient, scalable and affordable extraction processes, unproblematic toxicological and eco-toxicological profiles and novelty. Intellectual property protection is crucial for an engagement of industrial partners, which in turn is crucial for product development, registration and commercialization. Major challenges for the development of plant-derived plant protection products include registration hurdles, scalability of supply chains and extraction processes, compliance with the Nagoya protocol, closing of the gap between publicly funded research and commitment of industry, and economic competitiveness of products. Due to their unique properties to combat pests and diseases, plant-derived plant protection products will play an important role in future sustainable agriculture providing promising perspectives for bio-economy.

5 Where to look for further information

General information on plant diseases and their management in organic agriculture: Finckh et al. (2015).

Links to all regulations relevant for registration of pesticides and supporting information by the European Food Safety Authority (EFSA): https://www.efsa.europa.eu/de/applications/pesticides/regulationsandguidance.

International Biocontrol Manufacturer Association (IBMA): An association with the aim to promote biocontrol manufacturers and to facilitate market entry of 'green technologies': https://www.ibma-global.org/en/about-ibma.

European Patent Office Database on existing patents; legal texts and further information about patenting: https://www.epo.org/index.html.

Information on access and benefit sharing (Nagoya protocol): https://www.cbd.int/abs/nagoya-protocol.

6 References

Anon. 2004. Commission regulation (EC) No 2229/2004. Commission regulation (EC) No 2229/2004 of 3 December 2004 laying down further detailed rules for the implementation of the fourth stage of the programme of work referred to in Article 8(2) of Council Directive 91/414/EEC. European Commission.

Anon. 2006. REGULATION (EC) No 1907/2006. REGULATION (EC) No 1907/2006 of the European Parliament and of the Council of 18 December 2006 concerning the Registration, Evaluation, Authorisation and Restriction of Chemicals (REACH), establishing a European Chemicals Agency, amending Directive 1999/45/EC and repealing Council Regulation (EEC) No 793/93 and Commission Regulation (EC) No 1488/94 as well as Council Directive 76/769/EEC and Commission Directives 91/155/EEC, 93/67/EEC, 93/105/EC and 2000/21/EC. European Parliament and Council.

Anon. 2009a. 2009/128/EC. Directive 2009/128/EC of the European Parliament and of the Council of 21 October 2009 establishing a framework for Community action to achieve the sustainable use of pesticides (Text with EEA relevance). European Parliament and Council.

Anon. 2009b. (EC) No 1107/2009 Regulation (EC) No 1107/2009 of the European parliament and of the council of 21 October 2009 concerning the placing of plant protection products on the market and repealing Council Directives 79/117/EEC and 91/414/EEC. European Parliament and Council.

Anon. 2011. Nagoya protocol on access to genetic resources and the fair and equitable sharing of benefits arising from their utilization to the convention on biological diversity. United Nations Environmental Programme. Secretariat of the Convention on Biological Diversity. Available at: https://www.cbd.int/abs/doc/protocol/nagoya-protocol-en.pdf.

Anon. 2014. SANCO/11470/2012 - rev. 8. Guidance document on botanical active substances used in plant protection products. European Commission. Health & Consumer Protection Directorate-General. Available at: https://ec.europa.eu/food/sites/food/files/plant/docs/pesticides_ppp_app-proc_guide_doss_botanicals-rev-8.pdf.

Anon. 2018. IBMA white paper. New EU regulatory framework for bioprotection agents. IBMA Vision on how to improve regulation in the European Union. International Biocontrol Manufacturer Association (IBMA). Available at: https://www.ibma-global.org/upload/documents/ibmawhitepapergermanwebpages.pdf.

Bengtsson, M., Wulff, E., Lyngs Jørgensen, H. J., Pham, A., Lübeck, M. and Hockenhull, J. 2009. Comparative studies on the effects of a yucca extract and acibenzolar-S-methyl (ASM) on inhibition of *Venturia inaequalis* in apple leaves. *European Journal of Plant Pathology* 124(2), 187–98. doi:10.1007/s10658-008-9405-z.

Cohen, Y., Wang, W., Ben-Daniel, B. H. and Ben-Daniel, Y. 2006. Extracts of *Inula viscosa* control downy mildew of grapes caused by *Plasmopara viticola*. *Phytopathology* 96(4), 417–24. doi:10.1094/PHYTO-96-0417.

Coimbra, M., Isacchi, B., Van Bloois, L., Torano, J. S., Ket, A., Wu, X., Broere, F., Metselaar, J. M., Rijcken, C. J., Storm, G., Bilia, R. and Schiffelers, R. M. 2011. Improving solubility and chemical stability of natural compounds for medicinal use by incorporation into liposomes. *International Journal of Pharmaceutics* 416(2), 433–42. doi:10.1016/j.ijpharm.2011.01.056.

Cressey, D. 2014. Biopiracy ban stirs red-tape fears. *Nature* 514(7520), 14–5. doi:10.1038/514014a.

Dagostin, S., Formolo, T., Giovannini, O., Pertot, I. and Schmitt, A. 2010. *Salvia officinalis* extract can protect grapevine against *Plasmopara viticola*. *Plant Disease* 94(5), 575–80. doi:10.1094/PDIS-94-5-0575.

Dagostin, S., Schärer, H.-J., Pertot, I. and Tamm, L. 2011. Are there alternatives to copper for controlling grapevine downy mildew in organic viticulture? *Crop Protection* 30(7), 776–88. doi:10.1016/j.cropro.2011.02.031.

Disalvo, A. F. 1974. Antifungal properties of a plant extract I. Source and spectrum of antimicrobial activity. *Mycopathologia et Mycologia Applicata* 54(2), 215–9. doi:10.1007/bf02050042.

Ehlers, R.-U. (Ed.). 2011. *Regulation of Biological Control Agents*. Springer, Dordrecht, Heidelberg, London and New York.

European Commission. 2019. European pesticides database. Available at: https://ec.europa.eu/food/plant/pesticides/eu-pesticides-database/public/?event=homepage&language=EN (accessed on 17 October 2019).

Finckh, M. R., Van Bruggen, A. H. and Tamm, L. (Eds). 2015. *Plant Diseases and Their Management in Organic Agriculture*. APS Press, St. Paul, MN.

Gómez, A. M., Pereyra López, C. and Martinez de la Ossa, E. 1996. Recovery of grape seed oil by liquid and supercritical carbon dioxide extraction: a comparison with conventional solvent extraction. *The Chemical Engineering Journal and the Biochemical Engineering Journal* 61(3), 227–31. doi:10.1016/0923-0467(95)03040-9.

Güçlü-Üstündağ, O. and Mazza, G. 2007. Saponins: properties, applications and processing. *Critical Reviews in Food Science and Nutrition* 47(3), 231–58. doi:10.1080/10408390600698197.

Harkey, M. R., Henderson, G. L., Gershwin, M. E., Stern, J. S. and Hackman, R. M. 2001. Variability in commercial ginseng products: an analysis of 25 preparations. *The American Journal of Clinical Nutrition* 73(6), 1101–6. doi:10.1093/ajcn/73.6.1101.

Harris, E. S., Cao, S., Littlefield, B. A., Craycroft, J. A., Scholten, R., Kaptchuk, T., Fu, Y., Wang, W., Liu, Y., Chen, H., Zhao, Z., Clardy, J., Woolf, A. D. and Eisenberg, D. M. 2011. Heavy metal and pesticide content in commonly prescribed individual raw Chinese Herbal Medicines. *Science of the Total Environment* 409(20), 4297–305. doi:10.1016/j.scitotenv.2011.07.032.

Hasan, H. A., Raauf, A. M. R., Razik, B. M. A. and Hassan, B. 2012. Chemical composition and antimicrobial activity of the crude extracts isolated from *Zingiber officinale* by different solvents. *Pharmaceutica Analytica Acta* 3, 1–5.

Isman, M. B. 2000. Plant essential oils for pest and disease management. *Crop Protection* 19(8–10), 603–8. doi:10.1016/S0261-2194(00)00079-X.

Isman, M. B. 2006. Botanical insecticides, deterrents, and repellents in modern agriculture and an increasingly regulated world. *Annual Review of Entomology* 51, 45–66. doi:10.1146/annurev.ento.51.110104.151146.

Isman, M. B. 2014. Botanical insecticides: a global perspective. In: Gross, A., Coats, J. R. and Duke, S. O. (Eds), *Biopesticides: State of the Art and Future Opportunities*. ACS Symposium Series; vol. 1172. American Chemical Society, pp. 21–30.

Kaushik, N., Singh, B. G., Tomar, U., Naik, S., Vir, S., Bisla, S., Sharma, K., Banerjee, S. and Thakkar, P. 2007. Regional and habitat variability in azadirachtin content of Indian neem (*Azadirachta indica* A. Jusieu). *Current Science* 92, 1400–6.

Kuflik, T., Prodorutti, D., Frizzi, A., Gafni, Y., Simon, S. and Pertot, I. 2009. Optimization of copper treatments in organic viticulture by using a web-based decision support system. *Computers and Electronics in Agriculture* 68(1), 36–43. doi:10.1016/j.compag.2009.04.008.

Leahy, J., Mendelsohn, M., Kough, J., Jones, R. and Berckes, N. 2014. Biopesticide oversight and registration at the US Environmental Protection Agency. In: *Biopesticides: State of the Art and Future Opportunities*. ACS Publications, pp. 3–18.

Li, J., Jongsma, M. A. and Wang, C.-Y. 2014. Comparative analysis of pyrethrin content improvement by mass selection, family selection and polycross in pyrethrum [*Tanacetum cinerariifolium* (Trevir.) Sch. Bip.] populations. *Industrial Crops and Products* 53, 268–73. doi:10.1016/j.indcrop.2013.12.023.

Lomascolo, A., Uzan-Boukhris, E., Sigoillot, J. C. and Fine, F. 2012. Rapeseed and sunflower meal: a review on biotechnology status and challenges. *Applied Microbiology and Biotechnology* 95(5), 1105–14. doi:10.1007/s00253-012-4250-6.

Mendes, R. L., Nobre, B. P., Cardoso, M. T., Pereira, A. P. and Palavra, A. F. 2003. Supercritical carbon dioxide extraction of compounds with pharmaceutical importance from microalgae. *Inorganica Chimica Acta* 356, 328–34. doi:10.1016/S0020-1693(03)00363-3.

Michielin, E. M., Salvador, A. A., Riehl, C. A., Smânia Jr., A., Smânia, E. F. and Ferreira, S. R. 2009. Chemical composition and antibacterial activity of *Cordia verbenacea* extracts obtained by different methods. *Bioresource Technology* 100(24), 6615–23. doi:10.1016/j.biortech.2009.07.061.

Mukherjee, N. 1976. Antifungal activities of some oils and detergents. II. Effect on germ tube growth, sclerotici germination and growth of sclerotial plant pathogens/Pilzhemmende Wirkung einiger Öle und Detergentien. II. Einfluß auf Keimschlauchwachstum, Keimung der Sklerotien und Wachstum sklerotienbildender Phytopathogene. *Zeitschrift für Pflanzenkrankheiten und Pflanzenschutz/Journal of Plant Diseases and Protection* 83(6), 305–8.

Mulholland, D. A., Thuerig, B., Langat, M. K., Tamm, L., Nawrot, D. A., James, E. E., Qayyum, M., Shen, D., Heap, K., Jones, A., Hokkanen, H., Demidova, N., Izotov, D. and Schärer, H.-J. 2017. Efficacy of extracts from eight economically important forestry species against grapevine downy mildew (*Plasmopara viticola*) and identification of active constituents. *Crop Protection* 102, 104–9. doi:10.1016/j.cropro.2017.08.018.

Narain, A. 1972. Control of some fungal diseases by antibiotics and crude plant extracts. *Proceedings of the Indian National Science Academy* 37B, 332–40.

National Research Council. 1992. *Neem: A Tree for Solving Global Problems*. The National Academies Press, Washington DC.

OECD. 2017. Guidance document on botanical active substances used in plant protection products. ENV/JM/MONO(2017)6. Pesticide series No. 90. OECD, Paris.

Ostroukhova, L. A., Raldugin, V. A., Babkin, V. A., Onuchina, N. A. and Levchuk, A. A. 2012. Investigation of the chemical composition of larch wood resin. *Russian Journal of Bioorganic Chemistry* 38(7), 775–9. doi:10.1134/S1068162012070151.

Philogène, B. J., Regnault-Roger, C. and Vincent, C. 2008. Biopesticides d'origine végétale: bilan et perspectives. In: Philogène, B. J., Regnault-Roger, C. and Vincent, C. (Eds), *Biopesticides d'origine végétale*. Lavoisier, Paris, pp. 1–24.

Pietarinen, S. P., Willför, S. M., Ahotupa, M. O., Hemming, J. E. and Holmbom, B. R. 2006. Knotwood and bark extracts: strong antioxidants from waste materials. *Journal of Wood Science* 52(5), 436–44. doi:10.1007/s10086-005-0780-1.

Pollard, A. J., Reeves, R. D. and Baker, A. J. 2014. Facultative hyperaccumulation of heavy metals and metalloids. *Plant Science: an International Journal of Experimental Plant Biology* 217–218, 8–17. doi:10.1016/j.plantsci.2013.11.011.

Potterat, O. and Hamburger, M. 2006. Natural products in drug discovery-concepts and approaches for tracking bioactivity. *Current Organic Chemistry* 10(8), 899–920. doi:10.2174/138527206776894401.

Potterat, O. and Hamburger, M. 2014. Combined use of extract libraries and HPLC-based activity profiling for lead discovery: potential, challenges, and practical considerations. *Planta Medica* 80(14), 1171–81. doi:10.1055/s-0034-1382900.

Pullen, E. M. 2012. Biorational insecticide and fungicide and method of application. EP2123157 (A2)

Ramsey, E., Sun, Q., Zhang, Z., Zhang, C. and Gou, W. 2009. Mini-Review: green sustainable processes using supercritical fluid carbon dioxide. *Journal of Environmental Sciences* 21(6), 720–6. doi:10.1016/s1001-0742(08)62330-x.

Ramseyer, J., Thuerig, B., De Mieri, M., Schärer, H. J., Oberhänsli, T., Gupta, M. P., Tamm, L., Hamburger, M. and Potterat, O. 2017. Eudesmane sesquiterpenes from *Verbesina lanata* with inhibitory activity against grapevine downy mildew. *Journal of Natural Products* 80(12), 3296–304. doi:10.1021/acs.jnatprod.7b00868.

Sauter, H., Steglich, W. and Anke, T. 1999. Strobilurins: evolution of a new class of active substances. *Angewandte Chemie* 38(10), 1328–49. doi:10.1002/(SICI)1521-3773(19990517)38:10<1328::AID-ANIE1328>3.0.CO;2-1.

Scherf, A., Treutwein, J., Kleeberg, H. and Schmitt, A. 2012. Efficacy of leaf extract fractions of *Glycyrrhiza glabra* L. against downy mildew of cucumber (*Pseudoperonospora cubensis*). *European Journal of Plant Pathology* 134(4), 755–62. doi:10.1007/s10658-012-0051-0.

Schuster, C., Konstantinidou-Doltsinis, S. and Schmitt, A. 2010. *Glycyrrhiza glabra* extract protects plants against important phytopathogenic fungi. *Communications in Agricultural and Applied Biological Sciences* 75(4), 531–40.

Singh, U. P., Singh, H. B. and Singh, R. B. 1980. The fungicidal effect of neem (*Azadirachta indica*) extracts on some soil-borne pathogens of gram (*Cicer arietinum*). *Mycologia* 72(6), 1077–93. doi:10.1080/00275514.1980.12021288.

Soloway, S. B. 1976. Naturally occurring insecticides. *Environmental Health Perspectives* 14, 109–17. doi:10.1289/ehp.7614109.

Speiser, B., Schärer, H.-J. and Tamm, L. 2018. Direct plant protection in organic farming. In: *Improving Organic Crop Cultivation*. Burleigh Dodds Science Publishing, pp. 1–21. doi:10.19103/AS.2017.0029.12.

Thuerig, B., Ramseyer, J., Hamburger, M., Oberhänsli, T., Potterat, O., Schärer, H. J. and Tamm, L. 2016. Efficacy of a *Juncus effusus* extract on grapevine and apple plants against *Plasmopara viticola* and *Venturia inaequalis*, and identification of the major active constituent. *Pest Management Science* 72(9), 1718–26. doi:10.1002/ps.4199.

Thuerig, B., James, E. E., Schärer, H. J., Langat, M. K., Mulholland, D. A., Treutwein, J., Kleeberg, I., Ludwig, M., Jayarajah, P., Giovannini, O., Markellou, E. and Tamm, L. 2018a. Reducing copper use in the environment: the use of larixol and larixyl acetate to treat downy mildew caused by *Plasmopara viticola* in viticulture. *Pest Management Science* 74(2), 477–88. doi:10.1002/ps.4733.

Thuerig, B., Ramseyer, J., Hamburger, M., Ludwig, M., Oberhänsli, T., Potterat, O., Schärer, H.-J. and Tamm, L. 2018b. Efficacy of a *Magnolia officinalis* bark extract against grapevine downy mildew and apple scab under controlled and field conditions. *Crop Protection* 114, 97–105. doi:10.1016/j.cropro.2018.08.011.

Välimaa, A. L., Honkalampi-Hämäläinen, U., Pietarinen, S., Willför, S., Holmbom, B. and Von Wright, A. 2007. Antimicrobial and cytotoxic knotwood extracts and related pure compounds and their effects on food-associated microorganisms. *International Journal of Food Microbiology* 115(2), 235–43. doi:10.1016/j.ijfoodmicro.2006.10.031.

Wang, W., Ben-Daniel, B. H. and Cohen, Y. 2004. Control of plant diseases by extracts of *Inula viscosa*. *Phytopathology* 94(10), 1042–7. doi:10.1094/PHYTO.2004.94.10.1042.

Welch, E. W., Shin, E. and Long, J. 2013. Potential effects of the Nagoya Protocol on the exchange of non-plant genetic resources for scientific research: actors, paths, and consequences. *Ecological Economics* 86, 136–47. doi:10.1016/j.ecolecon.2012.11.019.

Wiens, M. N. 1978. Evaluation of phytoalexins and other plant natural products as protective fungicides and insect repellents. Theses (Dept. of Biological Sciences). Simon Fraser University.

Xue, J., Hao, L. and Peng, F. 2008. Residues of 18 organochlorine pesticides in 30 traditional Chinese medicines. *Chemosphere* 71(6), 1051–5. doi:10.1016/j.chemosphere.2007.11.014.

Zai-kang, T., Yan-ru, Z. and Jin-ping, S. 2002. Variation, heredity and selection of effective ingredients in *Magnolia officinalis* of different provenances. *Journal of Forestry Research* 13(1), 7–11. doi:10.1007/BF02857136.

Zhang, J., Wider, B., Shang, H., Li, X. and Ernst, E. 2012. Quality of herbal medicines: challenges and solutions. *Complementary Therapies in Medicine* 20(1–2), 100–6. doi:10.1016/j.ctim.2011.09.004.

Złotek, U., Mikulska, S., Nagajek, M. and Świeca, M. 2016. The effect of different solvents and number of extraction steps on the polyphenol content and antioxidant capacity of basil leaves (*Ocimum basilicum* L.) extracts. *Saudi Journal of Biological Sciences* 23(5), 628–33. doi:10.1016/j.sjbs.2015.08.002.

Zule, J., Cufar, K. and Tisler, V. 2015. Lipophilic extractives in heartwood of European Larch (*Larix decidua* Mill.). *Drvna Industrija* 66(4), 305–13. doi:10.5552/drind.2015.1442.

Index

www.ingramcontent.com/pod-product-compliance
Ingram Content Group UK Ltd.
Pitfield, Milton Keynes, MK11 3LW, UK
UKHW021826100926
8556IPUK00003B/175

9 781786 7635